I0036619

Smart Nanomaterials for Sensor Application

Edited by

Songjun Li

Jiangsu University
China

Yi Ge

Cranfield University
UK

He Li

University of Jinan
China

eBooks End User License Agreement

Please read this license agreement carefully before using this eBook. Your use of this eBook/chapter constitutes your agreement to the terms and conditions set forth in this License Agreement. Bentham Science Publishers agrees to grant the user of this eBook/chapter, a non-exclusive, nontransferable license to download and use this eBook/chapter under the following terms and conditions:

1. This eBook/chapter may be downloaded and used by one user on one computer. The user may make one back-up copy of this publication to avoid losing it. The user may not give copies of this publication to others, or make it available for others to copy or download. For a multi-user license contact permission@benthamscience.org

2. All rights reserved: All content in this publication is copyrighted and Bentham Science Publishers own the copyright. You may not copy, reproduce, modify, remove, delete, augment, add to, publish, transmit, sell, resell, create derivative works from, or in any way exploit any of this publication's content, in any form by any means, in whole or in part, without the prior written permission from Bentham Science Publishers.

3. The user may print one or more copies/pages of this eBook/chapter for their personal use. The user may not print pages from this eBook/chapter or the entire printed eBook/chapter for general distribution, for promotion, for creating new works, or for resale. Specific permission must be obtained from the publisher for such requirements. Requests must be sent to the permissions department at E-mail: permission@benthamscience.org

4. The unauthorized use or distribution of copyrighted or other proprietary content is illegal and could subject the purchaser to substantial money damages. The purchaser will be liable for any damage resulting from misuse of this publication or any violation of this License Agreement, including any infringement of copyrights or proprietary rights.

Warranty Disclaimer: The publisher does not guarantee that the information in this publication is error-free, or warrants that it will meet the users' requirements or that the operation of the publication will be uninterrupted or error-free. This publication is provided "as is" without warranty of any kind, either express or implied or statutory, including, without limitation, implied warranties of merchantability and fitness for a particular purpose. The entire risk as to the results and performance of this publication is assumed by the user. In no event will the publisher be liable for any damages, including, without limitation, incidental and consequential damages and damages for lost data or profits arising out of the use or inability to use the publication. The entire liability of the publisher shall be limited to the amount actually paid by the user for the eBook or eBook license agreement.

Limitation of Liability: Under no circumstances shall Bentham Science Publishers, its staff, editors and authors, be liable for any special or consequential damages that result from the use of, or the inability to use, the materials in this site.

eBook Product Disclaimer: No responsibility is assumed by Bentham Science Publishers, its staff or members of the editorial board for any injury and/or damage to persons or property as a matter of products liability, negligence or otherwise, or from any use or operation of any methods, products instruction, advertisements or ideas contained in the publication purchased or read by the user(s). Any dispute will be governed exclusively by the laws of the U.A.E. and will be settled exclusively by the competent Court at the city of Dubai, U.A.E.

You (the user) acknowledge that you have read this Agreement, and agree to be bound by its terms and conditions.

Permission for Use of Material and Reproduction

Photocopying Information for Users Outside the USA: Bentham Science Publishers grants authorization for individuals to photocopy copyright material for private research use, on the sole basis that requests for such use are referred directly to the requestor's local Reproduction Rights Organization (RRO). The copyright fee is US $25.00 per copy per article exclusive of any charge or fee levied. In order to contact your local RRO, please contact the International Federation of Reproduction Rights Organisations (IFRRO), Rue du Prince Royal 87, B-I050 Brussels, Belgium; Tel: +32 2 551 08 99; Fax: +32 2 551 08 95; E-mail: secretariat@ifrro.org; url: www.ifrro.org This authorization does not extend to any other kind of copying by any means, in any form, and for any purpose other than private research use.

Photocopying Information for Users in the USA: Authorization to photocopy items for internal or personal use, or the internal or personal use of specific clients, is granted by Bentham Science Publishers for libraries and other users registered with the Copyright Clearance Center (CCC) Transactional Reporting Services, provided that the appropriate fee of US $25.00 per copy per chapter is paid directly to Copyright Clearance Center, 222 Rosewood Drive, Danvers MA 01923, USA. Refer also to www.copyright.com

CONTENTS

EDITORS' BIOGRAPHIES

Professor Songjun Li

Professor Songjun Li is a distinguished professor of functional polymers in Jiangsu University, and currently the president of the Chinese Advanced Materials Society. He was the chairman of the 1st International Congress on Advanced Materials. Professor Li has published 50-plus papers in peer-reviewed journals and edited 5 books in prestigious publishers including Wiley-VCH, Elsevier, Bentham Science, Nova Science, and Research Signpost. He is also the PI for one EU Marie Curie FP7-IIF project and two Chinese National Science Funding projects. He sits on the editorial boards of "American Journal of Environmental Sciences", "Journal of Public Health and Epidemiology", "the Open Electrochemistry Journal", and "Journal of Computational Biology and Bioinformatics Research".

Professor Li was appointed an associate professor by Central China Normal University, in the wake of his PhD degree awarded by Chinese Academy of Sciences in 2005. As a post-doctoral associate, he joined the University of Wisconsin-Milwaukee (USA) in 2008, followed by his Marie Curie Fellowship in Cranfield University (UK), where he has worked with the world-renowned scientists Professor Anthony P.F. Turner and Professor Sergey A Piletsky during 2009-2011. He joined Jiangsu University as the distinguished professor in Dec 2011, where he is leading the research group of 'Molecular Imprinting and Catalytic Nanoreactors' in the School of Materials Science and Engineering.

Dr. Yi Ge

Currently is appointed as the Course Director and a Lecturer in Nanomedicine in Cranfield Health at Cranfield University. He obtained his bachelor's degree (1st Class Hons) in Biopharmaceutics and went on to an MPhil degree in Pharmaceutical Chemistry at Aston University. Afterwards, he moved to the University of Sheffield for a PhD in Chemistry. He was later employed as a research scientist in a UK pharmaceutical company and was then a postdoctoral research associate at Imperial College London, before joining Cranfield University in the group of advanced sensor and smart material in 2006. He is a member of

Royal Society of Chemistry and has served as an expert reviewer for the Engineering and Physical Sciences Research Council and Biotechnology and Biological Sciences Research Council. His activity in the field of nanotechnology was recognized by the Institute of Nanotechnology and he was admitted as a Professional Fellow of the Institute. He has been appointed as a Visiting Professor at the University of Jinan (China) and the Associate Editor for Advanced Materials Letters. One of his research interests and activities is on novel nanomaterials and advanced sensor and sensing technology, where he has extensive experiences in the design, synthesis, characterization and integration of nano-sized materials for various sensing and/or biological applications.

Dr. He Li

One of editors for the book "Biosensor Nanomaterials", is a professor of chemistry. He is currently the associate editor for the international principal journal "Advanced Materials Letters". He got his PhD degree in 2004 in Chinese Academy of Sciences. Subsequently, he joined the University of Jinan (China) and became an associate professor with research interests in Nanomaterials and their biomedical applications. He doubled also as chair of the Pharmaceutical Engineering Department during the period from 2007 to 2009. At present, he is working in the University of Wisconsin (USA) as a senior visiting scientist. In his personal database, he has published over 30 papers in international peer-reviewed journals. He has also been the invited reviewer for various grants and journals (beyond 40 times). His recent works are focused on designing and developing advanced functional materials for nanomedicine and biosensor application. Specifically, he is designing and synthesizing multifunctional nanocarriers (e.g., unimolecular micelles, polymer vesicles, functionalized inorganic nanoparticles) for the combined delivery of therapeutic and diagnostic agents in targeted cancer therapy and diagnosis, and fabricating biosensors (especially electrochemical biosensors) made of Nanomaterials to detect various biomolecules in the field of clinical diagnosis, bioaffinity assays and environmental monitoring.

FOREWORD

I welcome the timely publication of this e-Book on smart materials and their applications in sensors. The actual and potential impact of sensors and sensing systems is enormous and covers multifarious applications ranging from environmental monitoring and protection to pharmaceutical separation and analysis, and from defence and security to medicine and healthcare. It encompasses everything from electronics and materials to biotechnology and nanotechnology. The emerging challenges associated with public exposure to pollution and hazardous substances have fueled an urgent need for novel sensors and sensing systems to support toxicogenomic studies and protect populations. Over the past decades, scientists in this field have been working under pressure to meet these challenges. Sophisticated sensors and sensing materials are now available for evaluation and significant progress has been made in environmental protection, analytical technologies and the development of new materials. Prominent among them are smart nanomaterials, which combine the exciting properties of nanomaterials (such as electrical superconduction, paramagnetic properties and quantum optics) with enhanced diffusion properties and surface area to support a revolution in the design of sensors.

This e-Book compiles ten well-organized chapters related to the field of sensors and smart nanomaterials. The emphasis is on highlighting rapid, specific, sensitive, inexpensive, in-field, on-line and/or real-time detection by smart nanomaterials sensors. Singh *et al.* provide a review in the first chapter on smart nanomaterials and their applications in the field of sensors. Emerging trends and challenges for smart nanomaterials sensors are given in detail in this opening chapter. Diverging disciplines and fields, such as bionanoelectronics, nanotechnology, biotechnology and miniaturization, are exerting remarkable influences on the development of new sensing devices. The chapter by Marrazza is focused on metal nanoparticle-based biosensors. This is a developing field that combines nanoscale materials with biosensor technologies to achieve the direct wiring of enzymes onto electrode surfaces, and to promote electrochemical reactions, as well as incorporating nanobarcodes and signal amplification from biorecognition events. Studies have demonstrated that metal nanoparticles-based biosensors have important potential applications in the fields of environmental and medical analysis, due to their sensitivity, specificity, rapidity, simplicity and cost-effectiveness.

Molecular imprinting is one of the latest developments in the field of sensors. In chapter 3, Zhao *et al.* introduce molecularly imprinted nanomaterials-based sensors. Molecularly imprinted nanomaterials are classified as nanoparticles (including core-shell nanoparticles), nano-wires and tubes, and nanofilms. They review the working principles, methods used and binding events by molecularly imprinted nanosensors. In Chapter 4, Singh and Chen describe thermosensitive polymers and their applications for the prolonged delivery of contraceptive hormones to women. Salaimutharasan *et al.* review the recent progress made in nanosensors in Chapter 5. Nanosensors are used as the biological, chemical and surgical information sources to convey information. Current developments will allow the transfer of information from a nanoscale space to the macroscopic world. In the following chapter Banerjee *et al.* study the preparation of CdSe nanoparticles in the presence of self-assembled Abscisic acid (ABA). It appears possible that CdSe nanoparticles may lead to a new family of bio-imaging nanomaterials for cancer cell targeting and, in addition, provide a host for optoelectronic applications. Smith *et al.* describe the fabrication and optimization of a hydrogel delivery system for wound healing in Chapter 7. Zhu *et al.* introduce, in Chapter 8, the use of advanced carbon nanotubes and fibers in the field of sensors. Carbon nanotube-based biosensors demonstrate potential for the rapid diagnosis of life-threatening diseases. Singh *et al.* review selected gold nanoparticle-based biosensors. Giuliani and Ge describe 1D nanostructures and their application in sensors in Chapter 9. Nanowires, nanorods and nanotubes are known as the 1-D nanostructure prototype, which is characterized by cross sections as small as 1 micrometer and some microns in length. They can be generated by either adding gold nuclei to a growing solution or by using a template-based method. The 1-D nanostructure offers a significant potential in healthcare and safety.

It would take several extensive volumes to cover all the details of smart nanomaterials sensors. Thus, this e-Book can only hope to provide an overview and highlight some of the most important recent researches.

The editors have made an admirable attempt to include the most extensively studied areas, which should be of interest to a broad range of investigators and researchers.

I would like to thank both Bentham Science Publishers and the leading editor Dr. Songjun Li for their invitation to write this Foreword and to congratulate all the contributors for making this interesting e-Book possible. Thanks also should be expressed to the Research Directorate-General of European Commission and the National Science Foundation of China for supporting this work under both the Marie Curie Actions (No. PIIF-GA-2009-236799 and PIIF-GA-2010-254955) and the National Science Funding Project No. 21073068. It is my hope that this e-Book will provide valuable information to a broad range of researchers.

Professor Anthony P.F. Turner, PhD, DSC, FRSC

Professor of Biosensors & Bioelectronics
Editor-in-Chief, *Biosensors & Bioelectronics*
Chair, World Congress on Biosensors.
IFM-Linköping University, S-58183, Linköping
Sweden

PREFACE

There is tremendous implication in sensors and sensing systems, from environmental survey to protection, from separation to analysis, from electronics to materials, and from biotechnology to nanotechnology as well. Rising challenges in public exposure to pollution and hazardous substances have fueled a stringent requirement of developing novel sensors and sensing systems. Over the past decades, scientists with such backgrounds have been working under pressure to meet this stringent requirement. Sophisticated sensors and sensing materials are now assessable, leading to significant progress in environmental protection, analytic technologies and materials. Prominent among them are smart nanomaterials, which combine both the excellent properties of nanomaterials with smart functional materials, that have caused profound revolution in the understanding of the basic concept of 'sensors'.

Impressive progress has been made in this field due to the employment of novel preparation technologies and methods. The use of smart nanomaterials in the sensing applications enables to alter texture in conventional sensing models into controlled modes. The unique electronic, magnetic, acoustic and light properties of nanomaterials, coupled with smart materials capable of responsiveness to external stress, electric and magnetic fields, temperature, moisture and pH make accurate, real-time and modulated analysis possible. This e-Book summarized the main applications of smart nanomaterials in the field of sensors. The emphasis is to highlight the latest and significant progress made in this field. Other aspects including the use of functional materials into sensing systems, such as molecular device materials, bio-mimetic polymers, hybridized composites, supramolecular systems, information and energy-transfer materials, and environmentally friendly materials, are also described in this e-Book. When providing a relatively comprehensive profile on the current knowledge and technologies, we hope to provide insight into some new directions in this field. As such, this e-Book can be used not only as a textbook for advanced undergraduate and graduate students, but also as a reference e-Book for researchers in biotechnology, nanotechnology, biomaterials, medicine and bioengineering *etc*.

Several e-Books each composed of many chapters are probably not enough to cover all details of this field. Thus, it is very challenging to live up to the absolute and comprehensive summarization. Fortunately, all contributors because of their expert backgrounds have done their best while preparing their chapters. Because of the multidisciplinary nature of this subject, a large number of experts from different backgrounds have been invited to contribute their researches. Without doubt, if there was not participation of such a diverse group of experts, we would not have been able to accomplish our goal of developing a systematical e-Book in smart nanomaterials for sensor applications.

Songjun Li
Jiangsu University
China

Yi Ge
Cranfield University
UK

He Li
University of Jinan
China

List of Contributors

Ravindra Pratap Singh

Nanotechnology Application Centre,
University of Allahabad,
Allahabad- 211 002,
India

Joeng-Woo Choi

Department of Chemical & Biomolecular Engineering,
Interdisciplinary Program of Integrated Biotechnology,
Sogang University #1 Sinsoo-Dong, Mapo-Gu, Seoul 121-742,
Korea

Avinash Chandra Pandey

Nanotechnology Application Centre,
University of Allahabad,
Allahabad- 211 002,
India

Giovanna Marrazza

Università di Firenze, Dipartimento di Chimica,
Via della Lastruccia, 3; 50019 Sesto Fiorentino (Fi),
Italy

Shanshan Wang

Beijing National Laboratory for Molecular Sciences,
MOE Key Laboratory of Bioorganic Chemistry and Molecular Engineering,
College of Chemistry and Molecular Engineering,
Peking University,
Beijing, 100871,
China

Xiaocui Zhu,

Beijing National Laboratory for Molecular Sciences,
MOE Key Laboratory of Bioorganic Chemistry and Molecular Engineering,
College of Chemistry and Molecular Engineering,
Peking University,
Beijing, 100871,
China

Meiping Zhao

Beijing National Laboratory for Molecular Sciences,
MOE Key Laboratory of Bioorganic Chemistry and Molecular Engineering,
College of Chemistry and Molecular Engineering,
Peking University,
Beijing, 100871,
China

Songjun Li

School of Materials Science and Engineering,

Jiangsu University,
Zhenjiang 212013,
China

Priyanka Singh

University of North Dakota School of Medicine and Health Sciences,
1919 Elm Street,
Fargo, ND 58102,
USA

Sibao Chen

Purdue Pharmaceuticals L.P. 4701 Purdue Drive,
Wilson, NC 27893,
USA

Salaimutharasan Gnanamani

Department of Nanotechnology,
Faculty of Engineering and Technology, SRM University,
 Kattankulathur-603 203,
India

Siva Chidhambaram

Department of Nanotechnology,
Faculty of Engineering and Technology, SRM University,
Kattankulathur-603 203,
India

Mani Prabaharan

Department of Chemistry,
Faculty of Engineering and Technology, SRM University,
Kattankulathur-603 203,
India

Stephen H. Frayne

Fordham University, Department of Chemistry,
441 E. Fordham Road, Bronx, NY 10458,
USA

Stacey N. Barnaby

Fordham University, Department of Chemistry,
441 E. Fordham Road, Bronx, NY 10458,
USA

Areti Tsiola

Biology Department, Queens College, The City University of New York,
6530 Kissena Boulevard, Flushing, NY 11367,
USA

Karl R. Fath

Biology Department, Queens College, The City University of New York,
6530 Kissena Boulevard, Flushing, NY 11367,

USA

Evan M. Smoak

Fordham University, Department of Chemistry,
441 E. Fordham Road, Bronx, NY 10458,
USA

Ipsita A. Banerjee

Fordham University, Department of Chemistry,
441 E. Fordham Road, Bronx, NY 10458,
USA

Thomas J. Smith

Materials Research Institute,
Athlone Institute of Technology,
Dublin Rd, Athlone, Co. Westmeath,
Ireland

James E. Kennedy

Materials Research Institute,
Athlone Institute of Technology,
Dublin Rd, Athlone, Co. Westmeath,
Ireland

Clement L. Higginbotham

Materials Research Institute,
Athlone Institute of Technology,
Dublin Rd, Athlone, Co. Westmeath,
Ireland

Zhigang Zhu

Electrical Engineering Division,
Department of Engineering,
University of Cambridge,
9 JJ Thomson Avenue,
Cambridge, CB3 0FA,
UK

Andrew J Flewitt

Electrical Engineering Division,
Department of Engineering,
University of Cambridge,
9 JJ Thomson Avenue, Cambridge, CB3 0FA,
UK

William I Milne

Electrical Engineering Division,
Department of Engineering,
University of Cambridge,
9 JJ Thomson Avenue, Cambridge, CB3 0FA,
UK

Francis Moussy

Brunel Institute for Bioengineering,
Brunel University,
Uxbridge, Middlesex, UB8 3PH,
UK

Ram Singh

Department of Applied Chemistry,
Delhi Technological University,
Bawana Road, Delhi - 110 042,
India

Geetanjali

Department of Chemistry,
Kirori Mal College, University of Delhi,
Delhi - 110 007,
India

Vinita Katiyar

Civil Engineering Department,
Indian Institute of Technology,
Delhi - 110 016,
 India

S. Bhanumati

Department of Chemistry, Gargi College,
University of Delhi,
Srifort Road, New Delhi - 110 049,
India

Alessio Giuliani

SINTEA PLUSTEK Srl, Via E. Fermi 44, 20090 Assago (MI), Italy
Cranfield Health, Vincent Building,
Cranfield University,
Bedfordshire, MK43 0AL,
UK

Yi Ge

Cranfield Health, Vincent Building,
Cranfield University,
Bedfordshire, MK43 0AL,
UK

2

CHAPTER 1

Smart Nanomaterials for Biosensors, Biochips and Molecular Bioelectronics

Ravindra Pratap Singh[1,2*], Ashutosh Tiwari[3], Joeng-Woo Choi[2] and Avinash Chandra Pandey[1]

[1]*Nanotechnology Application Centre, University of Allahabad, Allahabad- 211 002, India;* [2]*Department of Chemical & Biomolecular Engineering, Interdisciplinary Program of Integrated Biotechnology, Sogang University #1 Sinsoo-Dong, Mapo-Gu, Seoul 121-742, Korea and* [3]*Cranfield Health, Vincent Building, Cranfield University, Cranfield, Bedfordshire, MK43 0AL, UK*

Abstract: The domain of biology has greatly been benefited by advances in other sciences leading to new levels of sensitivity, precision and resolution in biomolecular detection. The key driving force is the complementary length scale between biological structures that range from the 10's of nanometers (proteins, DNA, viruses) to the micron scale (cells and cellular assemblies) and capabilities of nanosystems to manipulate and control such feature sizes within our environment. Progress and development in biosensor development will inevitably focus upon the technology of the nanomaterials that promise to solve the biocompatibility and biofouling problems. The biosensors are integrated with new technologies in molecular biology, micro-fluidics, and smart nanomaterials, have applications in agricultural production, food processing, and environmental monitoring for rapid, specific, sensitive, inexpensive, in-field, on-line and/or real-time detection of pesticides, antibiotics, pathogens, toxins, proteins, microbes, plants, animals, foods, soil, air, and water. Thus, biosensors are excellent analytical tools for pollution monitoring, by which implementation of legislative provisions to safeguard our biosphere could be made effectively plausible. The current trends and challenges with smart nanomaterials for various applications have been the focuse in this chapter that pertains to biosensor development, bionanoelectronics, nanotechnology, biotechnology and miniaturization. All these growing areas will have a remarkable influence on the development of new ultra biosensing devices to resolve the severe pollution problems in the future that not only challenge the human health but also affect adversely other various comforts to living entities.

Keywords: Smart nonmaterials; biosensors; biochips; molecular bioelectronics.

1. INTRODUCTION

Richard Feynman in 1959 (the birth of nanotechnology) proposed for the first time about the possibility of manipulating the atoms for its application in data storage down to the scale of a single atom. When the characteristic length of the microstructures is in the 1-100 nm range, it becomes comparable with the critical length scales, *i.e.* size and shape effects on physical phenomena offer their usefulness in the devices using nano-structured materials, nano-tools, and nano-devices. Nano-devices such as Atomic Force Microscopes (AFM), Scanning Tunneling Microscopes (STM), Atomic-Layer-Deposition (ALD) and nanolithography tools, can manipulate matter at the atomic or molecular scale. In medical diagnostics and biosensors, lithography, Chemical Vapor Deposition (CVD), 3-D printing, and nano-fluidics are mostly used. Many other promising applications of nano tools and nanodevices are in the developmental phase such as nanoelectronic memory devices, nanosensors and drug delivery systems utilizing nanomaterials, semiconducting organic molecules, polymers and high purity chemicals and materials [1, 2].

A smart nanomaterial utilizes nano-scale engineering and system integration of existing materials to develop better materials and products. Various smart materials, including piezoelectric, thermoresponsive, shape memory alloys, polychromic, chromogenic and halochromic materials have become an integral part of our modern society. Smart materials exhibit properties that can be engineered in such a manner that their

---

*Address correspondence to Ravindra Pratap Singh:** Nanotechnology Application Centre, University of Allahabad, Allahabad-211 002, India; Tel: +91-0532-2460675; Fax: +91-0532-2460675; Mobile: +91-9451525764; E-mail: rpsnpl69@gmail.com

Songjun Li, Yi Ge and He Li (Eds)
All rights reserved - © 2012 Bentham Science Publishers

properties could be varied in a controlled manner under the influence of external stimuli such as temperature, force, moisture, electric charge, magnetic fields and pH. The piezoelectric materials produce voltage under stress or alter the shape under the influence of electric charge. Thermoresponsive materials, sometimes also known as shape memory alloys or shape memory polymers, alter their shape under the influence of the ambient temperature. Like thermoresponsive materials, magnetic shape memory alloys change shape due to changes in magnetic fields. Polychromic, chromogenic and halochromic materials change their color due to external influences like pH, temperature, light or electricity. Materials that change colour due to temperature are normally known as thermochromic materials and those that of light are a photo chromic materials. Applications of smart nanomaterials have made their presence strongly felt in various areas like healthcare, implants and prostheses; smart textiles, energy generation and conservation with energy generating materials and highly efficient batteries, defence, security, terrorism, and surveillance using smart dust and smart dust motes (nano-sized machines used in a range of sensors and wireless communication devices) due to their wherewithal of amplifying the signals employing biomarkers [3]. Fig. **1** shows the various kinds of nanomaterials which may be used to amplify biomarker signals.

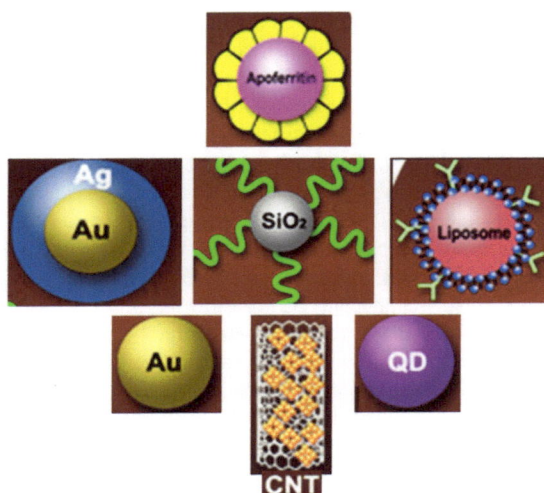

Figure 1. Various kinds of nanomaterials utilized for the amplification of biomarker signals.

Bionanomaterial's research has emerged as a new exciting field and the importance of DNA, RNA and peptides in designing the bionanomaterials for the fundamental development in biotechnology and nanomaterials have begun to be recognized as a new interdisciplinary frontier in the field of life science and material science. Great advances in nanobiochip materials, nanoscale biomimetic materials, nanomotors, nanocomposite materials, interface biomaterials, nanobiosensors and nano-drug-delivery systems have the enormous prospect in industrial, defense, and clinical medicine applications. Biomolecules assumes the very important role in Nanoscience and Nanotechnology, for example, Peptide Nucleic Acids (PNAs) replace DNA, and act as a biomolecular tool/probe in the molecular genetics diagnostics, cytogenetics, and also have enormous potentials in pharmaceutics for the development of sensors/arrays/chips besides many more applications. One of the current aspects related with PNA is the making of a new hot device for the commercial application, *e.g.* nanobiosensor arrays [4]. The integration of nanotechnology, micro fabrication techniques, and miniaturized devices with novel biochemical detection methodologies, leads to very sensitive and fast assays for the detection of desired biomolecules related with various commercial sectors. Nanotechnology involves the assembly of small molecules into complex architectures for improvised function by controlling the precise location of each atom in a 3-dimensional space. It forms larger functional elements and is being explored as the potential tool to fabricate nanometer size devices. [5, 6] Numerous reports are documented regarding the use of oligonucleotides for building nanostructures, which include DNA matrices based on subunits of fixed Holliday junctions, streptavidin-DNA fragment nanoparticle networks, DNA dendrimer formations for drug delivery, molecular tweezers (ssDNA based) and molecular switches. [7, 8] PNAs are promising connectors for the assembly of DNA based nanostructures with an exceptional ability to hybridize the sequences within the duplex

DNA by the strand invasion. High affinity binding by PNA has already been used for nanostructure assembly, with applications in labeling of DNA and strand invasion into DNA hairpins and tetra loop motifs. The bis-PNAs are two PNA sequences with a tethering spacer region. Amino acids were included in some of the spacer regions to increase the distance between the PNA sequences. The ability of bis-PNAs to assemble DNA with simple chemical modifications could be used in generating DNA: bis-PNA: DNA units for nanotechnology and DNA nanostructure assembly [9, 10].

Nanomaterials have attracted great attention in the research and development, due to their unique size dependent properties, originating from the small particle dimensions (10-100 nm) and size quantization effects. They are utilized broadly for optical, electrical, magnetic based nanodevices, especially for biomedical applications. Nanomaterials have the diverse range of applications such as a magnetic storage media, environment protection, sensors, catalysis, clinical diagnosis and treatment, *etc.* [11-16] Among the various types of nanomaterials, magnetic material of iron oxides (Fe_2O_3 and Fe_3O_4) are the most popular and promising materials, due to their many technological applications such as gas sensing material, heterogeneous catalyst, photo catalyst and pigments and anodes for electrolysis of water [17].

2. BIOSENSORS/BIOCHIPS

Biosensor consists of a biosensing material and a transducer that can be used for detection of biological and chemical agents. Biosensing materials, like enzymes, antibodies, nucleic acid probes, cells, tissues, and organelles selectively recognizes the target analytes, whereas transducers like electrochemical, optical, piezoelectric, thermal, and magnetic devices can quantitatively monitor the biochemical reaction [18]. Biosensors have become an emerging area of interdisciplinary research and Fig. **2** shows the process of biosensor and its various kinds.

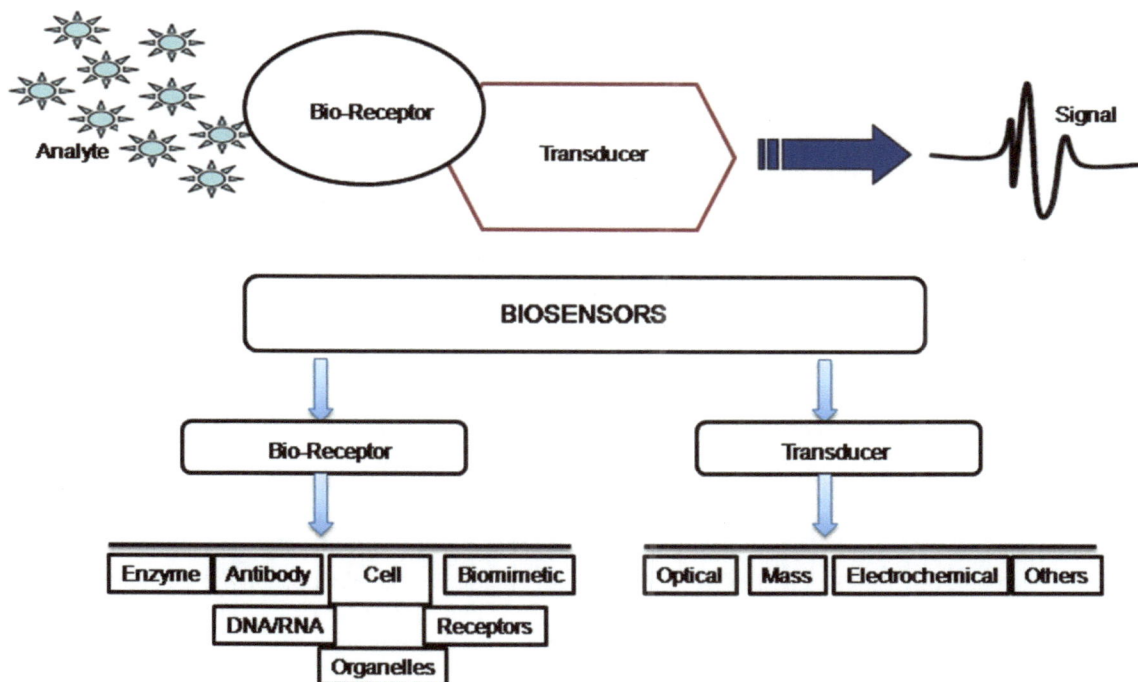

Figure 2. Concept of biosensor and its various kinds.

Various types of biosensors are used with inherent advantages and limitations in conjunction with different transducers forming the biosensing devices for the detection of various kinds of targeted biomolecules. Nucleic acid elements, including aptamers, DNAzymes, aptazymes, and PNA are widely used in nanobiotechnology (lab-on-a-chip, nanobiosensors array) [19-23]. In addition, the nano biosensor can be easily integrated into disposable polymer lab-on-a-chip for numerous applications in biochemical analysis and clinical diagnostics. The recent developments in nanobiosensors and their applications in biology, especially in medical diagnostics encompasses the concepts of coordinated nanobiosensors integrating the desirable properties of the individual components: protein machinery for sensitivity and specificity of binding, peptide or nucleic acid chemistry for aligning the various electron transducing units and the nanoelectrodes for enhancing sensitivity in electronic detection. Results from these systems focus on the potential advantages of use of nanoscale biosensor diligence, which will revolutionize biomedical diagnostics and treatments drastically. Thus the development and application of nanodevices in biology and medicine will have enormous implications for the benefit of society and human health [24, 25].

The biosensors integrated with the new technologies in molecular biology, microfluidics, and nanomaterials have applications in agricultural production, food processing, clinical care and environment for rapid, specific, sensitive, inexpensive, in-field, online and/or real time detection as well as monitoring of pesticides, antibiotics, pathogens, toxins, proteins, microbes, plants, animals, foods, soil, air, and water. Thus, ultrasensitive biosensors offer to be an excellent analytical tool for pollution monitoring enabling the surveillance of biosphere possible. The future emerging trends towards biosensor development in the context of bioelectronics, nanotechnology, miniaturization, and especially biotechnology seems to be all growing areas that will have a remarkable influence on the development of new biosensing platform to resolve our future clinical diagnostics and address the challenges relating to severe pollution problems concerning not only human health but also all living entities. To understand biological processes at a single molecule level and to implement them for the possible future applications in nanobiotechnology is of current interest. Studies relating to single molecule enabled probe have opened exciting avenues of research, especially in Nanoscience and Nanotechnology for the development of versatile biomolecule detection technology [26].

Nanomaterials have enabled the development of ultrasensitive biosensors, because of their high surface area, electronic properties and electrocatalytic activity as well as good biocompatibility. They have also been used to achieve direct wiring of enzymes to the electrode surface, to promote electrochemical reaction, to amplify a signal of biorecognition events. The uses of nanomaterials including nanoparticles, nanowire, nanoneedle, nanosheet, nanotube, nanorod, nanobelt, *etc.* for biosensing have been reported in research articles as well as review articles [27, 28]. Few important examples are: low potential detection of NADH using Carbon Nanotube (CNT) modified electrodes, gold nanoparticles for electrochemical immunosensors, carbon nanotube-based sensors [29-38], nanoparticles based biosensing [39-42] and nanowire as sensing materials [43].

Nanowires belong to a growing family of nano objects, which also includes nanotubes, nanoparticles, nanorods and many more. Nanowires can serve as the electrode or interconnects between micro and nanoelectronic devices. Their dimensions are sometimes at the same scale as biomolecules, which enables exciting possibilities for their interaction with biological species, such as cells, antibodies, DNA and other proteins. Chen *et al.* reported positively charged Ni-Al Layered Double Hydroxide Nanosheets (Ni-Al LDHNS) for the first time as matrices for immobilization of Horseradish Peroxidase (HRP) in order to fabricate enzyme electrodes for the purpose of studying direct electron transfer between the redox centers of proteins and underlying electrodes. The immobilized HRP in Ni-Al LDHNS on the surface of a Glassy Carbon Electrode (GCE) exhibited good direct electrochemical and electrocatalytic responses to the reduction of hydrogen peroxide and Trichloroacetic Acid (TCA). The linear detection results showed that Ni-Al LDHNS provides a novel and efficient platform for the immobilization of enzymes and the fabrication of third-generation biosensors. [44] ZnO nanosheet and other nanofibers and metal oxide were also reported by several investigators as sensing materials in electrochemical biosensors [45-48].

An ideal biosensing platform is pre-requisite and required to be not only miniaturized and cost efficient but also capable of simultaneous detection of multiple analytes. The biorecognition arrays (microarrays) have posed new technical challenges for the probes, transducers, and their detection apparatus. Microarrays can analyze large numbers of biological molecules using small amounts of material that are separated on a substrate (*e.g.* plastic, glass, and semiconductor). The biological materials deposited or synthesized on the chip surface include nucleic acids, proteins, peptides, antibody and carbohydrates. Fig. **3** shows biochips made by micro fabrication technology with the capability to sense a wide variety of analytes, including DNA, protein, antibodies, and small biological molecules, determined by most commonly fluorescence method.

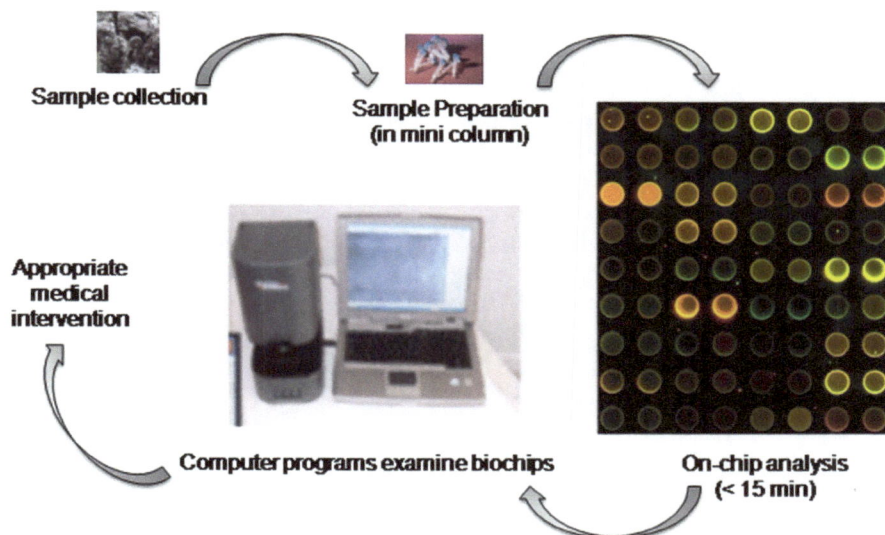

Figure 3. Biochips made by microfabrication technology, sense a wide variety of analytes.

The methods used for placing materials onto the chip surface rely on physical spotting, piezoelectric deposition, and *in situ* synthesis. Developments of DNA biosensors and DNA microarrays have progressed tremendously as reported by the several investigators [49]. DNA microarrays (gene chips, DNA chips, or biochips) utilize the preferential binding of complementary single stranded nucleic acid sequences. Unlike DNA biosensors, DNA microarrays are made onto the glass, plastic, or silicon supports consisting of tens to thousands of 10-100 μm reaction sites onto which individual oligonucleotide sequences have been immobilized [50]. The immobilization of a DNA probe in DNA biosensors is achieved directly onto a transducer surface. The exact number of DNA probes varies in accordance with the application. Contrary to DNA biosensors that allow single shot measurements, DNA microarrays allow multiple parallel detection and analysis of the patterns of expression of thousands of genes in a single experiment. The most common method for analyzing hybridization events on DNA microarrays is fluorescence. Additional methods have also been developed, such as surface plasmon resonance, atomic force microscopy, quartz crystal microbalance, and cantilevers. However, electrochemical detection offers several advantages over conventional fluorescence, such as portability, higher performance with lower background, fewer expensive components, and measurements could be made with turbid samples. Electrochemical transducers have often been used for detecting DNA hybridization due to their high sensitivity, small dimensions, low cost, and compatibility with micro manufacturing technology. There are numerous labeled electrochemical DNA biosensors where the tag can be an enzyme, ferrocene, an interactive electro active substance (a groove binder, such as Hoechst 33258, or an intercalator), or nanomaterials and label free electrochemical DNA biosensors [51].

Kumar and Dill reported applications of microarray in the development of electrochemical DNA microarray biosensors. Microelectrode arrays are composites of microelectrodes, where single microelectrodes are wired in parallel with each one, acting diffusionally independent; generating a signal, which is thousands times larger [52]. One of the microelectrode arrays is RAM^{TM} electrode where thousands of carbon wires are randomly sealed into epoxy resin producing random assemblies of micro disks [53]. Microfluidic environments add value to biosensing tasks because they consume lower amounts of probe molecules and target analyte [54]. A microelectrode array consisting of Boron Doped Diamond

(BDD) microelectrode disks a versatile electrode device with the advantages of both microelectrode arrays and of boron doped diamond as an electrode material was also developed. The BDD-microelectrode arrays are excellent substrates for the deposition of a range of metals, copper, silver and gold allowing a single electrode array to act as a template for a microelectrode array of many different electrode materials for a variety of analytical tasks, all of which can be carried out with a single BDD array after a suitable electrodeposition [55, 56].

Microfluidic technology consisting of microfluidic mixer, valves, pumps, channels, chambers in a single chip device have been established for detecting infectious particles (viruses and bacteria) in complex biological samples. These sensors are miniaturized arrays of individually addressable microelectrodes controlled by active Complementary Metal Oxide Semiconductor (CMOS) circuitry. The devices with capabilities of on-chip sample processing and detection provide a cost effective solution to direct sample-to-answer biological analysis for point-of-care genetic analysis, disease diagnosis, and in-field bio-threat detection [57, 58]. An integrated CMOS electrochemical sensor array capable of performing impedance spectroscopy, potentiometry, voltammetry, and ion-sensitive detection was reported by Hassibe and Lee [59]. The complete system is fabricated within a single chip and built in a standard digital 0.18 µm CMOS processor with no post processing requirements.

2.1. Nanoparticles in Nanobiodevices

The sensitivity and performance of devices are being improved using nanomaterials. Nanomaterials with at least one of their dimensions ranging in scale from 1 to 100 nm, display unique and remarkably different property as compared to its bulk because their nanometer size gives rise to high reactivity and other enhanced beneficial physical properties (electrical, electrochemical, optical and magnetic) owing to non-linearity after crossing the performance barrier threshold. Their applications can potentially translate into new assays that improve upon the existing methods of biomolecular detection. Nanoparticles have been widely used in biosensors for detection of nucleic acids, peptide nucleic acid and proteins [60-63]. Fig. **4** shows nanometer size scale of various kinds of nanoparticles materials.

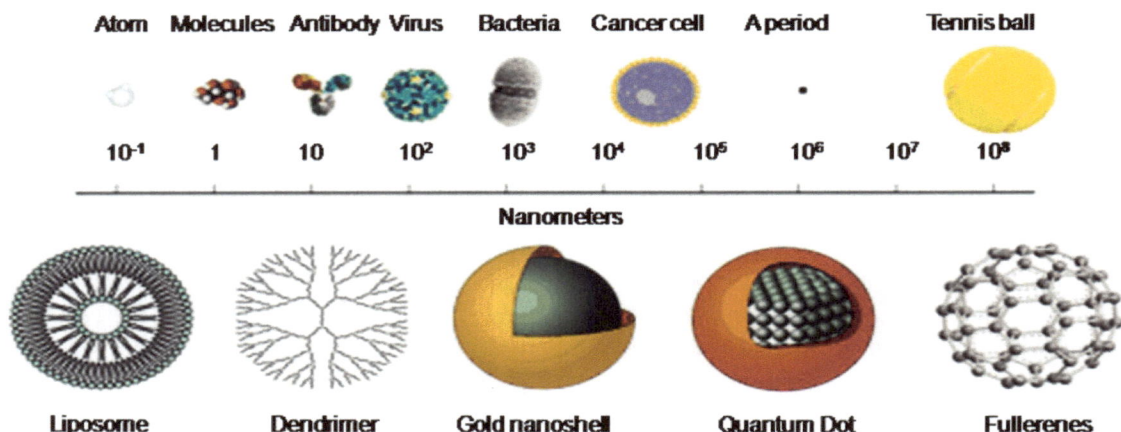

Figure 4. Nanometer size scale of various kinds of nanoparticles materials.

The enhancement in redox properties of gold nanoparticles coupled with silver have led to their widespread application as electrochemical labels in biosensor development with remarkable sensitivity [64, 65]. The gold nanoparticles coated with ferrocenyl hexathiol and streptavidin to monitor the DNA hybridization by using the ferrocene groups as the reporter molecule with a linear range between 7 and 150 pM for DNA implying that this approach requires no enzyme or enzyme substrate for amplification. Nanoparticles have also been coupled with magnetic particles to capture target DNA, which then hybridizes with a secondary probe DNA tagged to metal nanoparticle. These nanoparticles were chemically oxidized and detected by anodic stripping voltammetry. The detection limit of nm was reported using three different nanoparticle tags (ZnS, CdS, and

PdS) [66, 67]. A common problem with silver enhancement is a high background signal resulting from nonspecific precipitation of silver onto the substrate electrode and to overcome the setback various electrode surface treatments and electrochemically or enzymatically controlled deposition methods of silver have been reported. For reducing the silver related background signal and increasing the sensitivity, a new system of electrochemical detection of DNA hybridization based on stripping voltammetry of enzymatically deposited silver has been developed. The target DNA and a biotinylated DNA immobilized probe hybridize to a capture DNA probe tethered onto a gold electrode. Neutravidin (NA) conjugated alkaline phosphatase binds to the biotin of the detection probe on the electrode surface converting the nonelectroactive substrate to a reducing agent. The latter reduces the metal ions in solutions leading to the deposition of metal onto the electrode surface and DNA backbone [68, 69].

The application of underpotential deposition of Ag monolayer as a means of enhancing the electrochemical sensitivity of biomolecular reaction was first reported by K'Owino *et al.*, based on the development of a new DNA platform *via* metal-enhanced electrochemical detection (MED sensors) using immobilized metal layer, either as continuous film, particle, colloids or monolayer for the detection of parts per trillion (ppt) levels of anticancer drug cisplatin and PCBs. [70] A modified MED concept was also explored for the detection of two base pair mismatches using *Microcystis* as a model [71, 72] for monitoring DNA-cisplatin interactions. The detection may also be used to monitor other biomolecular reactions including DNA hybridization, mismatch detection, DNA-protein, antigen-antibody, and DNA-RNA reactions [73]. Aptamers, the artificial nucleic acid ligands consisting of single stranded DNA and/or RNA sequences and typically synthesized *in vitro* using systematic evolution of ligands by exponential enrichment (SELEX). Aptamer was utilized as a recognition element for biosensing applications for the detection of various analytes of interest [74-77] to enhance high sensitivity and selectivity using nanoparticles and quantum dot labels [78-81].

Semiconductor or conducting Polymer Nanowires (NW), are extremely attractive for designing high density protein arrays because of their high surface-to-volume ratio and electron transport properties as the electronic conductance is strongly influenced by minor surface perturbations (*e.g.*, binding on biomolecules). Hahm and Lieber reported the potential of functionalized NW for highly sensitive real time biodetection for monitoring DNA hybridization. [82] Patolsky *et al.* reported single viruses in connection with p-type silicon NW (SiNW) functionalized with PNA probes or antibodies for influenza [83]. Zheng *et al.* demonstrated the use of an antibody functionalized silicon nanowire sensor array for the multiplexed label free real time monitoring of cancer markers in undiluted serum samples. [84] The biosensing utility of conducting polymer NW or carbon nanotubes has also been reported [85, 86]. Recently, Tang *et al.* reported a simple and sensitive label free electrochemical immunoassay electrode for detection of Carcinoembryonic Antigen (CEA). CEA antibody (CEAAb) was covalently attached on a glutathione (GSH) monolayer modified gold nanoparticle (AuNP) and the resulting CEAAb-AuNP bioconjugates were immobilized on Au electrode by electrocopolymerization with ortho-aminophenol (OAP). Electrochemical impedance spectroscopy and cyclic voltammetry studies demonstrate that the formation of CEA antibody-antigen complexes increases the electron transfer resistance of $Fe\,[(CN)]_6{}^{3-/4-}$ redox pair at the poly-OAP/CEAAb-AuNP/Au electrode [87].

The micro fabricated Interdigitated Array Microelectrodes (IDAM) have been received great attention, there are one or multiple electrode pairs in an IDA, and the distance between finger electrodes can be in the range from micron to nanometer in order to probe the volume close to the electrodes. The integration of impedance technique with biosensor technology has led to the recent development of impedance biosensors that is expanding rapidly for bacteria detection. Impedance biosensors for bacteria detection are based on impedance analysis of the electrical properties of bacterial cells when they are attached to or associated with the electrodes [88]. The advances in micro fabrication technologies have enabled the use of micro fabricated microarray electrodes in impedance detection and the miniaturization of impedance microbiology into a chip format, which has shown great promise for rapid detection of bacterial growth. Yang and Bashir reviewed the electrical/electrochemical impedance for the rapid detection of food borne pathogenic bacteria, including microchip micro fabricated microelectrodes based (interdigitated array microelectrodes) and microfluidic based Faradic electrochemical impedance biosensors (microchips), non-Faradic impedance biosensors, and the integration of impedance biosensors with other techniques such as dielectrophoresis and electro-permeability. Magnetic particles are useful in separating target cells from a mixture of bacteria and food matrices and also help to concentrate separated cells into a very small volume with the help of a magnetic field and improve the sensitivity [89]. Recently, a microfluidic flow cell with embedded gold interdigitated array microelectrode was developed and integrated with magnetic Nanoparticle-Antibody Conjugates (MNAC) into an impedance biosensor to detect pathogenic bacteria in ground beef samples like *E. coli* O157:H7, which can produce toxins that damage the lining of the intestine, cause anemia, stomach cramps and bloody diarrhea, *etc.* [90-96] Cretich *et al.* reported a new, rapid and robust method for PDMS functionalization, based on chemisorptions of copolymer (DMA-NAS-MAPS), which provides an effective way to immobilize DNA fragments for the bacterial genotyping and food pathogen identification. Actually, PDMS bound surface of biomolecules can be applied to DNA, protein and peptide microarrays, biosensors and cell culturing. Furthermore, PDMS, with its attractive physico-chemical properties and the ease of patterning is the material of choice for the development of lab-on-chip devices. Liao *et al.* reported the first species-specific detection of bacterial pathogens in human clinical fluid samples using a micro fabricated electrochemical sensor array. Each of the 16 sensors in the array consisted of three single-layer gold electrodes: working, reference, and auxiliary [97]. Sadik and co-workers [98] describes the integration of a fully autonomous electrochemical biosensor (96-well-type electrodes array, DOX-dissolved oxygen sensor) with pattern recognition techniques for the detection and classification of the bacteria kingdom at subspecies and strain levels. The operation principle of a high-throughput 96-electrode DOX sensor is based on the measurement of the difference in the oxygen consumed by different bacteria classes and strains over a period of time and monitoring the oxygen reduction current at a fixed potential (-700 mV vs. gold), where the oxygen reduction current is maximized to ensure the highest sensitivity of the system. Additional selectivity is provided using pattern recognition, which allows better differentiation and classification of bacteria.

Clark and Lyons [99] proposed for the first time in 1962, the initial concept of glucose enzyme electrodes. Now, tremendous efforts directed towards the development of reliable devices for diabetes control have been developed like electrochemical glucose biosensors. [100, 101] Various nanomaterials, including gold nanoparticles or carbon nanotubes, have also been used as electrical connectors between the electrode and the redox center of glucose oxidase. Carbon nanotubes can be coupled to enzymes to provide a favorable surface orientation and thus can act as an electrical connector between their redox center and the electrode surface as molecular wires (nanoconnectors) between the underlying electrode and a redox enzyme [102].

Patolsky *et al.* demonstrated that the edge of Single Walled Carbon Nanotubes (SWCNT) can be linked to an electrode surface. Such enzyme reconstitution on the end of CNT represents an extremely efficient approach for plugging an electrode into glucose oxidase. Electrons were transported along distances higher than 150 nm with the length of the SWCNT controlling the rate of electron transport. An interfacial electron-transfer rate constant of 42 s^{-1} was estimated for 50 nm long SWCNT. [103] Efficient direct electrical connection to GOx was reported by Gooding and co-workers in connection to aligned SWCNT arrays. At present, activation of the bioelectrocatalytic functions of GOX by nanoparticles or CNT requires electrical overpotentials (beyond the thermodynamic redox potential of the enzyme redox center) [104]. Improving the contact between the nanomaterial and the electrode might decrease this overpotential. Subsequently, nanowires as sensing materials were also used for hydrogen peroxide and glucose sensors to overcome the overvoltage in the detection of H_2O_2 [105]. Gold nanowires were prepared by an electrodeposition strategy using nanopore polycarbonate (PC) membrane, with the average diameter of the nanowires of about 250 nm and length of about 10 nm. The nanowires were prepared and dispersed into chitosan (CHIT) solution and stably immobilized onto the glassy carbon electrode surface. The modified electrode allows low potential detection of hydrogen peroxide with a high sensitivity and fast response time. Glucose oxidase was adsorbed onto the nanowire surface to fabricate glucose biosensor as an application example. The detection of glucose was performed in the phosphate buffer (pH 6.98) at -0.2V [106, 107]. Similarly, platinum nanowires (PtNWs) prepared with the help of porous Anodic Aluminum Oxide (AAO) templates have been solubilized in chitosan together with Carbon Nanotubes (CNTs) to form a PtNW-CNT-CHIT organic-inorganic system. The PtNW-CNT-CHIT film modified electrode offered a significant decrease in the overvoltage for the hydrogen peroxide and was shown to be excellent amperometric sensors for hydrogen peroxide at -0.1 V over a wide range of concentrations with the sensitivity recorded being 260 μAmM^{-1} cm^{-2}. By linking glucose oxidase, an amplified biosensor toward glucose was prepared, which exhibits selective determination of glucose at -0.1V. The application of polypyrrole coated glucose oxidase nanoparticles (GOx/Ppy) in electrochemical biosensor design was also described, and the increase of Km by over 10 times was determined for polypyrrole-coated GOx, compared with native GOx. [108] Tao and co-workers [109] have developed a conductivity based glucose nanobiosensor based on conducting-polymer-based nanogap. Such nano junction-based sensor was formed by using polyaniline/glucose oxidase for bridging a pair of nanoelectrodes separated with a small gap (ca. 20-60 nm) [110]. Ultimately, one would like to eliminate the mediator and develop a reagent less glucose biosensor with a low operating potential, close to that of the redox potential of the enzyme. In this case, the electron is transferred directly from glucose to the electrode *via* the active site of the enzyme. The absence of mediators is the main advantage of such third generation biosensors, leading to a very high selectivity (owing to the very low operating potential). Jing and Yang reported oxidized boron-doped diamond electrodes suggesting some promise for mediator-free glucose detection based on direct electron transfer. [111] An ideal sensor would be one that provides a reliable real-time continuous monitoring of all blood glucose variations throughout the day with high selectivity and speed over extended periods under harsh conditions. Wilson reported continuous glucose monitoring, which addresses the deficiencies of the test strip based meters and provides the opportunity of making fast and optimal therapeutic interventions [112] and further Wilson and Gifford, reported a needle type multi electrode array for the hypodermic continuous glucose monitoring sensor using MEMS technology. The developed multielectrode sensor had four electrodes of two working (Pt) electrodes, one counter (Pt) electrode, and one reference (Ag/AgCl) electrode. [113] Jung *et al.* reported two working electrodes for the enzyme and non-enzyme electrodes, which measure glucose concentration and the background current. This would minimize short term crises and long-term complications of diabetes leading to improved quality and length of life for diabetic people. [114] Heller reported glucose biosensors based on closed loop glycemic control systems for regulating a person's blood glucose. The concept of closed loop (sense/release) systems is expected to have a major impact upon the treatment and management of other diseases revolutionizing the patient monitoring [115]. However, the challenges for meeting these demands include rejection of the sensor by the body, miniaturization, long term stability of the enzyme, *in vivo* calibration, short stabilization times, baseline drift, safety, and convenience. The sensor must be of a very tiny size and of proper shape to allow for easy implantation resulting in minimal discomfort. Alternative sensing sites, particularly the subcutaneous tissue, have thus received growing attention. The subcutaneous tissue is minimally invasive, and its glucose level reflects the blood glucose concentration. However, such subcutaneous implantation generates a wound site that experiences an intense local inflammatory reaction. This inflammatory response associated with the wound formation is characterized with the problems such as scar tissue formation accompanied by adhesion of bacteria and macrophage leading to distortion of the glucose concentration in the immediate vicinity of the sensor. Recent approaches for designing more biocompatible *in vivo* glucose sensors focused on preparing interfaces that resist biofouling [116-118]. Nitric oxide is an effective inhibitor of platelet and bacterial adhesion. Gifford *et al.* reported nitric oxide release glucose sensors fabricated by doping the outer polymeric membrane coating of needle type electrochemical sensors with suitable lipophilic diazeniumdiolate species or diazeniumdiolate-modified sol-gel particles. The utility and application of nanomaterials are targeted for the improved electrical contact between the redox center of GOx and electrode supports using genetically engineered GOx [119]. Recently, Andreescu and Luck, reported genetically engineered periplasmic glucose receptors as a biomolecular recognition element on gold nanoparticles (AuNPs) that holds promise for sensitive and reagent less electrochemical glucose biosensor. The receptors were immobilized on AuNPs by a direct sulfur-gold bond through a cysteine residue that was engineered in position 1 on the protein sequence. This finding will result in the advances in new painless *in vitro* testing, artificial (biomimetic) receptors for glucose, advanced biocompatible membrane materials, invasive monitoring with a compact insulin delivery system, new innovative approaches for non-invasive monitoring, and miniaturized long term implants [120].

3. BIOMATERIALS

The biomaterial is natural or man made that comprise whole or part of a living structure or biomedical device, which performs or replaces a natural function. They are used every day in dental applications, surgery, and drug delivery. A biomaterial may also be an auto graft, allograft or xenograft used as a transplant material. Silicates in algae and diatoms, carbonates in invertebrates, calcium phosphates and carbonates in vertebrates are few examples of biomaterials [121-126]. Self-assembly is described by the spontaneous aggregation of particles (atoms, molecules, colloids, micelles, *etc.*) without the influence of any external forces. Molecular self-assembly is found widely in biological systems and provides the basis of a wide variety of complex biological structures. Molecular crystals, liquid crystals, colloids, micelles, emulsions, phase separated polymers, thin films and self-assembled monolayers, all represent examples of highly ordered structures [127, 128].

In the biomedical field, artificial materials are used in the human body to measure, restore, improve physiologic function, and enhance survival and quality of life. Typically, inorganic (metals, ceramics, and glasses) and polymeric (synthetic and natural) materials have been used for such items as an artificial heart valves, (polymeric or carbon based), synthetic blood vessels, artificial hips (metallic or ceramic), medical adhesives, sutures, dental composites, and polymers for controlled slow drug delivery. The biocompatible materials are integrated into the biological environment as well as other tailored properties depending on the specific *in vivo* application [129]. The biomimetics may be described as the abstraction of good design from nature or the stealing of ideas from nature. The aim is to make materials for non biological uses under an inspiration from the natural world by combining them with man made, non biological devices or processes. The aim of the controlled release devices is to maintain the drug in the desired therapeutic range with just a single dose, reduces the need for follow-up care, preserves medications that are rapidly destroyed by the body, and increases patient comfort and/or improves compliance [130]. Biodegradability can therefore, be engineered into polymers by the addition of chemical linkages such as anhydride, ester, or amide bonds, among others. The mechanism for degradation is by hydrolysis or enzymatic cleavage resulting in a scission of the polymer backbone [131]. Microorganisms can eat and sometimes digest polymers, and initiate a mechanical, chemical, or enzymatic aging. Biodegradable polymers with hydrolysable chemical bonds are being researched extensively for biomedical, pharmaceutical, agricultural, and packaging applications. In order to be used in medical devices and controlled drug release applications, the biodegradable polymer must be biocompatible and should meet other criteria to be qualified as biomaterial-processable, sterilizable, and capable of controlled stability or degradation in response to biological conditions [132]. Chitosan is an environmentally biodegradable polymer that possesses a wide range of useful properties. Specifically, it is a biocompatible, antibacterial and environmentally friendly polyelectrolyte, thus lending itself to a variety of applications, including water treatment, chromatography, additives for cosmetics, textile treatment for antimicrobial activity, novel fibers for textiles, photographic papers, biodegradable films, biomedical devices, and microcapsule implants for controlled release in drug delivery [133].Poly Ethylene Oxide (PEO) is a polymer which shows biocompatibility, hydrophilicity, and versatility. The simple water soluble linear polymer can be modified by chemical interaction to form water insoluble but water swellable hydrogels retaining the desirable properties associated with the ethylene oxide part of the structure. Poly Ethylene Glycol (PEG) has been used increasingly for a variety of pharmaceutical applications. Multiblock copolymers of Poly Ethylene Oxide (PEO) and Poly Butylene Terephthalate (PBT) are under development process for the prosthetic devices and artificial skin. Ethylene-vinyl acetate copolymer is a widely used nondegradable polymer, and displays excellent biocompatibility, physical stability, biological inertness, and processability [134].

3.1. Nanocomposites/Bionanocomposites

Nanocomposites are materials with a nanoscale structure that improve the macroscopic properties of the products. They are clay, polymer or carbon, or a combination of these materials with nanoparticle building blocks and can greatly enhance the properties of materials. Fig. **5** shows SEM images of various kinds of nanostructured nanocomposites.

Magnetic multilayered materials are one of the most important aspects of nanocomposites as they have led to significant advances in storage media [135, 136]. Clay's polymer nanocomposites are among the most successful nanotechnological materials because they can simultaneously improve material properties without significant trade offs. Polymer layered silica nanocomposites and polymer/clay composites materials have improved mechanical stability, which contributes to an increased heat deflection temperature, enhance barrier and mechanical properties and are less flammable. Compression injection molding, melt intercalation, and co extrusion of the polymer with ceramic nanopowders can form nanocomposites. Often no solvent or mechanical shear is needed to promote intercalation [137, 138].

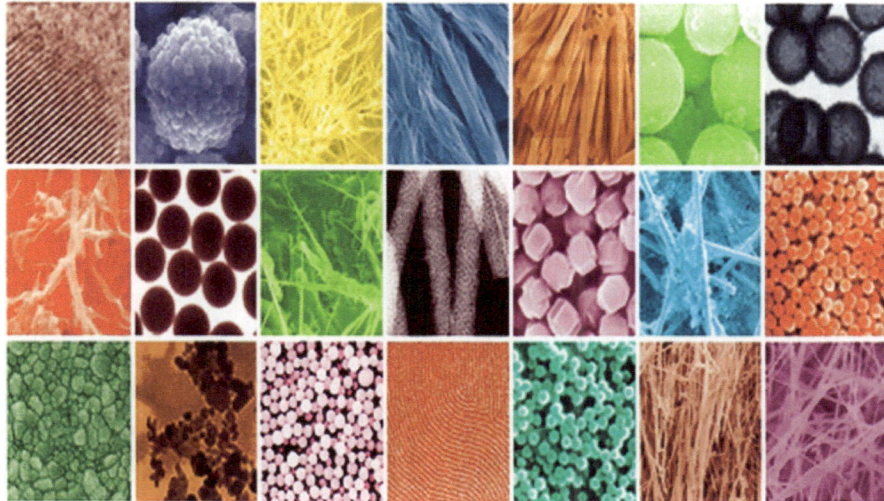

Figure 5. SEM images of various kinds of nanostructured nanocomposites.

The properties of nanocomposite materials depend not only on the properties of their individual parents but also on their morphology and interfacial characteristics. The nanocomposites have mechanical properties such as strength, modulus and dimensional stability, decreased permeability to gases, water and hydrocarbons, thermal stability and heat distortion temperature, flame retardancy and reduced smoke emissions, chemical resistance, surface appearance, electrical conductivity and optical clarity in comparison to conventionally filled polymers. The nanocomposites promise new applications in many fields such as mechanically re-inforced light weight components, non-linear optics, battery cathodes and ionics, nanowires, sensors and other systems [139]. The nanoparticulate/fibrous loading to confer significant property improvements with very low loading levels, resulting in significant weight reductions (better for military and aerospace applications), greater strength and increased barrier performance. The areas of nanocomposite applications are various that include drug delivery systems, anti-corrosion barrier coatings, UV protection gels, lubricants and scratch free paints, new fire retardant materials, new scratch/abrasion resistant materials, and superior strength fibers and films. Although nanocomposites are realizing many key applications in numerous industrial fields, a number of keys technical and economic barriers exist for its widespread commercialization. These include impact performance, the complex formulation relationships and routes to achieve and measure nanofiller dispersion and exfoliation in the polymer matrix [140].

Bionanocomposites from a fascinating interdisciplinary area that brings together biology, materials science, and nanotechnology. New bionanocomposites are impacting diverse areas, in particular, biomedical science. Generally, polymer nanocomposites are the results of the combination of polymers and inorganic/organic fillers at the nanometer scale. The extraordinary versatility of these new materials may spring from the large selection of biopolymers and fillers available to researchers. Few existing biopolymers, including (but not limited) polysaccharides, aliphatic polyesters, polypeptides, proteins, and polynucleic acids are predominantly reported in the literature along with fillers like clays, hydroxyapatite, and metal nanoparticles [141]. The interaction between filler components of nanocomposites at the nanometer scale enables them to act as molecular bridges in the polymer matrix. This is the basis for enhanced mechanical properties of the nanocomposites as compared to conventional microcomposites [142].

Bionanocomposites add a new dimension to these enhanced properties in that they are biocompatible and/or biodegradable materials. The biodegradable materials can be described as materials degraded and gradually absorbed and/or eliminated by the body, whether degradation is caused mainly by hydrolysis or mediated by metabolic processes [143]. Therefore, these nanocomposites are of immense interest to biomedical technologies such as tissue engineering, bone replacement/repair, dental application, and controlled drug delivery. Current opportunities for polymer nanocomposites in the biomedical arena arise from the multitude of applications and the vastly different functional requirements for each of these applications. For example, the screws and rods that are used for internal bone fixation bring the bone surfaces in close proximity to promote healing. This stabilization must persist for weeks to months without loosening or breaking. The modulus of the implant must be close to that of the bone for efficient load transfer [144]. The screws and rods must be non corrosive, non toxic, and easy to remove if necessary. Thus, a polymer nanocomposite implanted must meet certain design and functional criteria, including biocompatibility, biodegradability, mechanical properties, and, in some cases, aesthetic demands. [145, 146] Yamaguchi *et al.* synthesized and studied flexible chitosan-HAP nanocomposites [147].The matrix used for this study was chitosan (a cationic, biodegradable polysaccharide), which is flexible and has a high resistance against heating because of intramolecular hydrogen bonds formed between the hydroxyl and amino groups [148, 149]. The resulting nanocomposites, prepared by the co precipitation method is mechanically flexible and can be formed into any desired shape. Nanocomposites formed from gelatin and HAP nanocrystals are conducive to the attachment, growth, and proliferation of human osteoblast cells. [150] Collagen based polypeptidic gelatin has a high number of functional groups and is currently being used in wound dressings and pharmaceutical adhesives in clinics. [151, 152] The flexibility and cost effectiveness of gelatin can be combined with the bioactivity and osteoconductivity of HAP to generate potential engineering biomaterials. The traditional problem of HAP aggregation was overcome by precipitation of the apatite crystals within a polymer solution [153, 154]. The porous scaffold generated by this method exhibited well developed structural features and pore configuration to induce blood circulation and cell in growth. Such nanocomposites have the high potential for their use as hard tissue scaffolds. Three dimensional porous scaffolds from biomimetic HAP/chitosan-gelatin network composites with micro scale porosity have shown adhesion, proliferation, and expression of osteoblasts [155, 156]. Porosity is critical for tissue engineering applications because it enables the diffusion of cellular nutrients and waste, and provides support for cell movement [157, 158]. Polysaccharides, such as alginate, provide a natural polymeric sponge structure that has been used in tissue engineering scaffold design. The week, soft alginate scaffolds can be strengthened with HAP and have widespread applications [159, 160]. Composite membranes from HAP nanoparticles and chitosan/collagen sols have also been synthesized to study connective tissue reactions [161, 162]. Studies that target the nucleation of calcium phosphates and bone cell signaling within the matrix have used acidic macromolecules as the nanocomposite matrix [163]. Specifically, amino acids like aspartic acid and glutamic acid have been used as the matrix protein. Both amino acids are known to play an important role in intercellular communication and osteoblast differentiation that increases extracellular mineralization. Related studies have also highlighted the significance of aspartic acid in the treatment of osteoporosis and other bone dysfunctions [164].

Aliphatic polyester nanocomposites have been used for the predominant choice for materials in degradable drug delivery systems [165, 166]. Of the polyesters that show promise in biomedical fields, poly (L-lactic acid) (PLLA) is the most prevalent. It has widespread applications in sutures, drug delivery devices, prosthetics, scaffolds, vascular grafts, bone screws, pins, and plates for temporary internal fixation [167, 168]. Good mechanical properties and degradation into nontoxic products are the main reasons for such an array of applications [169, 170]. Poly Glycolic Acid (PGA) is aliphatic polyester with their applicability for biomedical use. However, unlike PLLA, PGA is readily soluble in water. Mechanical properties of self re-inforced PGA have been investigated and are found to worsen on exposure to distilled water. Thus, water solubility and its high melting point limit the use of PGA in bionanocomposites. (Poly ε-caprolactone) (PCL) is also a promising candidate for controlled release and soft tissue engineering. The range of properties can be furthered by copolymerization with other lactones such as glycolide, lactide, and Poly Ethylene Oxide (PEO) or by nanofiller incorporation. Nanocomposite polymer clay hydrogels have also been studied extensively by microscopy and scattering techniques [171, 172]. Recently, electrospinning has been used as an alternative scaffold fabrication technique in soft tissue transplantation and hard tissue regeneration. This method provides woven mats with individual fiber diameters ranging from 50 nm to a few microns. The interconnected, porous network so formed is desirable for drug delivery as well as biomedical substrates for tissue regeneration, wound dressing, artificial blood vessels, and other uses. Electro spinning helps tailor the mechanical, biological, and kinetic properties of the scaffolds by varying parameters such as polymer solution properties and processing conditions (*e.g.* electrical force, the distance between the electro spinning needle and the oppositely charged surface acting as ground, spinneret geometry, and solution flow rate) [173, 174]. However, some of the restrictions of this method, like controlling the pore size and softness of the electrospun mat, currently prevent it from being used for hard tissue applications. Polypeptides as matrices provide an additional array of opportunities in materials design and application in terms of a unique ability to adopt specific secondary, tertiary, and quaternary structures, a feature not available with synthetic polymers. Functionality can also be incorporated using natural and non natural amino acids with desired activity at specific sites along the polypeptide backbone. Materials like this, in addition to proteins, introduce the possibility of fibril incorporation into nanocomposites as re-enforcement with biodegradable matrices. Such unique nanocomposites combine the degradability and strength of the gel matrix with control over functionality and morphology of the fibrillar fillers. Potential applications of such a nanomaterials, include drug delivery matrices, tissue engineering scaffolds, and bioengineering materials. Nanocomposites from bioactive molecules and clays have also been reported. One such example is smectite nanocomposites that use the ability of smectites to induce specific cointercalation of purines and pyrimidines. [173, 174] Several bioactive compounds like DNA and pharmaceuticals have been incorporated within Layered Double Hydroxides LDH hosts [175, 176]. poly (Urethane Urea) (PUU) segmented block copolymers are common in ventricular assisted devices and total artificial hearts as blood sacs. One of the main disadvantages of PUUs in these devices is their relatively high permeability to air and water vapor, a result of the diffusion through the poly (tetramethylene oxide) soft segments that are present as the majority component of the copolymer. The use of organically modified layered silicates seems particularly attractive for the variety of approaches taken to reduce permeability while maintaining desired biocompatibility and mechanical properties. Nanocomposites used for various biomedical applications have different requirements, *e.g.* nanocomposites used for dental applications have unique necessities. To exemplify it further, thermoset methacrylate based composites are commonly used as dental restorative materials, because of relatively high cure efficiency by free radical polymerization and excellent aesthetic qualities. However, the demands of the oral environment and the masticatory loads encountered by dental restoratives require further property improvements in these materials. Specifically, nanocomposites with improved modulus, better efficiency of the free radical polymerization, low water sorption, improved processability, and low shrinkage are needed. Selective functionalization of the filler can lead to better interactions at the filler matrix interphase and has been used in studies in which silica nanoparticles were silanized to varying extents using two different modifiers and then mixed with a dimethacrylate resin. A few practical advantages of the dual silanization are improved workability of the composite paste, higher filler loadings leading to better modulus values, and nanocomposites with lower polymer shrinkage [177, 178]. The design of cardiovascular interfaces necessitates a combination of amphiphilicity and antithrombogenicity. Amphiphilicity ensures an optimal endothelial cell response at the vascular interface. Thrombogenicity refers to blood clot formation and can lead to early graft occlusion. Using reports of Polyhedral Oligomeric Silsequioxanes (POSS) acting as an amphiphile at the water air interface, researchers have explored the possibility of using POSS at the vascular interface. The strong intermolecular forces between constituent molecules and neighbors and the robust framework of shorter bond lengths make POSS nanocomposites more resistant to degradation. Initial work has shown that POSS nanocomposites are cytocompatible, making them potentially suitable for tissue engineering. Future efforts in this field are directed at assessing the thrombogenic potential of these nanocomposites, which would be critical in their application as cardiovascular interfaces [179, 180].

3.2. Metal and Ceramic/Bioceramic Materials

Ceramics are nonmetallic, inorganic materials typically formed as powders and often sintered into useful forms. They have a wide range of structures from amorphous to polycrystalline or even single crystals. They are typically very hard but rigid, and tend to break under mechanical stress. Since they are nonmetallic, they are usually electrically nonconductive or wide band-gap semiconductors but can be doped to decrease the band-gap to the semiconductor regime. They are often noted for their thermal stability at high temperatures [181]. Silicon carbide, for example, has excellent high-temperature characteristics but due to its brittleness has limited utility as a structural material. A micron thin coating of alumina (Al$_2$O$_3$) on SiC platelets gives a composite, which is 2-3 times stronger than traditional SiC ceramics. The result is a stress and oxidation resistant material for high temperature structural applications. Silicon carbide is also used to enhance the properties of other structural materials. Titanium composites inlaid with SiC monofilaments yield a stiff, strong, thermally stable structural material that can be used in aircraft components [182].

Alumina is a common and versatile structural ceramic due to its chemical, electrical, mechanical, and thermal properties. It possesses poor fracture resistance. The addition of yttrium stabilized zirconia to alumina significantly improves the fracture toughness of Al$_2$O$_3$. Engineering polymer is materials with exceptional mechanical properties such as stiffness, toughness, and low creep that make them valuable in the manufacture of structural products like gears, bearing, electronic devices, and auto parts. Typical engineering plastics include acetals, polyamides, poly(amide-imide)s, polyarylates, poly(ether-etherketone)s, poly(ether-imides)s, poly(phenylene oxide)s, poly(phenylene sulfide)s, and polysulfones [183, 184].

Polyamides are crystalline and have good impact strength, toughness, and abrasion resistance. Polyesters are often used with fillers like fiberglass, mica, and minerals to increase strength and stiffness. Sulfone containing polymers show high resistance to acids and alkalis. These thermally stable polymers are used in electronic connectors, circuit boards, sterilizable items, and appliance covers. Polyimides have outstanding thermal properties. Additionally, thermoplastics, thermosets, and rubbers have significant applications as engineering polymers [185,186].

Bioceramics can have structural functions as joint or tissue replacements, used as coatings to improve the biocompatibility of metal implants, function as resorbable lattices, which provide temporary structures, and a framework that is dissolved, replaced as the body rebuilds tissue. The thermal and chemical stability of ceramics, their high strength, wear resistance and durability all contribute in making the ceramics a good candidate for surgical implants. Some ceramics even feature drug delivery capability. Materials used for surgical implants, and medical devices are non toxic, bioinert, non interactive with biological systems, and durable that can undergo interfacial interactions with surrounding tissues; biodegradable, soluble, or resorbable (eventually replaced or incorporated into tissue) [187]. Due to their strength, flexibility, and biocompatibility, titanium alloys are often used for joint and bone implants. A chromium and cobalt alloy, as well as stainless steel is also, though somewhat less commonly, used for bone implants for similar reasons. Shape memory alloys have biomedical applications in angioplasty where they can prevent blood vessels from becoming reblocked. Nanocrystalline titanium powders used for bone implants are biocompatible, stronger and have the possibility of being used as hollow hip implants that more closely match the natural bone. The ceramic coatings that mimic the texture and appearance of natural teeth coated over metal supports provide a tool for prosthodonic tooth replacement. Ceramic coated biocompatible metals seem to offer an excellent compromise between the strength and flexibility of metals and the ability of ceramics to be incorporated into biological systems [188].

Much more work has been done on the interfacial reactions of biological systems with hydroxyapatite, a ceramic with chemical structure similar to the hard structure of bone. It is used as a coating for metal surgical implants. Ceramics are in a number of forms and compositions like stable SiO$_2$-Na$_2$O-CaO (SiO$_2$ content of 65 wt % or more). Although these 65 wt % silica glasses are extremely bioinert, weak, shatter easily *i.e.* brittle. The addition of P$_2$O$_5$ to the SiO$_2$-Na$_2$O-CaO matrix makes the glass extremely bioactive. Alumina and zirconia are among the bioinert ceramics that are used for prosthetic devices. Porous ceramics such as calcium phosphate based materials are used for filling bone defects. The ability to control porosity and solubility of some ceramic materials offers the possibility of use as drug delivery systems [189].

Devices that are to be used within the body must be able to withstand corrosion in a biological environment and endure use for years without undue wear (and without causing damage to surrounding tissues). The biomaterials discipline itself evolves the startling advances in genomics and proteomics in the last few years, in various high throughput cells processing techniques, in supramolecular and permutational chemistry, and in information technology and bioinformatics promise to support the quest for new materials with powerful analytic tools and insights of boundless energy and sophistication [190].

3.3. Carbon Nanotubes /Fullerenes

The discovery of fullerines in 1985 by Curl, Kroto, and Smalley, culminated in their Nobel Prize in 1996. Fullerines, or Buckminsterfullerenes, are named after Buckminster Fuller the architect and designer of the geodesic dome and are sometimes called bucky balls. The names derive from the basic shape that defines fullerines; an elongated sphere of carbon atoms formed by interconnecting six-member rings and twelve isolated five-member rings forming hexagonal and pentagonal faces. The first isolated and characterized fullerine, C_{60}, contains 20 hexagonal faces and 12 pentagonal faces just like a soccer ball and possess perfect icosahedral symmetry [191]. Magnetic nanoparticles (nanomagnetic materials) show great potential for high density magnetic storage media. Recent work has shown that C_{60} dispersed into ferromagnetic materials such as iron, cobalt, or cobalt iron alloy can form thin films with promising magnetic properties. A number of organometallic fullerine compounds have been synthesized, for example, a ferrocene-like C_{60} derivative and pair of fullerines bridged by a rhodium cluster. Carbon Nanotubes (CNTs) are hollow cylinders of carbon atoms. Their appearance is that of rolled tubes of graphite such that their walls are hexagonal carbon rings and are often formed in large bundles. The ends of CNTs are domed structures of six-membered rings capped by a five-membered ring. Generally speaking, there are two types of CNTs: Single Walled Carbon Nanotubes (SWNTs) and Multi Walled Carbon Nanotubes (MWNTs). As their names imply, SWNTs consist of a single, cylindrical graphene layer, whereas MWNTs consist of multiple graphene layers telescoped about one another [192-197]. Fig. **6** shows carbon allotropes and fullerines, various carbon wall nanotubes with closed graphene sheet structure.

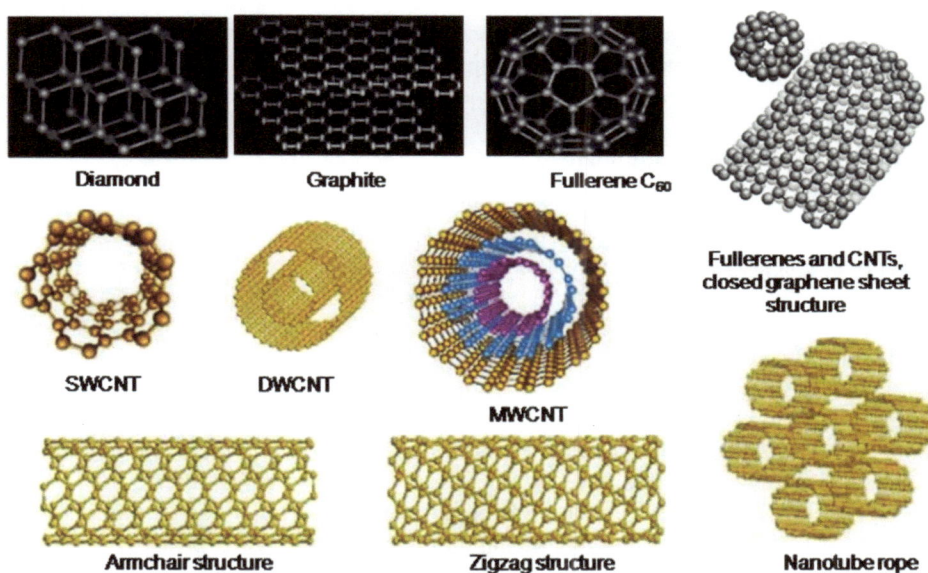

Figure 6. Carbon allotropes and fullerines, various carbon wall nanotubes with closed graphene sheet structure.

Carbon Nanotubes (CNTs) were first isolated and characterized by Ijima in 1991 [198]. The unique physical and chemical properties of CNTs, such as structural rigidity, flexiblity, strongness (about 100 times stronger, *i.e.* stress resistant than steel at an one-sixth the weight). CNTs can also act as either conductors or semiconductors depending on their chirality, possess an intrinsic superconductivity, are ideal thermal conductors, and can also behave as field emitters [199-202].

Carbon Nanotubes (CNTs) are exceptionally multifaceted nanomaterials with a wide range of applications such as electrodes, power cables, fibers, composites, actuators, sensors, biosensors, in the field emission-based flat panel displays, novel semiconducting devices, molecular electronics or computers and many other devices. CNTs can have metallic or variable semiconducting properties with energy gaps ranging from a few meV to a few tenths of an eV. Conductivity of single nanotubes has rectification effects for some nanotubes and ohmic conductance for others [203]. The individual CNT bundles within the brush-like end act as multi nanoelectrodes that facilitate the efficient capture and promotion of electron transfer reactions, as well as an increase the electroactive surface area for enzyme immobilization [204]. Owing to its small size, high electrochemical activity, excellent physical properties, low density and biocompatibility. The CNT fiber has a huge potential for implantable applications for continuous monitoring of clinically relevant analytes, including glucose (to aid the control of diabetes), lactate, antibodies and antigens. Other areas of interest include the analysis of analytes in bioreactors, veterinary and clinical chemistry, the food industry and environmental science [205].

Still, the carbon based nanomaterials are currently one of the most attractive nanomaterials with their different forms, such as fullerines, single and multiple walled carbon nanotubes, carbon nanoparticles, nanofibers, and so forth. Although carbon nanotubes are less toxic than carbon fibers and nanoparticles, the toxicity of carbon nanotubes increases significantly when a carbonyl (C=O), carboxyl (COOH), and/or hydroxyl (OH) groups are present on their surface [206]. The carbon based nanomaterials used into living systems have opened the way for the investigation of their potential applications in an emerging field of nanomedicine. A wide variety of different nanomaterials based on allotropic forms of carbon, such as nanotubes, nanohorns and nanodiamonds, are currently being explored towards different biomedical applications. The characteristics, the advantages, the drawbacks, the benefits and the risks associated with these novel biocompatible forms of carbon are very crucial. Especially, Carbon Based Nanomaterials (CBNs) are currently considered to be one of the key elements in nanotechnology. Thus, it is primordial to know the health hazards related to their exposure. Carbon based nanomaterials, due to their numerous and wide range applications and increasing real life usage, get nano toxicological attention. Toxicity of carbon based nanomaterials (nanotubes, nanofibers and nanowires) as a function of their aspect ratio and surface chemistry indicates that these materials are toxic while the hazardous effect is the size dependent [207].

3.4. Inorganic-Organic Hybrid Nanoparticles

Hybrid inorganic-organic composites are an emerging class of new materials that hold significant promise. Materials are being designed with the good physical properties of ceramics and the excellent choice of functional group chemical reactivity associated with organic chemistry. New silicon containing organic polymers, in general, and polysilsesquioxanes, in particular, have generated a great deal of interest because of their potential replacement for and compatibility with currently employed, silicon based inorganics in the electronics, photonics, and other materials technologies. Hydrolytic condensation of trifunctional silanes yields network polymers or polyhedral clusters having the generic formula $(RSiO_{1.5})_n$. Hence, they are known by the not quite on the tip of the tongue name silsesquioxanes. Each silicon atom is bound to an average of one and a half (sesqui) oxygen atoms and to one hydrocarbon group (ane). Typical functional groups that may be hydrolyzed or condensed include alkoxy or chlorosilanes, silanols, and silanolates [208-211]. Synthetic methodologies that combine pH control of hydrolysis/condensation kinetics, surfactant mediated polymer growth, and molecular templating mechanisms have been employed to control molecular scale regularity as well as external morphology in the resulting inorganic/organic hybrids from transparent nanocomposites, to mesoporous networks, to highly porous and periodic organosilica crystallites, all of which have the silsesquioxane (or $RSiO_{1.5}$) stoichiometry. These inorganic-organic hybrids offer a unique set of physical, chemical, and size dependent properties that could not be realized from just ceramics or organic polymers alone. Silsesquioxanes are therefore depicted as bridging the property space between these two component classes of materials. Many of these silsesquioxane hybrid materials also exhibit an enhancement in properties such as solubility, thermal and thermo mechanical stability, toughness, optical transparency, gas permeability, dielectric constant, and fire retardancy, *e.g.* beryllium silsesquioxane, poly (hydridosilsesquioxane) and polyhedral oligomeric silsesquioxane etc [212].

Nanostructured films, dispersions, large surface area materials, and supramolecular assemblies are the high utility intermediates to many products with improved properties such as solar cells and batteries, sensors, catalysts, coatings, and drug delivery systems. They have been fabricated using various techniques. Nanoparticles are obvious building blocks of nanosystems but require special techniques such as the self-assembly to properly align the nanoparticles. Recent developments have led to air resistant, room temperature systems for nanotemplates with features as small as 67 nm. More traditionally, electron beam systems are used to fabricate devices down to 40 nm [213].

3.5. Dendrimers: High Performance Nanostructures

Dendrimers are like organic nanoparticles, *i.e.* a new structural class of organic polymer macromolecules, nanometer sized, hyper branched materials having compact hydrodynamic volumes in solution and high, surface, functional group content. They may be water soluble but due to their compact dimensions, they do not show usual rheological thickening properties like many polymers in solution [214]. Dendrimers are defined by their three components: a central core, an interior dendritic structure (the branches), and an exterior surface (the end groups). Fig. **7** shows the structure of dendrimer.

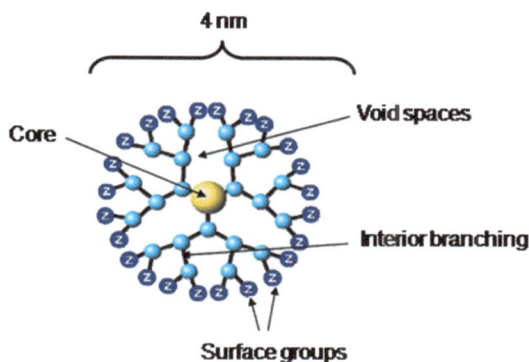

Figure 7. Structure of the dendrimer.

The characteristic features of dendrimers are nearly spherical structures, nanometer sizes, large numbers of reactive end group functionalities, shielded interior voids, and low systemic toxicity. These unique combinations of properties make them ideal candidates for potential nanotechnology applications in both biological and material sciences including a broad range of the fields, like materials engineering, industrial, pharmaceutical, biomedical, nanoscale catalysts, novel lithographic materials, rheology modifiers, targeted drug delivery systems, MRI contrast agents, and bioadhesives [215- 217].

Dendrimers's architectural component manifests a specific function and property, as they are grown generation by generation. The core may be thought of as the molecular information center from which size, shape, directionality, and multiplicity are expressed *via* the covalent connectivity to the outer shells. Within the interior, the branch cell amplification region, which defines the type and volume of interior void space that may be enclosed by the terminal groups as the dendrimer is grown. Branch cell multiplicity (Nb) determines the density and degree of amplification as an exponential function of generation (G). The interior composition and volume of solvent filled void space determine the extent and nature of guest host (endoreceptor) properties that are possible within a particular dendrimer family and generation. Finally, the surface consists of reactive or passive terminal groups that may perform several functions. With appropriate functionalization, they serve as a template polymerization region as each generation is amplified and covalently attached to the precursor generation. The surface groups may also function as passive or reactive gates controlling entry or departure of guest molecules from the dendrimer interior. These three architectural components (core, interior, and periphery) essentially determine the physical and chemical properties, as well as the overall size, shape, and flexibility of a dendrimer. The dendrimer diameters increase linearly as a function of shells or generations added, whereas the terminal functional groups increase exponentially as a function of generation. The lower generations are generally open, floppy structures, whereas higher generations become robust, less deformable spheroids, ellipsoids, or cylinders depending on the shape and directionality of the core [218-220].

Based on their systematic, size-scaling properties and electrophoretic and hydrodynamic behavior, they are referred to as artificial globular proteins. An overview of the hierarchical complexity that leads to precise, controlled nanostructures clearly illustrates the importance of quantized building blocks for viable bottom-up synthetic strategies. Fig. **8** shows various nanomolecules at nanoscales.

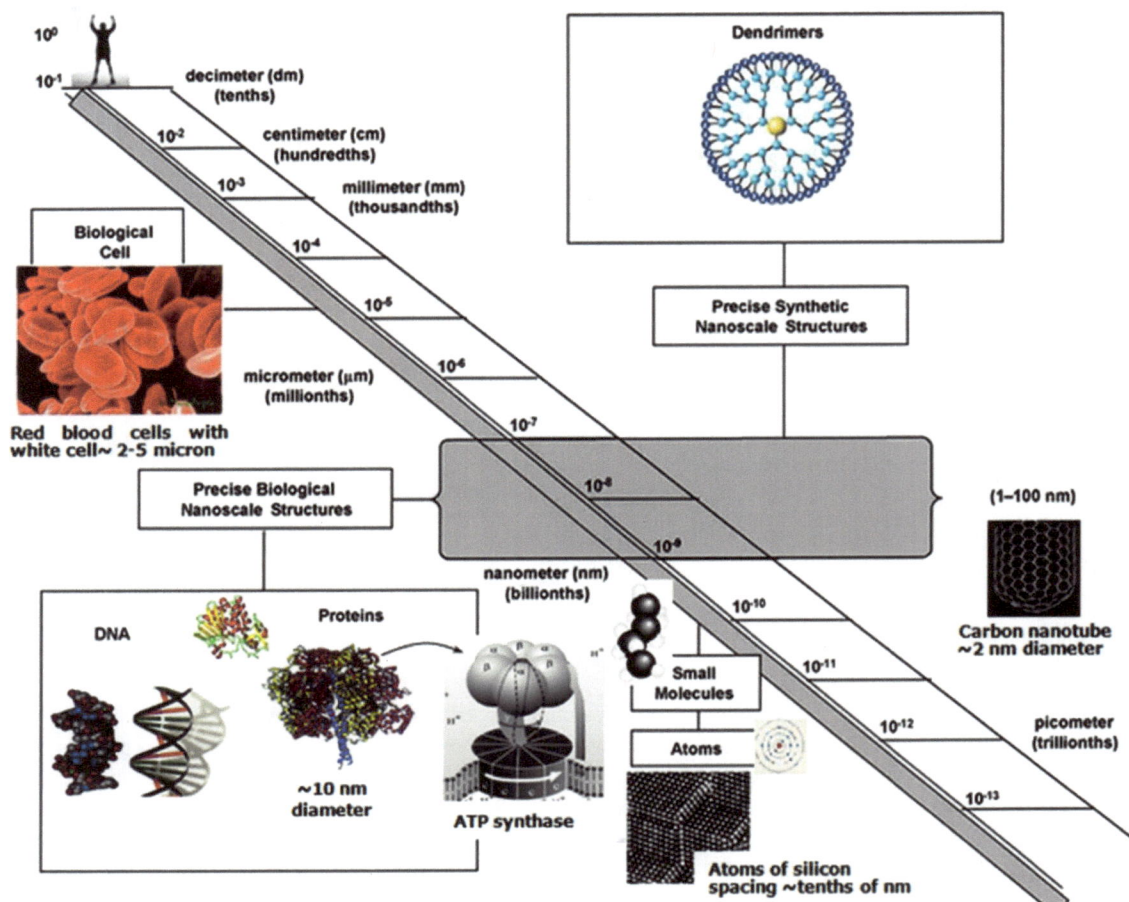

Figure 8. Various nanomolecules at nanoscales.

The future use of dendrimers as fundamental, reactive building blocks is expected to provide the enabling platform required for the routine synthesis of broad classes of well-defined synthetic organic, inorganic, and hybridized biomolecular nanostructures. The controlled shape, size, and differentiated functionality of dendrimers; their ability to provide both isotropic and anisotropic assemblies; their compatibility with many other nanoscale building blocks such as DNA, metal nanocrystals, and nanotubes; their potential for ordered self-assembly; ability to combine both organic and inorganic components; and their propensity to either encapsulated or be engineered into unimolecular functional devices make dendrimers uniquely versatile at existing nanoscale building blocks and materials. Dendrimers are spheroid or globular nanostructures, precisely engineered to carry molecules encapsulated in their interior void spaces or attached to the surface. Generation (shells) and chemical composition of the core, interior branching, and surface functionalities determine size, shape, and reactivity. Dendrimers are constructed through a set of repeating chemical synthesis procedures that build up from the molecular level to the nanoscale region. The dendrimer diameter increases linearly whereas the number of surface groups increases geometrically. Dendrimers are uniform with extremely low polydispersities, and are commonly created with dimensions incrementally grown in approximately nanometer steps from 1 to over 10 nm. The control over size, shape, and surface functionality makes dendrimers one of the smartest or customizable nanotechnologies commercially available. Dendrimers provide polyvalent interactions between surfaces and bulk materials for applications such as adhesives, surface coatings, or polymer cross-linking [221-226].

Dendrimers resemble covalent micelles characterized by well defined cavities responsible for their particular endoreceptor properties and terminal groups that define the solubility, the reactivity, and the exoreceptor characteristics of the molecule. Due to their intrinsic and exciting properties, dendrimers have been studied as an important new class of potential drug delivery system, protein models, molecular antennas, and catalytic materials. The dendrimers can also be used to modify an electrode surface due to their good biocompatibility and adequate functional groups for chemical fixation. It was reported that these materials are capable of increasing the concentration of hydrophobic molecules at the electrode solution interface, improving the sensitivity as well as the selectivity of certain specific electrochemical reactions [227-231].

Watanabe and Regen used an electrostatic layer-by-layer deposition technique to fabricate self- assembled films from alternating molecular layers of oppositely charged PAMAM dendrimers and the low molar mass compound K_2-$PtCl_4$ [232]. Tsukruk *et al.* Systematically studied assembled films of dendrimers in monolayers or multilayers on a solid surface [233-235]. Recently, Crooks and co-workers published a short communication reporting individual G4 and G8 PAMAM dendrimers adsorbed on the Au (111) substrate visualized by AFM. [236] Li *et al.* obtained AFM images of PAMAM dendrimers ranging from G5 to G10 ethylenediamine (EDA). The fourth generation polyamidoamine (PAMAM) polymers are of a particular interest because of its nanoscopic spherical structure and good biocompatibility. PAMAM dendrimer possesses 64 primary amine groups on the surface and has a globular shape with a diameter of about 4.5 nm. The mixture of proteins and PAMAM was cast on the electrode surface, forming protein-PAMAM films, and the direct electrochemistry of proteins was realized in the PAMAM films environment. [237-239] Kim and co-workers developed a glucose biosensor by alternate depositions of periodate-oxidized glucose oxidase (GOX) and PAMAM or ferrocenyl-tethered PAMAM. They also constructed a biosensor in which PAMAM functionalized with ferrocenyl and biotin analogues were assembled layer-by-layer on the gold electrode. There were also some works reporting that PAMAM dendrimers form Self-Assembled Monolayers (SAM) on a gold support for development biosensors. [240] Hianik reported the suitability of the polyamidoamine (PAMAM) dendrimer G1 for the development of the glucose biosensor. G1 was adsorbed on a gold support together with hexadecanethiol, and the properties of the sensor depending on the method of immobilization of GOx by cross linking with glutaraldehyde were studied. He has also been reported that an enzyme biosensor based on acetylcholinesterase (AChE) and Choline Oxidase (ChO) immobilized on the supported monomolecular layer composed of poly(amidoamine) (PAMAM) dendrimers of the fourth generation (G4) mixed with 1-hexadecanethiol (HDT) [241, 242]. Zhu *et al.* studied the suitability of the fourth-generation (G4) PAMAM dendrimers for the development of the DNA biosensor using self-assembly and covalent immobilization technique. Due to the presence of spherical G4 PAMAM on the electrode, the surface area and the density of amino groups on the electrode increased obviously, and this would result in the increase of the immobilized DNA probe. Experiments carried out with these novel materials not only showed an improved DNA attachment quantity on the dendrimers modified electrodes when compared to DNA sensors with oligonucleotides directly immobilized on Au electrodes, but also exhibited a high selectivity, sensitivity and stability for the measurement of DNA hybridization [218].

Immobilization of polyamidoamine (PAMAM) dendrimers, star-like structures, onto solid surfaces has been involved in a variety of applied and basic sciences [243]. Several immobilization strategies have been developed for modification of solid surfaces with dendrimers including chemical and physical methods, such as covalent binding and Layer-by-Layer (LbL) assembly techniques. [244, 245] Recently, Crooks group reported a direct immobilization of hydroxyl terminated PAMAM dendrimers on Glassy Carbon Electrodes (GCEs). They utilized multiple hydroxyl terminals of the PAMAM dendrimers for oxidative coupling of the dendrimers on GCEs *via* ether bond [246-249].

4. MICRO AND NANOELECTRONICS

As the electronic industry continues to move towards smaller and smaller tools and devices, researchers must continually strive for technologies to deposit electronic materials with precision at ever-smaller scales. This trend has led to a variety of techniques used in the microelectronics industry to layer materials with precision thickness, sometimes down to even the atomic level. During the manufacturing silicon semiconductor devices, layers of barrier material are deposited to separate interconnecting metal from silicon and prevent their diffusion. Transition metal nitrides (TaN, TiN, and NbN) or metal nitride-silicates are widely used barrier materials in ultra-large scale integration of microelectronic devices between copper/aluminum and silicon [250]. They prevent efficient diffusion of metals and silicon through the barrier layer, not react with the interconnecting metal or silicon, are stable during the manufacturing and operation of the device, have a low resistivity, good adhesion to the underlying material. Titanium nitride (TiN) has widely been studied barrier material with several limitations and will not meet future demanding requirements for device integrity. Recently, Tantalum nitride has been received extensive interest as a barrier material along with other materials like WN_x, $TiSi_xN_y$, WSi_xN_y, and WB_xN_y [251]. Tantalum nitride provides superior physical properties in comparison to titanium nitride. It has a high melting point, very hard, highly conductive, and thermodynamically stable with respect to Cu because it does not form copper-tantalum or copper-nitride compounds [252]. Several different techniques are used for deposition of barrier layers, Chemical Vapor Deposition (CVD) method is a main technique currently used for deposition of metal nitride layers. The films made by CVD, exhibit much better conformity than the films deposited by physical vapor deposition methods [253]. The other variant of the deposition process, Atomic Layer Deposition (ALD) and its several enhancement, allows very good control of layer thickness and uniformity, which provides a good step coverage, and relatively low deposition temperature. It is worthy to notice that thin-film transition metal nitrides have a potential in other applications besides diffusion barriers. In microelectronics, TaN, for example, has been used as a thin-film resistor and acts as a passivation layer against copper oxidation. It has been considered as an electrode material for those Dynamic Random Access Memories (DRAM) storage capacitors where tantalum (V) oxide (Ta_2O_5,) is used as a dielectric material [254]. Tantalum nitride has also been investigated as a nonmagnetic interlayer in NiFeCo/TaN/NiFeCo nonvolatile magnetic random access memories [255]. In addition to microelectronics, TaN films have also been examined for high-temperature ceramic pressure sensors due to its good stability and piezoresistive properties [256]. Chemical precursors were used in barrier deposition processes, including several classes of materials such as metals, metal halides, metal carbonyls, and metal amides [257].

As integrated circuit design rules shrink, the elements of the circuit must be miniaturized. This relentless miniaturization has led to enhance chip performance in terms of processing speed and expanded memory capacities at diminishing unit costs. The limitation of silica (SiO_2) is in its ability to behave as an electrical insulator (dielectric barrier). The dielectric strength of a material is expressed in terms of its dielectric constant, κ (for SiO_2, $\kappa \approx 4$ and for SiO_xN_y, $\kappa = 5$-6). Electrical current will leak across the dielectric, and the capacitor will discharge. In an effort to remove this obstacle to miniaturization, researchers in semiconductor device architecture and manufacturing have considered alternative materials featuring high κ values for such critical dielectric films in both the gate electrodes and in DRAM (Dynamic Random Access Memory) storage capacitors, few examples are transition metal oxides and their silicates (eg., titanium, zirconium and hafnium oxides, TiO_2, $\kappa = 60$, ZrO_2, $\kappa = 23$; HfO_2, $\kappa = 20$ or tantalum oxide, Ta_2O_5, $\kappa = 25$), multi-metallic oxides (*e.g.,* Barium Strontium Titanate, BST, $\kappa = 300$), *etc.* The properties of any new high κ material include high permittivity or dielectric constant, barrier properties to prevent tunneling, stability in direct contact with silicon, having good film interface quality. The dielectric materials are able to demonstrate its compatibility with the gate material, process compatibility, and high reliability [258, 259].

The absolute limit for oxide/nitride films appears to be 7 Å, but 12 Å may represent the practical limit, since effects of gate leakage or reliability will likely prevent further improvement in device performance. The replacement of SiO_2 with a high permittivity (high-κ) gate dielectric (*e.g.* $HfSi_xO_y$, $ZrSi_xO_y$, Si_3N_4) presents a formidable challenge to the semiconductor industry. To effectively replace silica, a high κ gate dielectric must exhibit low leakage current with good capacitance, able to be integrated into a full process flow, showed sufficient reliability at operating conditions. Zirconium and hafnium oxides were emerged as a possible candidate for gate materials. According to the ITRS (International Technology Roadmap for Semiconductor) data the best candidate for gate material should possess κ in the 10-25 range. Material for memory capacitors should have been much higher κ value. In order to enable the development of this critical new technology, the semiconductor industry needs a source for a large selection of suitable molecular precursors (such as early transition metals and lanthanide alkyls, amides, alkoxides, β-diketonates, and nitrates) [260-262].

The robust route toward the synthesis of highly effective luminescent semiconductor nanocrystals with sizes ranging from 1.5-8 nm has been established. By controlling the particle growth of the II-VI semiconductors the size, crystal structure and lattice parameters of the nanocrystals can be accurately tailored [263-266].The ability to control these parameters has a profound impact in material properties, engineering assemblies of nanometer scale units with novel characteristics [267-270]. Quantum dots of II-VI semiconductors have, in particular, gained recent attention because of their ease to synthesize in the size range that leads to quantum confinement of the semiconductor nanocrystals. These materials are of potential interest in non-linear optics and in fast optical switching. Nanocrystalline CdS, CdTe, CdSe, ZnSe and PbS have been synthesized by a variety of methods, including precipitation, sputtering, electrochemical deposition and inverse micelles [271]. A reduction in the particle size strongly influences the crystallinity, melting point and structural stability. Semiconductor nanoparticles of II-VI are highly unstable, and in the absence of trapping media or some other form of encapsulation, they agglomerate or coalesce quickly, due to this reason bonding of capping agents to nanoparticles is necessary to provide surface passivation, and also to improve the surface state which significantly influenced the optoelectronic properties of nanoparticles [272-276].

Semiconducting material has attracted much interest owing to their unique electronic and optical properties, and their potential applications in solar energy conversion, photo conducting cells, non-linear optics and heterogeneous photo catalysis, [277-279] spintronics, photonics, optoelectronics, Light Emitting Diodes (LEDs), laser diodes, cosmetics, antibiotic material and gas sensors. II-VI semiconductor materials at nanometer scale have the potential in various applications such as piezoelectric transducers, gas sensors, light-emitting devices, photonic crystals, nanoelectronics photo detectors, photodiodes, optical waveguides, transparent conductive films, solar cells, transparent UV protection films, chemical sensors and biological (drug delivery, bio-imaging, *etc.*) systems. ZnO is an important II-VI semiconductor with a wide and direct band gap (3.37 eV), equivalent to that of GaN. ZnO has a wide range of applications in piezoelectric transducers, gas sensors, photonic crystals, light-emitting devices, photo-detectors, photodiodes, optical waveguides, transparent conductive films, varistors, solar cell windows, and transparent UV protection films, biological (drug delivery system) and chemical sensors photonics, nanoelectronics, optoelectronics and data storage [280-284]. CdSe Quantum Dots (QDs) have been studied and used as biological labels for the multiplexed biological detection, imaging and in molecular cell biology [285-301].

The development of electronics demands a continuous decrease of the element sizes. The purpose of this motif is not only increasing the integration level but also increase the operation speed. II-VI nanocrystals (NCs) are receiving considerable attention for their exploitation in appealing applications in optoelectronics and photonics, for example, optical switches, sensors, electro-luminescent devices, lasers and biomedical tags [302-306]. Nanometer size quantum dots exhibit a wide range of electrical and optical properties that depends sensitively on the size of the nanocrystals and are of both fundamental and technological interest. The control and improvement of the luminescence properties of Quantum Dots (QDs) has been a major goal in synthetic nanochemistry and related preparative procedures. Since the electronic energy levels of semiconductor nanocrystals are radically affected by their sizes, there is a large demand for reproducible material scientists to find some convenient, economical, less energy consuming and environmentally friendly synthesized routes to these semiconductor nanocrystals. The surface chemistry is an efficient tool not only to organize and immobilize the NCs but also to effectively modify the emission properties. The possible of performing the manipulation of the prepared NCs by properly engineering the surface by biomolecules enables the NCs to be placed in almost any chemical environment, being soluble in organic solvent [307-309].

Biological systems can control mineralization and synthesis of various nanocrystals in the exact shapes and sizes with high reproducibility and accuracy. Dameron *et al.* had biosynthesized CdS quantum semiconductor crystallines in yeasts *Condida glabmta* and *Schizosacharomyces pombe*. Recently, biomimetic synthesis of inorganic nanocrystals has attracted more and more attention. Biomoleculars, such as DNA, peptides, virus and bacterial rhapidosomes have been utilized for the direct deposition, assembly and patterning of iron oxide and semiconducting nanocrystals. Biomimetic syntheses of CdS nanocrystals mediated by DNA, peptides and polysaccharides have also been reported [310-317].

Molecular electronic device has advanced to overcome the limit of a current electronic device to develop molecular scale diode, organic thin film transistor, organic wire like nanowire, and organic photovoltaic cell using the organic materials. Various concepts for molecular information storage have been proposed [318-320]. Fig. **9** shows the schematic view of bioelectronic concepts and its application.

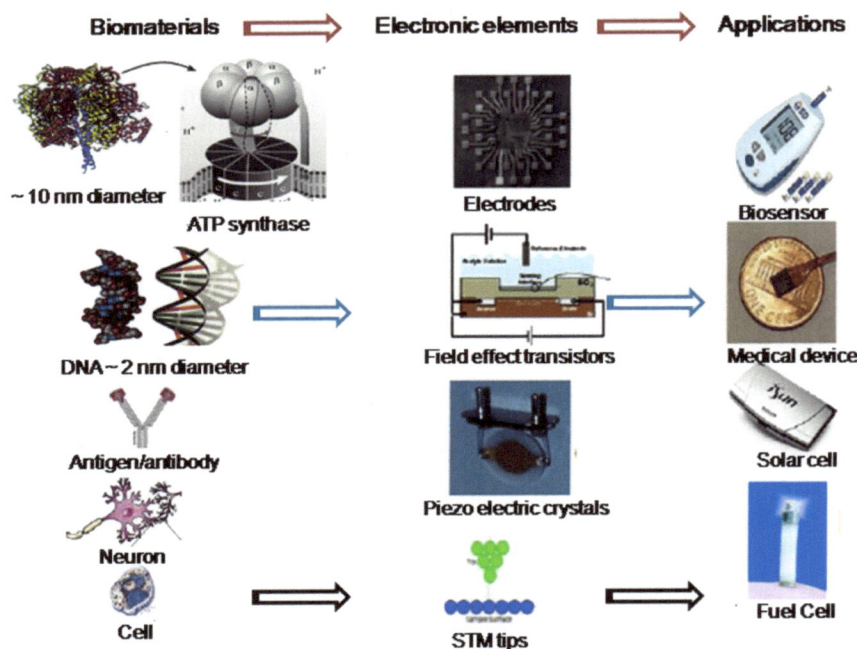

Figure 9. Schematic view of bioelectronic concepts and its application.

Hopfield *et al.* proposed the concept of shift register memory. The memory elements are based on a chain of electron transfer molecules incorporated on a very large scale integrated substrate, and the information is shifted by photo-induced electron-transfer reactions [321]. Hersam and co-workers proposed the probing charge transport at the single molecule level on silicon. It means that a single molecule can be applied to silicon based molecular electronic devices [322]. Bocian and co-workers investigated redox kinetics of the redox active molecule attached to the Si (100) surface. They demonstrated that porphyrin based molecules bound to Si (100), which exhibit redox behavior useful for information storage [323]. In the field of biomolecular electronics, many researchers have been investigated to develop the biomolecular information device. Willner and co-workers proposed the information storage logic that was composed of DNA or enzyme. The possibility of encoding information in the base sequences of DNA was shown by manipulating DNA by enzymes [324, 325]. According to *in vitro* studies, azurin is a very useful biomaterial in the field of bioelectronics. This metalloprotein with its redox property can function as an electron donor or acceptor. However, generally the protein is unstable and fragile in the solid state. Therefore, efficient film fabrication technology is strongly required in order to use the protein as a component of a bioelectronics device. Given that artificially controlling the arrangement of the biomolecule in the solid state, a nanoscale bioelectronics device can be developed with high performance. In a recent decade, Self-Assembly (SA) technique has been studied, which offers a useful method to make a thin layer onto the metal substrate for various applications, such as biosensor, Field-Effect Transistor (FET), Single Electron Transistor (SET), and so on. The most general system of self-assembly is to use alkanethiols as chemical linker. In the long hydrocarbon chains structure, one side is reacted with a solid substrate and another side is reacted with target protein. It has been found that sulfur compounds to coordinate very strongly to various metal surfaces, such as Au, Ag, Cu, and Pt. In most work till a date, a Au surface has been used for the Self-Assembly Monolayer (SAM) formation of alkanethiols, because gold can not be oxidize easily. Therefore, it can be handled in the ambient conditions. However, alkanethiols based self-assembly method has a drawback for the application to bioelectronics, because it can be a function as an insulator [326, 327].

Kim *et al.* suggested a novel immobilization technique of cupredoxin azurin on the Au surface. A recombinant azurin with cysteine residue by using Site-Directed Mutagenesis (SDM) was designed and then directly immobilized on a Au surface without chemical linker. Surface plasmon resonance (SPR) and Scanning Tunneling Microscopy (STM) confirmed the immobilization of the functionalized protein. The electrochemical property of the fabricated thin film was investigated by the Cyclic Voltammetry (CV). Thus, the direct immobilization of the recombinant azurin modified by site directed mutagenesis for the development of an efficient bioelectronics device has been reported [328-330].

There is significant interest in the development of molecular devices with molecules for their nanostructured capabilities and potential applications in nanotechnology. Nanoelectronic devices based on organic molecules, and biomolecules are currently being explored as a long term alternative to silicon based devices. Several researchers have investigated the molecular approach towards information storage, and molecular switches using biomolecules as the active component have also been developed. Redox functioning proteins for molecular electronic devices such as biosensors and biomemory devices have also been introduced with various applications [331-340]. Cysteine residue has been introduced into *Pseudomonas aeruginosa* azurin and their physiological and physiochemical properties have been extensively studied to improve the uniformity of the protein monolayer onto the gold. Therefore, azurin can be applied directly to make a new concept of memory function with biomaterial associated with electronic devices; due to the charge transfer and trap function as well as thermal and chemical stability. The basic principle in applying recombined azurin in a biomemory device is shown in Fig.**10**.

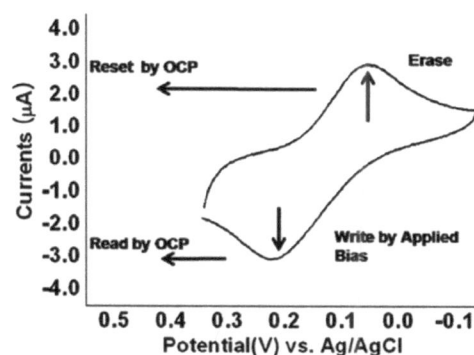

Figure 10. Schematic representation of cyclic voltammogram of azurin assembled layer on gold working electrode, which shows the memory function upon application of proper bias potentials.

The redox property of azurin is controlled upon application of an external potential to show the memory function of the developed biomemory device. When an oxidation potential is applied to the fabricated bioelectrode, the protein in the monolayer loses an electron to the inorganic base, and thereby entrapment of positive charge occurs in the protein film, as shown in Fig. **11a.** This trap process of positive charge inside the protein layer corresponds to the function of storage (writing) of information in a conventional memory device. The trapped charge (written information) in the protein film is measured (read) when an Open-Circuit Potential (OCP) is applied to the fabricated electrode. When a reduction potential is applied to the fabricated electrode after the initial step of oxidation the inorganic base gives back the electron to the protein monolayer, as shown in Fig.**11b.**

Thus, the initial charge trapped in the biofilm during oxidation, is neutralized (erasing). The sequential application of oxidation potential, OCP and reduction potential makes the developed biomemory device write, read and erase the information as a conventional inorganic memory device. However, efficient combination techniques of azurin in conventional electronic devices have not yet been reported; nevertheless, it is a key process to sustain and control the electrochemical properties of the azurin to mimic the biological memory system. Therefore, efficient immobilization of azurin onto the Au electrode is introduced in order to increase the electrochemical properties of the azurin active layer to achieve the suggested basic memory functions using a protein based biomemory device. The specific region of sequence of native azurin was modified to possess cysteine residue by using DNA recombinant technology to maximize electron transfer between a protein film and inorganic electrodes. This fabrication technique of a protein film on a gold surface was characterized by a surface plasmon resonance (SPR) technique followed by scanning tunneling microscopy (STM) to optimize the concentration and study the topography, respectively. Electrochemical measurements were carried out for the investigation of the memory function of the developed biomemory device [341].

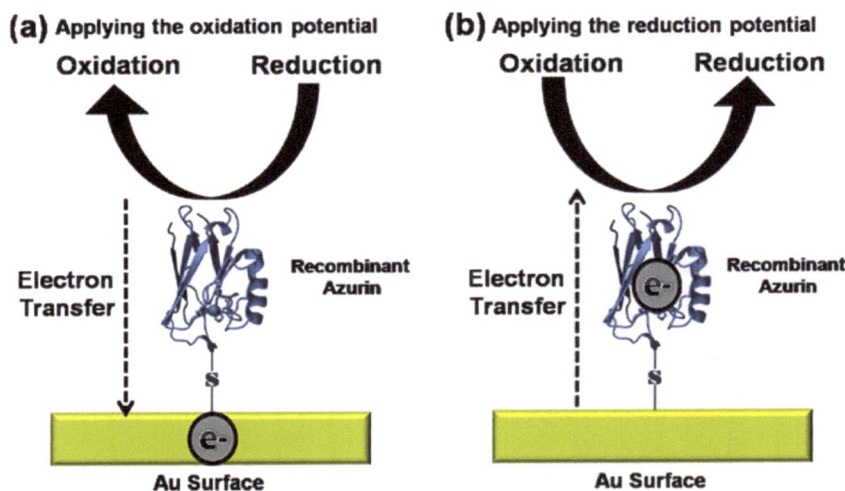

Figure 11. Schematic diagram shows an electron transfer mechanism of recombinant azurin on Au electrode. **(a)** Application of oxidation voltage causes the electron to move from protein to Au electrode. **(b)** Application of reduction voltage sends the electron back to the protein.

4.1. The Memory Function of the Protein Based Device

Faradaic currents were obtained when the working electrode received an oxidation potential of 0.247 V vs. Ag/AgCl, which caused the protein layers to become oxidized; the corresponding currents were measured as shown in Fig. **12**.

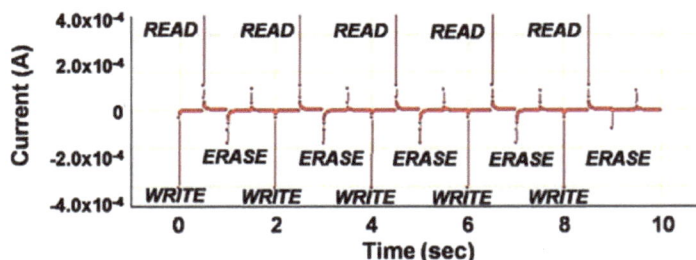

Figure 12. Memory functions of the protein-based biomemory device. Continuous pulses of oxidation (write), open-circuit (read), and reduction (erase) potentials are applied. Each has a pulse width of 20 m/s duration, and corresponding charging currents were measured for a total duration of 10 s.

Now, the electrode was left open for a small duration of time, and applied the Open Circuit Potential (OCP) of 0.2 V vs. Ag/AgCl and observed small-amplitude of currents. The charge stored in the protein molecules can be read by connecting the counter electrode at the OCP and subsequently, applying the oxidation voltage (writing step). Applying the open circuit voltage is the reading step with respect to the measurement of currents. Finally, the application of the reduction potential erases all the stored charge. On this basis, potentials were applied to the protein film for the duration of 10 s, and the corresponding charging currents were measured. We observed clear transient currents for the charge for the write, read, and erase functions that are requisite for a molecular memory storage device [342].

Molecular electronics with bio or organic molecules have been considered as a potential alternative technology that may overcome the problems conventional technologies currently face in commercial fields. To increase memory density, it would be imposed a multistate approach, wherein the charge storage element contains molecules with multiple redox states. This was developed by using a ferrocene-porphyrin conjugate that bears a single thiol tether. Recently, the blue copper azurin was demonstrated for the biomemory device. Storing and reading out the multi state information using proteins azurin with a simple fabrication method for the electronic device has not yet been reported. Blue copper azurin (about 14.6 kDa) is responsible for electron transfer in the respiratory system of several bacteria. The two anchoring sites of recombinant azurin containing cysteine residues derived from *Pseudomonas aeruginosa*, improve the uniformity of the immobilized protein monolayer on the gold surface when compared with the wild type form of the azurin (Fig. **13a)**. The wild type azurine was adsorbed onto the gold surface *via* its exposed disulfide moiety was an inconsistent and not uniform manner (Cys3-Cys26).

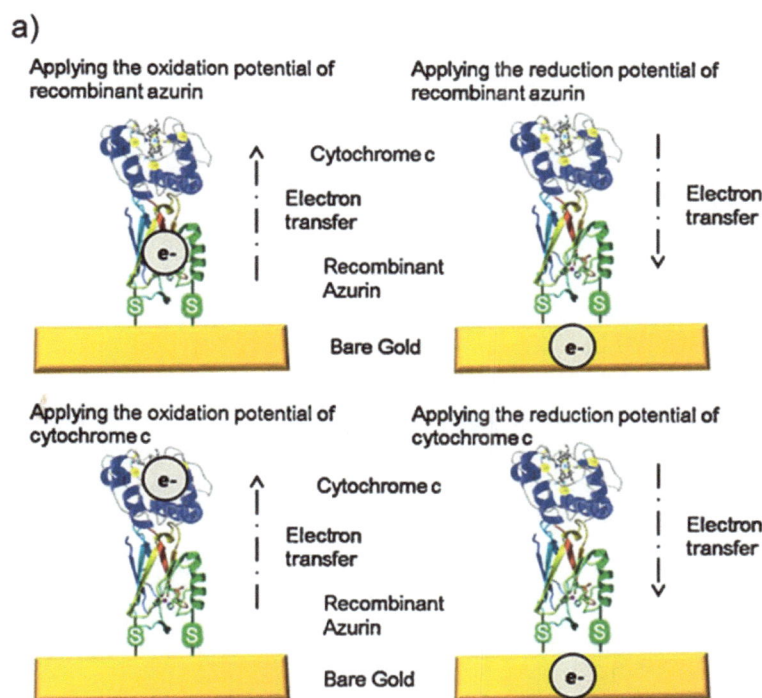

Figure 13a. The electron-transfer mechanism of a cytochrome c/recombinant azurin layer on a Au surface.

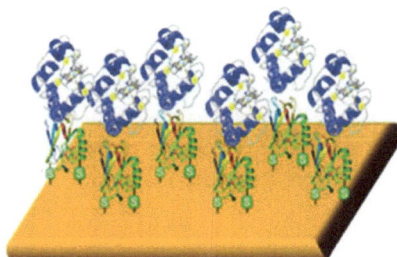

Figure 13b. Schematic diagram of protein immobilization, recombinant azurin immobilized onto gold surface, and then, cytochrome c adsorbed on the self-assembled azurin layer by electrostatic interactions.

Since azurin contains copper, which is a key element in the electron transfer mechanisms, it can be used as an electron acceptor in the development of molecular electronic devices. Therefore, due to its charge transfer and trap function as well as thermal and chemical stability, azurin was directly applied in the development of a novel biomemory device. Recombinant azurin containing cysteine residues was produced by site-directed mutagenesis and immobilized directly onto an Au surface; then, cytochrome c was adsorbed onto the immobilized azurin layer by electrostatic bonding (Fig. **13b**).

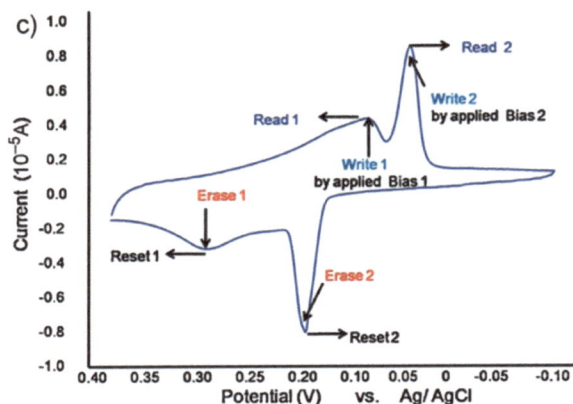

Figure 13c. Schematic representation of a cyclic voltammogram for heterolayers that consist of recombinant azurin/cytochrome c on a gold surface.

This direct immobilization enhanced the efficiency of electron transfer between the azurin layer and an Au surface and then between the cytochrome c layer and an Au surface due to the well-oriented protein structure, which resulted from using the Au-S bond without any chemical linker. This heterolayer consisting of recombinant azurin and cytochrome c, was capable of storing the two pairs of information, which is referred to as the multistate memory as shown in the Fig. **13c**. [243-344].

5. CONCLUSIONS

The current trends and challenges for smart nanomaterials for various applications are the focus of this chapter. Nanoscale materials have assumed importance in areas of cancer diagnostics, detection of pathogenic organisms, food safety, environmental measurements and clinical applications. Microfluidic technology consisting of microfluidic mixer, valves, pumps, channels, chambers in a single chip device have been commercialized. Aptamer-based microarrays for the quantitation of multiple protein analytes as well as a metal-enhanced electrochemical detection concept had been demonstrated for the sensitive detection of Ab-Ag, DNA-DNA, DNA-drug and DNA-toxin interactions. The key issues to be addressed in the future is the increasing demand for higher sensitivity and selectivity that will allow molecules to be monitored in real-time at minimal cost in the analysis of complex clinical and environmental samples. In addition, this book chapter comprehensively surveys the past, present and future of fast expanding and burgeoning research activities, the impact of nanotechnology on biosensors, biochips and micro/nanoelectronics including biomemory. It backed up by the rapid strides of nanotechnology, these research are making a progress in multidisciplinary research, which are beginning to make their way to the market place. Focusing on the salient developments with the viewpoints of novel smart materials, device structures and functionalities introduced, the chapter highlights the significant milestones achieved and elucidate further the emerging future prospects in this area.

ACKNOWLEDGEMENTS

R.P.Singh, A.T., J.W.Choi and A.C.Panday. contributed equally for this work and also the work was supported by Nano/Bioscience & Technology Program (M10536090001-05N3609-00110) of the Ministry of Education, Science and Technology (MEST), Korea Science and Engineering Foundation (KOSEF) grant of Korean government (M10644080003-06N4408-00310), Ministry of Knowledge Economy (MKE) & Korea Industrial Technology Foundation (KOTEF) through the Human Resource Training Project for Strategic Technology, South Korea and Nano-Mission (SR/NM/NS-87/2008), Department of Science and Technology, Government of India.

REFERENCES

[1]   Feynman, R.P. There's plenty of room at the bottom. *Eng. Sci.*, **1960**, *23*, 22-36.

[2]   Drexler, K.E. Molecular engineering: An approach to the development of general capabilities for molecular manipulation. *Proc. Natl. Acad. Sci.*, **1981**, *78*, 5275-5278.

[3]   Sailor, M. J.; Link, J. R. Smart dust: nanostructured devices in a grain of sand. *Chem. Commun.*, **2005**, *21*, 1375-1383.

[4]   Rothemund, P. W. K. Folding DNA to create nanoscale shapes and patterns. *Nature*, **2006**, *440*, 297-302.

[5]   Yurke, B.; Turberfield, A. J.; Mills, A. P. J.; Simmel, F. C.; Neumann, J. L. A DNA-fuelled molecular machine made of DNA. *Nature*, **2001**, *406*, 605-608.

[6]   Choi, J. W.; Kim Y. J.; Kim, S. U.; Min, J.; Oh, B. K. The fabrication of functional biosurface composed of iron storage protein, Ferritin. *Ultramicroscopy*, **2008**, *108*, 356-359.

[7]   Park, S. J.; Taton, T. A.; Mirkin C. A. Array-based electrical detection of DNA with nanoparticle probes. *Science*, **2002**, *295*, 1503-1506.

[8]   Seeman, N. C. Nanomaterials based on DNA. *Ann. Rev. Biochem.*, **2010**, *79*, 65-87.

[9]   Seeman, N. C. DNA Nanotechnology: Novel DNA constructions. *Ann. Rev. Biophys. Biomol. Struct.*, **1998**, *27*, 225-248.

[10]   Singh, R. P.; Choi, J. W. Bio-nanomaterials for versatile bio-molecules detection technology. *Adv. Mater. Letts.*, **2010**, *1*, 83-84.

[11]   Wang, R.; Chen,Y.; Fu.Y.; Zhang, H.; Kisielowski, C. Bicrystalline hematite nanowires. *J. Phys. Chem. B*, **2005**, *109*, 12245-12249.

[12]   Tenga, X.; Yang, H. Effects of surfactants and synthetic conditions on the sizes and self-assembly of monodisperse iron oxide nanoparticles. *J. Mater. Chem.*, **2004**, *14*, 774-779.

[13]   Chauhan, P.; Annapoorni, S.; Trikha, S. K. Humidity-sensing properties of nanocrystalline haematite thin films prepared by sol-gel processing. *Thin Solid Films*, **1999**, *346*, 266-268.

[14]   Bate, G.; Wohlforth, E. D. *Ferromagnetic Material*, North-Holland, Amsterdam; The Netherlands, **1980**.

[15]   Cornell, R. M, Schwertmann, U. *The iron oxides: Structure, Properties, Reactions, Occurence and Uses,* VCH, Weinhein, **1996**.

[16]   Lindgren, T.; Wang, H.; Beermann, N.; Vayssieres, L.; Hagfeldt, A.; Lindquist, S. E. Aqueous photoelectrochemistry of hematite nanorod array. *Solar Energy Mater Solar Cells*, **2002**, *71*, 231-243.

[17]   Monica, S. A.; Grabias, D.; Tarabasanu, M.; Diamandescu, L.; Influence of cobalt and nickel substitutions on populations, hyperfine fields, and hysteresis phenomenon in magnetite. *J. Appl. Phys.*, **2002**, *91*, 8135-8138.

[18]   Singh, R. P.; Kang, D. Y.; Oh, B. K.; Choi, J. W. Polyaniline based catalase biosensor for the detection of Hydrogen peroxide and Azide. *Biotech. Bioproc. Eng.*, **2009**, *14*, 443-449.

[19]   Singh, R. P.; Oh, B. K.; Koo, K. K.; Jyoung, J. Y.; Jeong, S.; Choi, J. W. Biosensor arrays for environmental pollutants detection. *Biochip J.,* **2008**, *2*, 223-234.

[20]   Singh, R. P.; Kim, Y. J.; Oh, B. K.; Choi, J. W. Glutathione-s-transferase based electrochemical biosensor for the detection of Captan. *Electrochem. Commun.,* **2009**, *11*, 181-185.

[21]   Kim, H.; Kang, D. Y.; Goh, H. J.; Oh, B. K.; Singh, R. P.; Oh, S. M.; Choi, J. W. Analysis of direct immobilized recombinant protein G on a gold surface. *Ultramicroscopy*, **2008**, *108*, 1152-1156.

[22]   Tombelli, S.; Minunni, M.; Mascini, M. Analytical applications of aptamers. *Biosens. Bioelectron.*, **2005**, *20*, 2424-2434.

[23]   Seeman, N. C. DNA engineering and its application to nanotechnology. *Treds. Biotech.*, **1999**, *17*, 437-443.

[24]   Singh, R. P.; Kang, D. Y.; Choi, J. W. Nanofabrication of bio-self assembled monolayer and its electrochemical property for toxicant detection. *J. Nanosci. Nanotech.*, **2011**, *11*, 408-412.

[25]   Singh, R. P.; Kang, D. Y.; Choi, J. W. Electrochemical DNA biosensor for the detection of sanguinarine in adulterated mustard oil. *Adv. Mater. Letts.*, **2010**, *1*, 48-54.

[26] Jianrong, C.; Yuqing, M.; Nongyue, H.; Xiaohua, W.; Sijiao, L. Nanotechnology and biosensors. *Biotech. Adv.*, **2004**, *22*, 505-518.

[27] Xiao, Y. H.; Li, C. M. Nanocomposites: from fabrications to electrochemical bioapplications. *Electroanalysis*, **2008**, *20*, 648-662.

[28] Pumera, M.; Sanchez, S.; Ichinose, I.; Tang, J. Electrochemical nanobiosensors. *Sens. Actuatos. B*, **2007**, *123*, 1195-1205.

[29] Agui, L.; Yanez-Sedeno, P.; Pingarron, J. M. Role of carbon nanotubes in electroanalytical chemistry: A review. *Anal. Chim. Acta*, **2008**, *622*, 11-47.

[30] Gooding, J. J. Nanostructuring electrodes with carbon nanotubes: A review on electrochemistry and applications for sensing. *Electrochim. Acta*, **2005**, *50*, 3049-3060.

[31] He, P. A.; Xu, Y.; Fang, Y. Z. Applications of carbon nanotubes in electrochemical DNA biosensors. *Microchim. Acta*, **2006**, *152*, 175-186.

[32] Kumar, S. A.; Chen, S. M. Nanostructured Zinc Oxide Particles in Chemically Modified Electrodes for Biosensor Applications. *Anal. Letts.*, **2008**, *41*, 141-158.

[33] Merkoci, A.; Pumera, M.; Llopis, X.; Perez, B.; Valle, M. D.; Alegret, S. New materials for electrochemical sensing VI: Carbon nanotubes. *Trac. Treds. Anal. Chem.*, **2005**, *24*, 826-838.

[34] Rivas, G. A.; Rubianes, M. D.; Rodriguez, M. C.; Ferreyra, N. E.; Luque, G. L.; Pedano, M. L.; Miscoria, S. A.; Parrado, C. Carbon nanotubes for electrochemical biosensing. *Talanta*, **2007**, *74*, 291-307.

[35] Sirivisoot, S.; Webster, T. J. Multiwalled carbon nanotubes enhance electrochemical properties of titanium to determine *in situ* bone formation. *Nanotechnology*, **2008**, *19*, 295101-295113.

[36] Wang, J. Nanomaterial-based amplified transduction of biomolecular interactions. *Small*, **2005**, *1*, 1036-1043.

[37] Wildgoose, G. G.; Banks, C. E.; Leventis, H. C.; Compton, R. G. Chemically modified carbon nanotubes for use in electroanalysis. *Microchim. Acta*, **2006,** *152*, 187-214.

[38] Yogeswaran, U.; Chen, S. M. Recent Trends in the Application of Carbon Nanotubes-Polymer Composite Modified Electrodes for Biosensors: A Rev. *Anal. Letts.*, **2008**, *41*, 210-243.

[39] Katz, E.; Willner, I. Biomolecule-Functionalized carbon nanotubes: applications in nanobioelectronics. *Chem. Phys. Chem.*, **2004**, *5*, 1084-1104.

[40] Luo, X. L.; Morrin, A.; Killard, A. J.; Smyth, M. R. Application of nanoparticles in electrochemical sensors and biosensors. *Electroanalysis*, **2006**, *18*, 319-326.

[41] Merkoci, A.; Aldavert, M.; Marin, S.; Alegret, S. New materials for electrochemical sensing V: Nanoparticles for DNA labeling. *Trends Anal. Chem.*, **2005**, *24, 341*-349.

[42] Pingarron, J. M.; Yanez-Sedeno, P.; Gonzalez-Cortes, A. Gold nanoparticle-based electrochemical biosensors. *Electrochim. Acta*, **2008**, *53*, 5848-5866.

[43] Yogeswaran, U.; Chen, S. M. A review on the electrochemical sensors and biosensors composed of nanowires as sensing material. *Sensors*, **2008**, *8*, 290-313.

[44] Chen, X.; Fu, C.; Wang, Y.; Yang, W.; Evans, D. G. Direct electrochemistry and electrocatalysis based on a film of horseradish peroxidase intercalated into Ni-Al layered double hydroxide nanosheets. *Biosens. Bioelectron.*, **2008**, *24,* 356-361.

[45] Lu, X.; Zhang, H.; Ni, Y.; Zhang, Q.; Chen, J. Porous nanosheet-based ZnO microspheres for the construction of direct electrochemical biosensors. *Biosens. Bioelectron.*, **2008**, *24*, 93-98.

[46] Tang, C. S.; Schmutz, P.; Petronis, S.; Textor, M.; Keller, B.; Voros, J. Locally Addressable Electrochemical Patterning Technique (LAEPT) applied to poly (L-lysine)-graft-poly (ethylene glycol) adlayers on titanium and silicon oxide surfaces. *Biotech. Bioeng.*, **2005**, *91*, 285-295.

[47] Vamvakaki, V.; Hatzimarinaki, M.; Chaniotakis, N. Biomimetically synthesized silica-carbon nanofiber architectures for the development of highly stable electrochemical biosensor systems. *Anal. Chem.*, **2008**, *80*, 5970-5975.

[48] Zhou, M.; Shang, L.; Li, B.; Huang, L.; Dong, S. Highly ordered mesoporous carbons as electrode material for the construction of electrochemical dehydrogenase and oxidase-based biosensors. *Biosens. Bioelectron.*, **2008**, *24*, 442-447.

[49] Audrey, S.; Leca, B.; Beatrice, D.; Blum, L. J. DNA Biosensors and microarrays. *Chem. Rev.*, **2008**, *108*, 109-139.

[50] Hahn, S.; Mergenthaler, S.; Zimmermann, B.; Holzgreve, W. Nucleic acid based biosensors: The desires of the user. *Bioelectrochemistry,* **2005,** *67*, 151-154.

[51]   Daniel, S.; Rao, T. P.; Rao, K. S.; Rani, S. U.; Naidu, G. R. K.; Lee, H. Y.; Kawai, T. A review of DNA functionalized/grafted carbon nanotubes and their characterization. *Sens. Actuatos. B*, **2007**, *122*, 672-682.

[52]   Kumar, A. Dill, K. Published by IVD Technology, **2005**. http://www.devicelink.com/ivdt/archive/05/05/ 003.html.

[53]   Davies, T. J.; Ward, J. S.; Banks, C. E.; Campo, F. J. D.; Mas, R.; Munoz, F. X.; Compton, R. G. The cyclic and linear sweep voltammetry of regular arraysof microdisc electrodes: Fitting of experimental data. *J. Electroanal. Chem.*, **2005**, *585*, 51-62.

[54]   Ordeig, O.; Banks, C. E.; Davies, T. J.; Campo, J.D.; Mas, R.; Munoz, F. X.; Compton, R. G. Regular arrays of microdisc electrodes: simulation quantifies the fraction of dead electrodes. *Analyst*, **2006**, *131*, 440-445.

[55]   Fletcher, S.; Horne, M. D. Random assemblies of microelectrodes (RAMy electrodes) for electrochemical studies. *Electrochem. Commun.*, **1999**, *1*, 502-512.

[56]   Wang, J. Electrochemical biosensors: Towards point-of-care cancer diagnostics. *Biosens. Bioelectron.*, **2006**, *21*, 1887-1892.

[57]   Liu, R. Integrated microfluidic biochips for immunoassay and DNA bioassays. *Conf. Proc. IEEE. Eng. Med. Biol. Soc.*, **2004**, *7*, 5394.

[58]   Simm, A. O.; Banks, C. E.; Ward-Jones, S.; Davies, T. J.; Lawrence, N. S.; Jones, T. G. J.; Li, J.; Compton, R.G. Boron-doped diamond microdisc arrays: electrochemical characterization and their use as a substrate for the production of microelectrode arrays of diverse metals (Ag, Au, Cu) *via* Electrodeposition. *Analyst*, **2005**, *130*, 1303-1311.

[59]   Hassibi, A.; Lee, T. H. Programmable 0.18 µ CMOS Electrochemical Sensor Microarray for Biomolecular Detection. *IEEE Sens. J.*, **2006**, *6*, 1380-1388.

[60]   Singh, R. P.; Oh, B. K.; Choi, J. W. Application of peptide nucleic acid towards development of Nanobiosensor arrays. *Bioelectrochemistry*, **2010**, *79*, 153-161.

[61]   Cai, W.; Peck, J.; Weide, D.; Hamers, R. Direct electrical detection of hybridization at DNA-modified silicon surfaces. *Biosens. Bioelectron.*, **2004**, *19*, 1013-1019.

[62]   Cai, H.; Wang, Y.; He, P.; Fang, Y. Electrochemical detection of DNA hybridization based on silver-enhanced gold nanoparticle label. *Anal. Chim. Acta*, **2004**, *469*, 165-172.

[63]   Kerman, K.; Morita, Y.; Takamura, Y.; Ozsoz, M.; Eiichi, T. Modification of *Escherichia coli* single-stranded DNA binding protein with gold nanoparticles for electrochemical detection of DNA hybridization. *Anal. Chim. Acta*, **2004**, *510*, 169-174.

[64]   Wang, J.; Li, J.; Baca, A.; Hu, J.; Zhou, F.; Yan, W.; Pang, D.W. Amplified voltammetric detection of DNA hybridization *via* oxidation of ferrocene caps on gold nanoparticle/streptavidin conjugates. *Anal. Chem.*, **2003**, *75*, 3941-3945.

[65]   Ozsoz, M.; Erdem, A.; Kerman, K.; Ozlam, D.; Tugrui, B.; Topcuoglu, N.; Ekem, H.; Taylam, M. Electrochemical genosensor based on colloidal gold nanoparticles for the detection of factor V Leiden mutation using disposable pencil graphite electrodes. *Anal. Chem.*, **2003**, *75*, 2181-2187.

[66]   Wang, J.; Liu, G.; Rivas, G. Encoded beads for electrochemical identification. *Anal. Chem.*, **2003**, *75*, 4667-4671.

[67]   Wang, J.; Xu, D.; Polsky, R. Magnetically-induced solid-state electrochemical detection of DNA hybridization. *J. Am. Chem. Soc.*, **2002**, *124*, 4208-4209.

[68]   Tansil, N. C.; Gao, Z. Nanoparticles in biomolecular detection. *Nano Today*, **2006**, *1*, 28-37.

[69]   Hwang, S.; Kim, E.; Kwak, J. Electrochemical detection of DNA hybridization using biometallization. *Anal. Chem.*, **2005**, *77*, 579-585.

[70]   K'Owino I. O.; Agarwal, R.; Sadik, O. A. Novel electrochemical detection scheme for DNA binding interactions using monodispersed reactivity of silver ions. *Langmuir*, **2003**, *19*, 4344-4350.

[71]   Aluoch, A. O.; Sadik, O. A.; Bedi, G. Development of an oral biosensor for salivary amylase using a monodispersed silver for signal amplification. *Anal. Biochem.*, **2005**, *340*, 136-144.

[72]   K'Owino I. O.; Mwilu, S.K.; Sadik, O. A. Metal enhanced biosensor for genetic mismatch detection. *Anal. Biochem.*, **2007**, *369*, 8-17.

[73]   Ngundi, M. *Rational design of chemical and biochemical sensors*, Chemistry Department, Binghamton University, Binghamon, New York, **2003**.

[74]   Lee, J. O.; So, H. M.; Jeon, E. K.; Chang, H.; Won, K.; Kim, Y. H. Aptamers as molecular recognition elements for electrical nanobiosensors. *Anal. Bioanal. Chem.*, **2008**, *390*, 1023-1032.

[75] Song, S. P.; Wang, L. H.; Li, J.; Zhao, J. L.; Fan, C. H. Aptamer-based biosensors. *Trac. Treds. Anal. Chem.*, **2008**, *27*, 108-117.

[76] Tombelli, S.; Minunni, M.; Mascini, M. Aptamers-based assays for diagnostics, environmental and food analysis. *Biomol. Eng.*, **2008**, *24*, 191-200.

[77] Willner, I.; Zayats, M. Electronic Aptamer-Based Sensors. *Angew Chem. Int. Ed.*, **2007**, *46*, 6408-6418.

[78] Hansen, J. A.; Wang, J.; Kawde, A. N.; Xiang, Y.; Gothelf, K. V.; Collins, G. Quantum-dot/aptamer-based ultrasensitive multi-analyte electrochemical biosensor. *J. Am. Chem. Soc.,* **2006**, *128*, 2228-2229.

[79] He, P. L.; Shen, L.; Cao, Y. H.; Lia, D. F. Ultrasensitive electrochemical detection of proteins by amplification of aptamer-nanoparticle bio barcodes. *Anal. Chem.*, **2007**, *79*, 8024-8029.

[80] Li, B. L.; Wang, Y. L.; Wei, H.; Dong, S. J. Amplified electrochemical aptasensor taking AuNPs based sandwich sensing platform as a model. *Biosens. Bioelectron.,* **2008**, *23*, 965-970.

[81] Numnuam, A.; Chumbimuni, Torres, K.; Xiang, Y.; Bash, R.; Thavarungkul, P.; Kanatharana, P.; Pretsch, E.; Wang, J.; Bakker, E. Aptamer-based potentiometric measurements of proteins using ion-selective microelectrodes. *Anal. Chem.*, **2008**, *80*, 707-712.

[82] Hahm, J.; Lieber, C. M. Direct Ultrasensitive electrical detection of DNA and DNA sequence variations using nanowire nanosensors. *Nano Letts.*, **2004**, *4*, 51-54.

[83] Patolsky, F.; Zheng, G. F.; Hayden, O.; Lakadamyali, M.; Zhuang, X. W.; Lieber, C. M. Electrical detection of single viruses. *Proc. Nat. Acad. Sci. USA.,* **2004**, *101*, 14017-14022.

[84] Zheng, G. F.; Patolsky, F.; Cui, Y.; Wang, W.U.; Lieber, C. M. Multiplexed electrical detection of cancer markers with nanowire sensor arrays. *Nat. Biotechnol.*, **2005**, *23*, 1294-1301.

[85] Katz, E.; Willner, I.; Wang, J. Electroanalytical and Bioelectroanalytical Systems Based on Metal and Semiconductor Nanoparticles. *Electroanalysis*, **2004**, *16*, 19-44.

[86] Ramanathan, K.; Bangar, M. A.; Yun, M.; Chen, W.; Myung, N. V.; Mulchandani, A. Bioaffinity sensing using biologically functionalized conducting-polymer nanowire. *J. Am. Chem. Soc.*, 2005, *127*, 496-497.

[87] Tang, H.; Chen, J. H.; Nie, L. H.; Kuang, Y. F.; Yao, S. Z. A label-free electrochemical immunoassay for carcinoembryonic antigen (CEA) based on gold nanoparticles (AuNPs) and nonconductive polymer film. *Biosens. Bioelectron.*, **2007**, *22*, 1061-1067.

[88] Yang, L. J.; Li, Y. B.; Erf, G. F. Interdigitated Array Microelectrode-based electrochemical impedance immunosensor for detection of Escherichia coli O157:H7. *Anal. Chem.,* **2004**, *76*, 1107-1113.

[89] Yang, L.; Bashir, R. Electrical/electrochemical impedance for rapid detection of foodborne pathogenic bacteria. *Biotech. Adv.*, **2008**, *26*, 135-150.

[90] Gomez-Sjoberg, R.; Morisette, D. T.; Bashir, R. Impedance microbiology-on-a-chip: microfluidic bioprocessor for rapid detection of bacterial metabolism. *J. Microelectromech. Syst.,* **2005**, *14*, 829-838.

[91] K'Owino, I. O.; Sadik, O. A. Impedance Spectroscopy: A powerful tool for rapid biomolecular screening and cell culture Monito. *Electroanalysis*, **2005**, *17*, 2101-2113.

[92] Radke, S. A.; Alocilja, E. C. A high density microelectrode array biosensor for detection of *E. coli* O157:H7. *Biosens. Bioelectron.,* **2005**, *20*, 16620-1667.

[93] Varshney, M.; Li, Y. B. Interdigitated array microelectrode based impedance biosensor coupled with magnetic nanoparticle-antibody conjugates for detection of *Escherichia coli* O157:H7 in food samples. *Biosens. Bioelectron.,* **2007**, *22*, 2408-2414.

[94] Varshney, M.; Li, Y. B.; Srinivasan, B.; Tung, S. A label-free, microfluidics and interdigitated array microelectrode-based impedance biosensor in combination with nanoparticles immunoseparation for detection of *Escherichia coli* O157:H7 in food samples. *Sens. Actuatos. B*, **2007**, *128,* 99-107.

[95] Yang, W.; Butler, J.; Russel, J.; Hamers, R. Interfacial electrical properties of DNA-modified diamond thin films: Intrinsic response and hybridization-induced field effects. *Langmuir*, **2004**, *20*, 6778-6787.

[96] Cretich, M. S. V.; Damin, F.; DiCarlo, G.; Oldani, C.; Chiari, M. Functionalization of poly (dimethylsiloxane) by chemisorption of copolymers: DNA microarrays for pathogen detection. *Sens. Actuatos. B*, **2008**, *132*, 258-264.

[97] Liao, J. C.; Mastali, M.; Gau, V.; Suchard, M. A.; Moller, A. K.; Bruckner, D. A.; Babbitt, J. T.; Li, Y.; Gornbein, J.; Landaw, E. M.; McCabe, E. R. B.; Churchill, B. M.; Haake, D. A. Use of electrochemical DNA biosensors for rapid molecular identification of uropathogens in clinical urine specimens. *J. Clin. Microbiol.*, **2006**, *44*, 561-570.

[98] Karasinski, J.; White, L.; Zhang, Y. C.; Wang, E.; Andreescu, S.; Sadik, O. A.; Lavine, B. K.; Vora, M. Detection and identification of bacteria using antibiotic susceptibility and a multi-array electrochemical sensor with pattern recognition. *Biosens. Bioelectron.*, **2007**, *22*, 2643-2649.

[99]   Clark, L. C.; Lyons, C. Electrode systems for continuous monitoring in cardiovascular surgery. *Ann. N. Y. Acad. Sci.*, **1962**, *102*, 29-45.

[100]   Wang, Y.; Xu, H.; Zhang, J. M.; Li, G. An electrochemical sensor for clinic analysis. *Sensors*, **2008**, *8*, 2043-2081.

[101]   Wang, J.; Musameh, M. Carbon nanotube screen printed electrochemical sensors. *Analyst*, **2004**, *129*, 1-2.

[102]   Liu, J. Q.; Chou, A.; Rahmat, W.; Paddon-Row, M. N.; Gooding, J. J. Achieving direct electrical connection to glucose oxidase using aligned single walled carbon nanotube arrays. *Electroanalysis*, **2005**, *17*, 38-46.

[103]   Patolsky, F.; Weizmann, Y.; Willner, I. Long range electrical contacting of redox enzymes by SWCNT connectors. *Angew Chem. Int. Ed.*, **2004**, *43*, 2113-2117.

[104]   Gooding, J. J.; Wibowo, R.; Liu, J.; Yang, W.; Losic, D.; Orbons, S.; Mearns, F. J.; Shapter, J. G.; Hibbert, D. B. Protein electrochemistry using aligned carbon nanotube arrays. *J. Am. Chem. Soc.*, **2003**, *125*, 9006-9007.

[105]   Cusma, A.; Curulli, A.; Zane, D.; Kauhs, S.; Padeletti, G. Feasibility of enzyme biosensors based on gold nanowires. *Mater. Sci. Eng. C*, **2007**, *27*, 1158-1161.

[106]   Lu, Y. H.; Yang, M. H.; Qu, F. L.; Shen, G. L.; Yu, R. Q. Enzyme-functionalized gold nanowires for the fabrication of biosensors. *Bioelectrochemistry*, **2007**, *71*, 211-216.

[107]   Qu, F. L.; Yang, M. H.; Shen, G. L.; Yu, R. Q. Electrochemical biosensing utilizing synergic action of carbon nanotubes and platinum nanowires prepared by template synthesis. *Biosens. Bioelectron.*, **2007**, *22*, 1749-1755.

[108]   Ramanavicius, A.; Kausaite, A.; Ramanaviciene, A. Polypyrrole-coated glucose oxidase nanoparticles for biosensor design. *Sens. Actuatos. B*, **2005**, *111*, 532-539.

[109]   Li, C. Z.; He, H. X.; Tao, N. J. Quantized tunneling current in the metallic nanogaps formed by electrodeposition and etching. *Appl. Phys. Letts.*, **2000**, *77*, 3995-3997.

[110]   Forzani, E. S.; Zhang, H. Q.; Nagahara, L.; Amlani, I.; Tsue, R.; Tao N. J. Conducting polymer nano junction sensor for glucose detection. *Nano Letts.*, **2004**, *4*, 1785-1788.

[111]   Jing, W.; Yang, Q. Mediator free amperometric determination of glucose based on direct electron transfer between glucose oxidase and an oxidized boron-doped diamond electrode. *Anal. Bioanal. Chem.*, **2006**, *385*, 1330-1335.

[112]   Wilson, M. S. An electrochemical immunosensor for the simultaneous detection of two tumor markers. *Anal. Chem.*, **2005**, *77*, 1496-1502.

[113]   Wilson, G. S.; Gifford, R. Biosensors for real time *in vivo* measurements. *Biosens. Bioelectron.*, **2005**, *20*, 2388-2403.

[114]   Jung, M. W.; Kim, D. W.; Jeong, R. A.; Kim, H. C. *In: engineering in medicine and biology society.* IEMBS 04, 26th annual international conference of the IEEE. San Francisco, USA, **2004**; *1*, pp. 1987-1989.

[115]   Heller, A. An integrated medical feed-back systems for drug delivery. *AIChE J.*, **2005**, *51*, 1054-1066.

[116]   Frost, M.; Meyerhoff, M. E. *In Vivo* Chemical Sensors: Tackling Biocompatibility. *Anal. Chem.*, **2006**, *78*, 7370-7377.

[117]   Oh, B. K.; Robbines, M. E.; Nablo, B. J.; Schoenfisch, M. H. Miniaturized glucose biosensor modified with a nitric oxide-releasing xerogel microarray. *Biosen. Bioelectron.*, **2005**, *21*, 749-757.

[118]   Shin, J. H.; Marxer, S. M.; Schoenfisch, M. H. Nitric oxide releasing sol-gel particles poly-urethane glucose biosensors. *Anal. Chem.*, **2004**, *76*, 4543-4549.

[119]   Gifford, R.; Batchelor, M. M.; Lee, Y.; Gokulrangan, G.; Meyerhoff, M. E.; Wilson, G. S. Mediation of *in vivo* glucose sensor inflammatory response *via* nitric oxide release. *J. Biomed. Mater. Res.*, **2005**, *75A*, 755-766.

[120]   Andreescu, S.; Luck, L. A. Studies of the binding and signaling of surface-immobilized periplasmic glucose receptors on gold nanoparticles: A glucose biosensor application. *Anal. Biochem.*, **2008**, *375*, 282-290.

[121]   Berg, J. M.; Tymoczko, J. L.; Stryer, L. *Biochemistry*, 5th ed.; W.H. Freeman & Co.: New York, **2002**.

[122]   Arnott, H. J. *In: The mechanisms of biomineralization in animals and plants*, Omori, M.: Watabe, N., eds.; Tokai University Press: Tokyo, **1980**, *pp* 211-218.

[123]   Perry, C. C. Silicification: The Processes by Which Organisms Capture and Mineralize Silica. *Rev. Miner. Geochem.*, **2003**, *54*, 291-327.

[124]   Weiner, S.; Lowenstam, H. A. *On biomineralization*, Oxford University Press: New York, **1989**.

[125]   Mann, S. *Biomineralization*, Oxford University Press: New York, **2005**.

[126]   Sarikaya, M. Biomimetics: materials fabrication through biology. Proc. Natl. Acad. Sci. U SA., **1999**, *96*, 14183-14185.

[127]   Whitesides, G. M.; Mathias, J. P.; Seto, C. T. Molecular Self-Assembly and Nanochemistry: A chemical strategy for the synthesis of nanostructures. *Science,* **1991**, *254*, 1312-1319.

[128] Ariga, K.; Hill, Lee, J. P.; M. V.; Vinu, A.; Charvet, R.; Acharya, S. Challenges and breakthroughs in recent research on self-assembly. *Sci. Technol. Adv. Mater.,* **2008,** *9,* 14109.

[129] Williams, D. F. *The Williams dictionary of biomaterials,* Liverpool University Press: Liverpool; UK, **1999.**

[130] Heller, J.; Sparer, R. V.; Zenter, G. M. *Poly (ortho esters) In biodegradable polymers as drug delivery systems,* Chasin, M.: Langer, R., eds.; Marcel Dekker: New York, **1990.**

[131] Ron, E.; Langer, R. *Erodible systems: In Treatise on controlled drug delivery,* Kydonieus, A. ed.; Marcel Dekker: New York, **1992,** pp. 199-224.

[132] Shalaby, S. W. *Biomedical polymers,* Hanser: New York, **1994.**

[133] Park, K.; Shalaby, W. S. W.; Park, H. *Biodegradable Hydrogels for drug delivery,* Technomic publishing Co. Inc.: Lancaster PA, **1993.**

[134] Stevens, M. P. *Polymer Chemistry: An Introduction,* 3rd ed.; Oxford University Press: New York, **1999.**

[135] Kresge, C. T.; Leonowicz, M. E.; Roth, W. J.; Vartuli, J. C.; Beck, J. S. Ordered mesoporous molecular sieves synthesized by a liquid-crystal template mechanism. *Nature,* **1992,** *359,* 710-712.

[136] Inagaki, S.; Guan, S.; Fukushima, Y.; Ohsuna, T.; Terasaki, O. Novel mesoporous materials with a uniform distribution of organic groups and inorganic oxide in their frame works. *J. Am. Chem. Soc.,* **1999,** *121,* 9611-9614.

[137] Plass, R.; Last, J. A.; Bartelt, N. C.; Kellogg, G. L. Nanostructures: self-assembled domain patterns. *Nature,* **2001,** *412,* 875-875.

[138] Jager, E. W. H.; Smela, E.; Inganäs, O. Microfabricating conjugated polymer actuators. *Science,* **2000,** *290,* 1540-1545.

[139] Ruiz-Hitzky, E.; Darder, M.; Aranda, P. Functional biopolymer nanocomposites based on layered solids. *J. Mater. Chem.,* **2005,** *15,* 3650-3662.

[140] Alivisatos, A. P. Semiconductor clusters, nanocrystals, and quantum dots. *Science,* **1996,** *271,* 933-937.

[141] Daniels, A. U.; Chang, M. K. O.; Andriano, K. P. Mechanical properties of biodegradable polymers and composites proposed for internal fixation of bone. *J. Appl. Biomater.,* **1990,** *1,* 57-58.

[142] Bradley, G. W.; McKenna, G. G.; Dunn, H. K.; Daniels, A. U.; Statton. W. O. Effects of flexural rigidity of plates on bone healing. *J. Bone Joint Surg.,* **1979,** *61A,* 866-872.

[143] Terjesen, T.; Apalset, K. The influence of different degrees of rigidity of fixation plates on experimental bone healing. *J. Orthop. Res.,* **1988,** *6,* 293-299.

[144] Gillett, N.; Brown, S. A.; Dumbleton, J. H.; Pool, R. P. The use of short carbon fibre reinforced thermoplastic plates for fracture fixation. *Biomater.,* **1985,** *6,* 113-121.

[145] Uchida, A.; Araki, N.; Shinto, Y.; Yoshikawa, H.; Kurisaki, E.; Ono, K. The use of calcium hydroxyapatite ceramic in bone tumour surgery. *J. Bone Joint Surg. Br.,* **1990,** *72-B,* 298-302.

[146] Cooke, F. W. Ceramics in Orthopedic Surgery. *Clin. Orthop. Relat. Res.,* **1992,** *276,* 135-146.

[147] Yamaguchi, I.; Tokuchi, K.; Fukuzaki, H.; Koyama, Y.; Takakuda, K.; Monma, H.; Tanaka, J. Preparation and microstructure analysis of chitosan/hydroxyapatite nanocomposites. *J. Biomed. Mater. Res.,* **2001,** *55,* 20-27.

[148] Ogawa, K.; Hirano, S.; Miyanishi, S.; Yui, T.; Watanabe, T. A new polymorph of chitosan. *Macromol.,* **1984,** *17,* 973-975.

[149] Okuyama, K.; Noguchi, K.; Hanafusa, Y.; Osawa, K.; Ogawa, K. Structural study of anhydrous tendon chitosan obtained *via* chitosan/acetic acid complex. *Int. J. Biol. Macromol.,* **1999,** *26,* 285-293.

[150] Kim, H. W.; Kim, H. E.; Salih, V. Stimulation of osteoblast responses to biomimetic nanocomposites of tissue engineering scaffolds. *Biomater.,* **2005,** *26,* 5221-5230.

[151] Mann, S.; Ozin, G. A. Synthesis of inorganic materials with complex form. *Nature,* **1996,** *382,* 313-318.

[152] Zerwekh, J. E.; Kourosh, S.; Scheinberg, R. Fibrillar collagen biphasic calcium phosphate composite as a bone graft substitute for spinal fusion. *J. Orthop. Res.,* **1992,** *10,* 562-572.

[153] Zerwekh, J. E.; Kourosh, S.; Scheinberg, R.; Kitano, D. T.; Edwards, M. L.; Shin, D.; Selby, D. K. *Biomater.,* **2002,** *23,* 3227-3234.

[154] Drury, L.; Mooney, D. J. Hydrogels for tissue engineering: scaffold design variables and applications. *Biomater.,* **2003,** *24,* 4337-4351.

[155] Lin, H. R.; Yeh, Y. J. Porous alginate/hydroxyapatite composite. *J. Biomed. Mater. Res., Part B*: *Appl. Biomater.,* **2004,** *71B,* 52-65.

[156] Ito, M.; Hidaka, Y.; Nakajima, M.; Yagasaki, H.; Kafrawy, A. H. Effect of hydroxyapatite content on physical properties and connective tissue reactions to a chitosan-hydroxyapatite composite membrane. *J. Biomed. Mater. Res.,* **1999,** *45,* 204-208.

[157] Du, C.; Cui, D. F. Z.; Feng, Q. L.; Zhu, X. D.; Groot, K. D. Tissue response to nano-hydroxyapatite/collagen composite implants in marrow cavity. *J. Biomed. Mater. Res.,* **1998**, *42*, 540-548.

[158] Mullermai, C. M.; Stupp, S. I.; Voigt, C.; Gross, U. Nanoapatite and organoapatite implants in bone histology and ultrastructure of the interface. *J. Biomed. Mater. Res.,* **1995**, *29*, 9-18.

[159] Boanini, E.; Torricelli, P.; Gazzano, M.; Giardino, R.; Bigi, A. Nanocomposites of hydroxyapatite with aspartic acid and glutamic acid and their interaction with osteoblast-like cells. *Biomater.,* **2006**, *27*, 4428-4433.

[160] Edlund, U.; Albertsson, A. C. Degradable polymer microspheres for controlled drug delivery. *Adv. Polym. Sci.,* **2002**, *157*, 67-112.

[161] Kronenthal, R. L. *Polymer medicine and surgery*, Plenum Press: New York, **1975**; *pp.* 119.

[162] Tsuji, H.; Ikada, Y. Properties and morphology of poly (L-lactide): Effects of structural parameters on long-term hydrolysis of poly (L-lactide) in phosphate-buffered solution. *Polym. Degrad. Stab.,* **2002**, *179*, 179-189.

[163] Bleach, N. C.; Nazhat, S. N.; Tanner, K. E.; Kellomaki, M.; Tormala, P. Effect of filler content on mechanical and dynamic mechanical properties of particulate biphasic calcium phosphate-polylactide composites. *Biomater.,* **2002**, *23*, 1579-1585.

[164] Alexander, H.; Langrana, N.; Massengill, J.; Weiss, A. Development of new methods for phalangeal fracture fixation. *J. Biomech.,* **1981**, *14*, 377-387.

[165] Krikorian, V.; Pochan, D. J. Poly (L-Lactic Acid)/layered silicate nanocomposite: fabrication, characterization, and properties. *Chem. Mater.,* **2003**, *15*, 4317-4324.

[166] Krikorian, V.; Pochan, D. J. Unusual crystallization behavior of organoclay reinforced poly (L-lactic acid) nanocomposites. *Macromolecule,* **2004**, *37*, 6480-6491.

[167] Krikorian, V., Pochan, D. J. Crystallization behavior of poly (l-lactic acid) nanocomposites: Nucleation and growth probed by infrared spectroscopy. *Macromolecule,* **2005**, *38*, 6520-6527.

[168] Zhang, J.; Tsuji, H.; Noda, I.; Ozaki, Y. Structural changes and infrared correlation spectroscopy. *Macromolecule,* **2004**, *37*, 6433-6439.

[169] Zhang, J.; Tsuji, H.; Noda, I.; Ozaki, Y. J. Weak intermolecular interactions during the melt crystallization of poly (L-lactide) investigated by two-dimensional infrared correlation spectroscopy. *Phys. Chem. B,* **2004**, *108*, 11514-11520.

[170] Loizou, E.; Butler, P.; Porcar, L.; Kesselman, E.; Talmon, Y.; Dundigalla, A.; Schmidt, G. Large Scale Structures in Nanocomposite Hydrogels. *Macromolecule,* **2005**, *38*, 2047-2049.

[171] Zong, X.; Kim, K.; Fang, D.; Ran, S.; Hsiao, B. S.; Chu, B. Structure and process relationship of electrospun bioabsorbable nanofiber membranes. *Polymer,* **2002**, *43*, 4403.

[172] Alexander, M.; Dubois, P. Polyactite/montmor-illonite nanocomposite: Study of the hydrolytic degradation. *Mater. Sci. Eng. R,* **2000**, *28*, 1-63.

[173] Lee, Y. H.; Lee, J. H.; An, I. G.; Kim, C.; Lee, D. S.; Lee, Y. K.; Nam, J. D. Electrospun dual-porosity structure and biodegradation morphology. *Macromolecule,* **2005**, *26*, 3165-3172.

[174] Hule, R. A.; Pochan, D. J. Poly (L-Lysine) and clay nanocomposite with desired matrix. *J. Polym. Sci., Part B: Polym. Phys.,* **2007**, *45*, 239-252.

[175] Karikoriani, V.; Kuriani, M.; Galvini, M. E.; Nowak, A. P.; Deming, T. J.; Darrin, J.; Pochani, J. Polypeptide-based nanocomposite: structure and properties of poly (L-lysine)/Na-Montmorillonite. *J. Polym. Sci. Part B: Polym. Phys.,* **2002**, *40*, 2579-2586.

[176] Ozbas, B.; Kretsinger, J.; Rajagopal, K.; Schneider, J. P.; Pochan, D. J. Salt-triggered peptide folding and consequent self-assembly into hydrogels with tunable modulus. *Macromolecule,* **2004**, *37*, 7331-7337.

[177] Lailach, G. E.; Brindley, G. W. Specific co-absorption of purines and montmorillonite (clay-organic studies XV). *Clays Clay Miner.,* **1969**, *17*, 95-100.

[178] Wei, M.; Yuan, Q.; Evans, D. G.; Wang, Z.; Duan, X. Layered solids as a molecular container for pharmaceutical agents: L-tyrosine-intercalated layered double hydroxides. *J. Mater. Chem.,* **2005**, *15*, 1197-1203.

[179] Wilson, K. S.; Zhang, K.; Antonucci, J. M.; Systematic variation of interfacial phase reactivity in dental nanocomposites. *Biomater.,* **2005**, *26*, 5095-103.

[180] Kannan, R. Y.; Salacinski, H. J.; Butler, P. E.; Seifalian, A. M. Polyhedral oligomeric silsesquioxane nanocomposites: the next generation material for biomedical applications. *Acc. Chem. Res.,* **2005**, *38*, 879-884.

[181] Williams, J. C. *The Production, behavior and application of Ti alloys in high performance materials in Aerospace.* Flower H.M., ed.; Chapman & Hall: London, **1995**, *pp* 85-134.

[182] Alger, M. S. M. *Polymer Science Dictionary*, ed.; Elsevier Science: New York, **1989**.

[183] Seymour, R. B.; Carraher, C. E. Jr. *Polymer Chemistry: An Introduction*, 3rd ed.; Marcel Dekker: New York, **1992**; *p* 243.

[184] Odian, G. *Principles of polymerization*, 3rd ed.; John Wiley & Sons: **1991**, *pp.* 100.

[185] Stevens, M. P. *Polymer Chemistry: An Introduction,* 2nd ed.; Oxford University Press: New York, **1990**, *pp.* 32.

[186] Saunders, K. J. *Organic Polymer Chemistry*, 2nd ed.; Chapman and Hall: New York, **1988**, *pp.* 281-285.

[187] Wicks, Z. W.; Jr, Jones, F. N.; Pappas, S. P. *Organic Coatings: Science and Technology*, 2nd ed.; Wiley-Interscience: New York; **1999**.

[188] Hench, L. L. *Bioactive Ceramics, in Ducheyne,* Lemons, P.; Lemons, J. *Bioceramics: Material characteristics versus in vivo behavior,* eds.; New York Academy of Sciences: New York; **1988**.

[189] Hench, L. L. *Bioceramics and the Origins of Life in Oonishi H ed. Bioceramics: proceedings of the first international bioceramic symposium,* Ishyaku/EuroAmerica Inc.: Tokyo; **1990**, *pp.* 5-1.

[190] Howden, A. *Tissue reaction to the Bioceramic Synthos in Hastings, G., ed.; Mechanical properties of biomaterials,* John Wiley and Sons: **1980**, *pp.* 445-456.

[191] Kroto, H. W.; Heath, J. R.; O'Brien, S. C.; Curl, R.F.; Smalley, R. E. C_{60}: Buckminsterfullerene. *Nature*, **1985**, *318*, 162-163.

[192] Lingyi, A.; Zheng, B. M.; Lairson, E. V.; Shull, B. R. D. Formation of nanomagnetic thin films by dispersed fullerenes. *Appl. Phys. Letts.*, **2000**, *77*, 3242.

[193] Zheng, L. A.; Barrera, E. V. Formation and stabilization of nanosize grains in ferromagnetic thin films by dispersed C_{60}. *J. Appl. Phys.*, **2002**, *92,* 523.

[194] Sawamura, M.; Kuninobu, Y.; Toganoh, M.; Matsuo, Y.; Yamanaka, M.; Nakamura, E. Hybrid of ferrocene and fullerene. *J. Am. Chem. Soc.*, **2002**, *124*, 9354-9355.

[195] Lee, K.; Song, H.; Kim, B.; Park, J. T.; Park, S.; Choi, M. G. The first fullerene-metal sandwich complex: An unusually strong electronic communication between two C_{60} cages. *J. Am. Chem. Soc.*, **2002**, *124*, 2872-2873.

[196] Margadonna, S.; Aslanis, E.; Prassides, K. Ammoniated alkali fullerides (ND_3) $xNaA_2C_60$: Ammonia specific effects and superconductivity. *J. Am. Chem. Soc.*, **2002**, *124*, 10146-10156.

[197] Schön, J. H.; Kloc, C.; Batlogg, B. High-temperature superconductivity in lattice-expanded C_{60}, *Science*, **2001**, *293,* 2432-2434.

[198] Iijima, S. Helical microtubules of graphitic carbon. *Nature*, **1991**, *354,* 56-58.

[199] Overney, G.; Zhong, W.; Tománek, D. Structural rigidity and low frequency vibrational modes of long carbon tubules. *Z. Phys. Chem.*, **1993**, *27*, 93.

[200] Mintmire, J. W.; Dunlap, B. I.; White, C. T. Are fullerene tubules metallic? *Phys. Rev. Lett.,* **1992**, *68*, 631-634.

[201] Hamada, N.; Sawada, S.; Oshiyama, A. New one-dimensional conductors: Graphitic microtubules. *Phys. Rev. Letts.*, **1992**, *68*, 1579-1581.

[202] Saito, R.; Fujita, M.; Dresselhaus, G.; Dresselhaus, M. S. Electronic structure of graphene tubules based on C_{60}. *Phys. Rev. B*, **1992**, *46*, 1804-1811.

[203] Kociak, M.; Kasumov, A. Y.; Gueron, S.; Reulet, B.; Khodos, I. I.; Gorbatov, Y. B.; Volkov, V. T.; Vaccarini, L.; Bouchiat, H. Superconductivity in ropes of single walled carbon nanotubes. *Phys. Rev. Letts.*, **2001**, *86*, 2416-2419.

[204] Rinzler, A. G.; Hafner, J. H.; Nikolaev, P.; Nordlander, P.; Colbert, D. T.; Smalley, R. E.; Lou, L.; Kim, S. G.; Tománek, D. Unraveling nanotubes: field emission from an atomic wire. *Science.,* **1995**, *269*, 1550-1553.

[205] Cantor, B.; Allen, C. M.; Burkowski, R. D.; Green, M. H.; Hutchinson, J. L.; Q' Reilly, K. A. Q.; Long, A. K. P.; Schumacher, P.; Sloan, J.; Warren, P. J. Applications of nanocomposites. *Scripta Mater.,* **2001**, *44*, 2055-2059.

[206] Tombler, T. W.; Zhou, C.; Alexseyev, L.; Kong, J.; Dai, H.; Liu, L.; Jayanthi, C. S.; Tang, M.; Wu, S. Y. Reversible electromechanical characteristics of carbon nanotubes under local-probe manipulation. *Nature*, **2000**, *405*, 769-771.

[207] Maiti, A.; Andzelm, J.; Tanpipat, N.; Allmen, P. V. Effect of adsorbates on field emission from carbon nanotubes. *Phs. Rev. Letts.,* **2001**, *87*, 155502.

[208] Haubold, T.; Bohn, R.; Birringer, R.; Gleiter, H. Nanocrystalline intermetallic compounds structure and mechanical properties. *Mater. Sci. Eng. A*, **1992**, *153*, 679-683.

[209] Hranisavlijevic, J. Photoinduced charge separation reactions of J-aggregates coated on silver nanoparticles. *J. Am. Chem. Soc.,* **2002**, *124*, 4536-4537.

[210] Baney, R. H.; Itoh, M.; Sakakibara, A.; Suzuki, T. Silsesquioxanes. *Chem. Rev.,* **1995**, *95*, 1409-1430.

[211] Lichtenhan, J. D. *Silsesquioxane based polymers in the polymeric materials encyclopedia: Synthesis, properties and applications,* ed.; CRC Press: Boca Raton, Florida, **1996**, *pp.* 7768.

[212]   Karian, H. G. *Handbook of polypropylene and polypropylene composites*, Dekker: NewYork, **1999**.

[213]   Shea, K. J.; Loy, D. A. Bridged Polysilsesquioxanes Molecular Engineered Hybrid Organic-Inorganic Materials. *Chem. Mater., * **2001**, *13*, 3306-3319.

[214]   Zhao, M.; Sun, L.; Crooks, R. M. Preparation of Cu Nanoclusters within dendrimer templates. *J. Am. Chem. Soc.,* **1998**, *120*, 4877-4878.

[215]   Balogh, L.; Tomalia, D. A. Poly (Amidoamine) dendrimer templated nanocomposites: Synthesis of zero valent copper nanoclusters. *J. Am. Chem. Soc.*, **1998**, *120*, 7355-7356.

[216]   De, B. E. M. M.; Nijenhuis, A. G.; Borggreve, R. *Star polycondensates: Large scale synthesis, rheology and material properties*, Polymer News, **1997**.

[217]   Wilbur, D. S.; Pathare, P. M.; Hamlin, D. K.; Buhler, K. R.; Vessella. R. L. Biotin reagents for antibody pretargeting: Synthesis, radioiodination, and evaluation of biotinylated star burst dendrimers. *Bioconjugate. Chem.,* **1998**, *9*, 813-825.

[218]   Zhu, N.; Gu, U.; Chang, Z.; He, P.; Fang, Y. Dendrimers based DNA biosensors for electrochemical detection of DNA hybridization. *Electroanalysis,* **2006**, *18*, 2107-2114.

[219]   Bosman, A. W.; Janssen, H. M.; Meijer, E. W. About dendrimers: Structure, physical properties, and applications. *Chem. Rev.,* **1999**, *99*, 1665-1688.

[220]   Tully, D. C.; Frechet, J. M. J. Dendrimers at surfaces and interfaces: Chemistry and applications. *Chem. Commun.,* **2001**, 1229-1239.

[221]   Patri, A. K.; Majoros, I. J.; Baker, J. R. Dendritic polymer macromolecular carriers for drug delivery. *Curr. Opin. Chem. Biol.,* **2002**, *6*, 466-471.

[222]   Lee, S. C.; Parthasarathy, R.; Duffin, T. D.; Botwin, K.; Zobel, J; Beck, T.; Lange, G.; Kunneman, D.; Janssen, R.; Rowold, E.; Voliva, C. F. Recognition properties of antibodies to PAMAM dendrimers and their use in immune detection of dendrimers. *Biomed. Microdev.*, **2001**, *1*, 53-59.

[223]   Balzani, V.; Campagna, S.; Denti, G.; Juris, A.; Serroni, S.; Venturi, M. Designing dendrimers based on transition-metal complexes. Light-harvesting properties and predetermined redox patterns. *Acc. Chem. Res.*, **1998**, *31*, 26.

[224]   Newkome, G. R.; Moorefield, C. N.; Vogtle, F. *Dendritic Molecules: Concepts, Syntheses and Perspectives*, VCH: Weinheim, Germany, **1996**.

[225]   Borkowski, B. J.; Awad, J.; Wasgestian, F. Reactions of Chromium (III) and Cobalt (III) amine complexes with Starburst (PAMAM) dendrimers. *J. Inclusion Phenom. Macrocyclic Chem.,* **1999**, *35*, 355-359.

[226]   Balogh, L.; Tomalia, D. A. Poly (Amidoamine) dendrimer templated nanocomposites. 1. Synthesis of zerovalent copper nanoclusters. *J. Am. Chem. Soc.,* **1998**, *120*, 7355-7356.

[227]   Jansen, J. F. G. A.; De, B. D.V.; Berg, E. M. M.; Meijer, E. W. Encapsulation of guest molecules into a dendritic box. *Sci.,* **1994**, *266,* 1226-1229.

[228]   Zimmerman, S. C.; Wendland, M. S.; Rakow, N. A.; Zharov, I.; Suslick, K. S. Synthetic hosts by monomolecular imprinting inside dendrimers. *Nature*, **2002**, *418*, 399-403.

[229]   Archut, A.; Azzellini, G. C.; Balzani, V.; Cola, L. D.; Vögtle, F. Toward photo switchable dendritic hosts: Interaction between azobenzene-functionalized dendrimers and eosin. *J. Am. Chem. Soc.*, **1998**, *120*, 12187-12191.

[230]   Yeung, L. K.; Crooks, R. M. Heck heterocoupling within a dendritic nanoreactor. *Nano Letts.* **2001**, *1*, 14-17.

[231]   Ledesma-Garcia, J.; Manriquez, J.; Gutierrez-Granados, S.; Godinez, L. A. Dendrimer modified thiolated gold surfaces as sensor devices for halogenated alkyl-carboxylic acids in aqueous medium: A promising new type of surfaces for electroanalytical applications. *Electroanalysis*, **2003**, *7,* 657.

[232]   Watanabe, S.; Regen, S. L. Dendrimers as building blocks for multilayer construction. *J. Am. Chem. Soc.*, **1994**, *116*, 8855-8856.

[233]   Tsukruk, V. V.; Rinderspacher, F.; Bliznyuk. V. N. Self-assembled multilayer films from dendrimers. *Langmuir*, **1997**, *13*, 2171-2176.

[234]   Bliznyuk, V. N.; Rinderspacher, F.; Tsukruk, V. V. On the structure of polyamidoamine dendrimer monolayers. *Polymer*, **1998**, *39*, 5249-5252.

[235]   Tsukruk, V. V. Dendritic macromolecules at interfaces. *Adv. Mater.*, **1998**, *10*, 253-257.

[236]   Hierlemann, A.; Campbell, J. K.; Baker, L. A.; Crooks, R. M.; Ricco, A. J. Structural Distortion of Dendrimers on Gold Surfaces: A tapping mode AFM investigation. *J. Am. Chem. Soc.*, **1998**, *120*, 5323-5324.

[237]   Li, J.; Piehler, L. T.; Qin, D.; Baker, J. R.; Tomalia, J. D. A. Visualization and characterization of poly (amidoamine) dendrimers by atomic force microscopy. *Langmuir*, **2000**, *16*, 5613-5616.

[238]   Shen, L.; Hu, N. Heme protein films with polyamidoamine dendrimer: Direct electrochemistry and electrocatalysis. *Biochim. Biophys. Acta, Bioenerg.*, **2004**, *1608*, 23-33.

[239]   Yoon, H. C.; Kim, H. S. Multilayered assembly of dendrimers with enzymes on gold: Thickness controlled biosensing interface. *Anal. Chem.*, **2000**, *72*, 922-926.

[240]   Yoon, H. C.; Cox, M. Y. Preparation of multilayered nanocomposites of polyoxometalates and poly (Amidoamine) dendrimers. *Electrochem. Commun.*, **2001** *3*, 285-289.

[241]   Svobodova, L.; Nejda´rkova, M. S.; Hianik, T. Properties of glucose biosensors based on dendrimer layers effect of enzyme immobilization. *Anal. Bioanal. Chem.*, **2002**, *373*, 735-741.

[242]   Snejdarkova, M.; Svobodova, L.; Nikolelis, D. P.; Wang, J.; Hianik, T. Acetylcholine biosensor based on dendrimer layers for pesticides detection. *Electroanalysis*, **2003**, *15*, 1185-1191.

[243]   Lu, X.; Imae, T. Size-controlled *in situ* synthesis of metal nanoparticles on dendrimer modified carbon nanotubes. *J. Phys. Chem. C*, **2007**, *111*, 2416-2420.

[244]   Koda, S.; Inoue, Y.; Iwata, H. Gene transfection into adherent cells using electroporation on a dendrimer modified gold electrode. *Langmuir,* **2008**, *24*, 13525-13531.

[245]   Humenik, M.; Pohlmann, C.; Wang, Y.; Sprinzl, M. Enhancement of electrochemical signal on gold electrodes by polyvalent esterase-dendrimer clusters. *Bioconjug. Chem.*, **2008**, *19*, 2456-2461.

[246]   Kim J. M.; Ju, H.; Choi, H. S.; Lee, J.; Kim, K. J. J.; Kim, H. D.; Kim, J. Self-assembled dendrimer/polyelectrolyte layers on indium tin oxide electrodes as a matrix for immobilization of gold nanoparticles and fluorophores. *Bull. Kor. Chem. Soc.*, **2010**, *31*: 491-494.

[247]   Oh, S. K.; Kim, Y. G.; Ye, H.; Crooks, R. M. Synthesis, characterization, and surface immobilization of metal nanoparticles encapsulated within bifunctionalized dendrimers. *Langmuir*, **2003**, *19*, 10420-10425.

[248]   Sun, L; Crooks, R. M. Dendrimer mediated immobilization of catalytic nanoparticles on flat, solid supports. *Langmuir*, **2002**, *18*, 8231-8236.

[249]   Ye, H.; Crooks, R. M. Electrocatalytic O_2 reduction at glassy carbon electrodes modified with dendrimer-encapsulated Pt nanoparticles. *J. Am. Chem. Soc.*, **2005**, *127*, 4930-4934.

[250]   Kaloyeros, A. E.; Eisenbraun, E. Ultrathin diffusion bsrriers/liners for gigascale copper metallization. *Annu. Rev. Mater. Sci.*, **2000**, *30*, 363-385.

[251]   Min, K. H.; Chun, K. C.; Kim, K. B. Comparative study of tantalum and tantalum nitrides as a diffusion barrier for Cu metallization. *J. Vac. Sci. Technol. B*, **2009**, *14*, 3263-3269.

[252]   Chuang, J. C.; Chen, M. C. Passivation of Cu by sputter deposited Ta and reactively sputter-deposited Ta-nitride layers. *J. Electrochem. Soc.*, **1998**, *145*, 3170-3177.

[253]   Danek, M. *Metal Barriers for Advanced Cu Interconnect Application,* 11th Dielectric and CVD Metallization Symposium, San Diego, **2000**.

[254]   Ramberg, C. E.; Blanquet, E.; Pons, M.; Bernard, C.; Madar, R. Application of equilibrium thermodynamics to the development of diffusion barriers for copper metallization. *Microelectron. Eng.*, **2000**, *50*, 357-368.

[255]   Lide, D. *CRC Handbook of Chemistry and Physics*, 71th ed.; CRC Press: Boston, **1990-1991**, *pp.*4-109.

[256]   Winter, C. H. The Chemical vapor deposition of metal Nitride films using modern metalorganic precursors. *Aldrichim. Acta*, **2000**, *33*, 3-8.

[257]   Tsai, M. H.; Sun, S. C.; Chiu, H. T.; Tsai, C. E.; Chuang, C. H. Metal organic chemical vapor deposition of tantalum nitride by tertbutylimidotris (diethylamido) tantalum for advanced metallization. *Appl. Phys. Letts.*, **1995**, *67*, 1128.

[258]   Chiu, H. T.; Chang, W. P. Deposition of Tantalum Nitride thin films from ethylimido-tantalum complex. *J. Mater. Sci. Lett.*, **1992**, *11*, 96.

[259]   Chaneliere, C.; Autran, J. L.; Devine, R. A. B.; Balland, B. Tantalum pentoxide (Ta_2O_5) thin films for advanced dielectric applications. *Mater. Sci. Eng. Rep.*, **1998**, *R22*, 269-322.

[260]   Ayerdi, I.; Castaño, E.; García-Alonso, A.; Gracia, J. High-temperature ceramic pressure sensor. *Sens. Actuatos. B*, **1997**, *A60*, 72c-75.

[261]   Vogel E. M, *High-k Gate Dielectrics*, MRS Workshop, New Orleans, Loss Angeles, **2000**.

[262]   Gordon, R. G.; *et al. Atomic layer deposition-conference*, AVS, Monterey, California, **2001**.

[263]   Murray, C. B.; Norris, D. J.; Bawendi, M. G. J. Synthesis and characterization of nearly monodisperse CdE (E = sulfur, selenium, tellurium) semiconductor nanocrystallites. *J. Am. Chem. Soc.*, **1993**, *115*, 8706.

[264]   Murray, C. B.; Kagan, C. R.; Bawendi, M. G. Synthesis and characterization of monodisperse nanocrystals and close packed nanocrystal assemblies. *Annu. Rev. Mater. Sci.*, **2000**, *30*, 545.

[265] Peng, X.; Manna, L.; Yang, W.; Wickham, J.; Scher, E.; Kadavanich, A.; Alivisatos, A. P. Shape control of CdSe nanocrystals. *Nature*, **2000**, *404*, 59-61.

[266] Manna, L.; Scher, E. C.; Alivisatos, A. P. Synthesis of soluble and processable rod-, arrow-, teardrop, and tetrapod-shaped CdSe nanocrystals. *J. Am. Chem. Soc.*, **2000**, *122*, 12700-12706.

[267] Welser, J. J.; Tiwari, S.; Rishton, S.; Lee, K. Y.; Lee, Y. Room temperature operation of a quantum-dot flash memory. *IEEE Electron. Device Letts.*, **1997**, *18*, 278-280.

[268] Murray, C. B.; Kagan, C. R.; Bawendi, M. G. Self organization of CdSe nanocrystallites into three dimensional quantum dot superlattices. *Science*, **1995**, *270*, 1335-1338.

[269] Huyhn, W.; Peng, X.; Alivisatos, A. P. CdSe nanocrystal rods/poly (3-hexylthiophene) composite photovoltaic devices. *Adv. Mater.*, **1999**, *11*, 923-927.

[270] Vlasov, Y. A.; Yao, N.; Norris, D. J. Synthesis of photonic crystals for optical wavelengths from semiconductor quantum dots. *Adv. Mater.*, **1999**, *11*, 165-169.

[271] Klimov, V. I. Mechanisms for photogeneration and recombination of multiexcitons in semiconductor nanocrystals: implications for lasing and solar energy conversion. *J. Phys. Chem. B*, **2006**, *110*, 16827-16845.

[272] Efros, A. L.; Rosen, M. The electronic structure of semiconductor nanocrystals. *Annu. Rev. Mater. Sci.*, **2000**, *30*, 475-521.

[273] Nirmal, M.; Brus, L. Luminescence photophysics in semiconductor nanocrystals. *Acc. Chem. Res.*, **1999**, *32*, 407-414.

[274] Winiarz, J. G.; Zhang, L.; Lal, M.; Friend, C. S.; Prasad, P. N. Photogeneration charge transport, and photoconductivity of a novel PVK/CdS-nanocrystal polymer composite. *Chem. Phys.*, **1999**, *245*, 417-428.

[275] Qi, L.; Colfen, H.; Antonietti, M. Synthesis and characterization of CdS nanoparticles stabilized by double-hydrophilic block copolymers. *Nano Letts.*, **2001**, *1*, 6-65.

[276] Hu, K.; Brust, M.; Bard, A. J. Characterization and surface charge measurement of self-assembled CdS nanoparticle films. *Chem. Mater.*, **1998**, *10*, 1160-1165.

[277] Brus, L. Electronic wave functions in semiconductor clusters: experiment and theory. *J. Phys. Chem.*, **1986**, *90*, 2555-2560.

[278] Henglein, A. Small-particle research: physicochemical properties of extremely small colloidal metal and semiconductor particles. *Chem. Rev.*, **1989**, *89*, 1861-1873.

[279] Amma, B. S.; Manzoor, K.; Ramakrishnaa, K.; Pattabi, M. Synthesis and optical properties of CdS/ZnS coreshell nanoparticles. *Mater. Chem. Phys.*, **2008**, *112*, 789-792.

[280] Verma, P.; Gali, S. M.; Pandey, A. C. Organic capping effect and mechanismin of Mn-doped CdS nanocomposites. *Physica. B*, **2010**, *405*, 1253-1257.

[281] Glaspell, G.; Dutta, P.; Manivannan, A. Room temperature and microwave synthesis of M-doped ZnO (M Co, Cr, Fe, Mn & Ni). *J. Cluster. Sci.*, **2005**, *16*, 523-536.

[282] Cong, C. J.; Hong, J. H.; Zhanga, K. L. Effect of atmosphere on the magnetic properties of the Co-doped ZnO magnetic semiconductors. *Mater. Chem. Phys.*, **2009**, *113*, 435-440.

[283] Singh, S.; Kumar, E. S.; Rao, M. S. R. Microstructural, optical and electrical properties of Cr-doped ZnO. *Scr. Mater.*, **2008**, *58*, 866-869.

[284] Thota, S.; Dutta, T.; Kumar, J. On the sol-gel synthesis and thermal, structural, and magnetic studies of transition metal (Ni, Co, Mn) containing ZnO powders. *J. Phys. Condens. Matter.*, **2006**, *18*, 2473.

[285] Chan, W. C. W.; Nie, S. Quantum dot bioconjugates for ultrasensitive nonisotopic detection. *Science*, **1998**, *281*, 2016-2018.

[286] Alivisatos, A. P. Semiconductor clusters, nanocrystals, and quantum dots. *Science*, **1996**, *271*, 933-937.

[287] Dahan, M.; Laurence, T.; Pinaud, F.; Chemla, D. S.; Alivisatos, A. P.; Sauer, M. Time gated biological imaging by use of colloidal quantum dots. *Opt. Letts*, **2001**, *26*, 825-827.

[288] Bruchez, M.; Moronne, M.; Gin, P.; Weiss, S.; Alivisatos, A. P. Semiconductor nanocrystals as fluorescent biological labels. *Science*, **1998**, *281*, 2013-6.

[289] Sharma, P. K.; Kumar, M.; Singh, P. K.; Pandey, A. C.; Singh, V. N. Properties of Sol-gel derived YAG: Eu3^{+} hierarchical nanostructures with their time evolution studies. *J. Appl. Phys.*, **2009**, *105*, 034309.

[290] Sharma, P. K.; Pandey, A. C.; Zolnierkiewicz, G.; Guskos, N.; Rudowicz, C. Relationship between oxygen defects and photoluminescence property of ZnO nanoparticles: A spectroscopic view. *J. Appl. Phys.*, **2009**, *106*, 094314.

[291] Yadav, R. S.; Mishra, P.; Pandey, A. C. Growth mechanism and optical property of ZnO nanoparticles synthesized by sonochemical method. *Ultrason. Sonochem.*, **2008**, *15*, 863.

[292] Sharma, P. K.; Dutta, R. K.; Kumar, M.; Singh, P. K.; Pandey, A. C. Luminescence studies and formation mechanism of symmetrically dispersed ZnO quantum dots embedded in SiO$_2$ matrix. *J. Lumin.*, **2009**, *129*, 605.

[293] Parashar, V.; Parashar, R.; Sharma, B.; Pandey, A. C. Parthenium Leaf Extract Mediated Synthesis of Silver Nanoparticles: A Novel Approach towards Weed Utilization. *Dig. J. Nanomater. Biostruct.*, **2009**, *4*, 4550.

[294] Mishra, P.; Yadav, R. S.; Pandey, A. C. Starch assisted sonochemical synthesis of flower-like ZnO nanostructure. *Dig. J. Nanomater. Biostruct.*, **2009**, *4*, 193-198.

[295] Dutta, R. K.; Sharma, P. K.; Pandey, A. C. Surface enhanced Raman spectra of *Escherichia Coli* Cells using ZnO nanoparticles. *Dig. J. Nanomater. Biostruct.*, **2009**, *4*, 83-87.

[296] Nakamura, S. The roles of structural imperfections in InGaN based blue light-emitting diodes and laser diodes. *Science*, **1998**, *281*, 956-961.

[297] Mirkin, C. A. Tweezers for the Nanotool Kit. *Science*, **1999**, *286*, 2095-2096.

[298] Karpina, V. A.; Lazorenko, V. I.; Lashkarev, C. V.; Dobrowolski, V. D.; Kopylova, L. I.; Baturin, V. A.; Lytuyn, S. A.; Ovsyannikov, V. P.; Mauvenko, E. A. Zinc oxide-analogue of GaN with new perspective possibilities. *Cryst. Res. Technol.*, **2004**, *39*, 980-992.

[299] Honma, H. S.; Yamada, K.; Bae, J. M. Synthesis of organic/inorganic nanocomposites protonic conducting membrane through sol-gel processes. *Solid State Ionics*, **1999**, *118*, 29-36.

[300] Phely-Bobin, T. S.; Muisener, R. J.; Koberstein, J. T.; Papadinmitrakopoulos, F. Site-specific self-assembly of Si/SiOx nanoparticles on micropatterned poly (dimethylsiloxane) thin films. *Synth. Met.*, **2001**, *116*, 439-443.

[301] Sharma, P. K.; Dutta, R. K.; Kumar, M.; Singh, P. K.; Pandey, A. C. Luminescence studies and formation mechanism of symmetrically dispersed ZnO quantum dots embedded in SiO$_2$ matrix. *J. Lumin.*, **2009**, *129*, 605-610.

[302] Alivisatos, A. P. Semiconductor clusters, nanocrystals, and quantum dots. *Science*, **1996**, *271*, 933-937.

[303] Tessler, N.; Medvedev, V.; Kazes, M.; Kan, S.; Banin, U. Efficient near-infrared polymer nanocrystal light-emitting diodes. *Science*, **2002**, *295*, 1506-1508.

[304] Klimov, V. L.; Mikhailowsky, A. A.; Xu, S.; Malko, A.; Hallingsworth, J. A.; Leatherdale, C. A.; Eisler, H. J.; Bawendi, M. G. Optical Gain and Stimulated Emission in Nanocrystal Quantum Dots. *Science*, **2002**, *290*, 314-317.

[305] Han, M.; Gao, X.; Su, J. Z.; Nie, S. Quantum-dot-tagged microbeads for multiplexed optical coding of biomolecules. *Nat. Biotechnol.*, **2001**, *19*, 631-635.

[306] Alivisatos, A. P. J. Perspectives on the physical chemistry of semiconductor nanocrystals. *Phys. Chem.B*, **1996**, *100*, 13226-13239.

[307] Peng, X.; Wickham, J.; Alivisatos, A. P. J. Kinetics of II-VI and III-V colloidal semiconductor nanocrystal growth: focusing of size distributions. *J. Am. Chem. Soc.*, **1998**, *120*, 5343-5344.

[308] Peng, X.; Manna, L.; Yang, W. D.; Wickham, J.; Scher, E.; Kadavanich, A.; Alivisatos, A. P. Shape control of CdSe nanocrystals. *Nature*, **2000**, *404*, 59-61.

[309] Qu, L.; Peng, X. J. Control of photoluminescence properties of CdSe nanocrystals in growth. *Am. Chem. Soc.*, **2002**, *124*, 2049-2055.

[310] Dameron, C. T.; Reese, R. N.; Mehra, R. K.; Kortan, A. R.; Carroll, P. J.; Steigerwald, M. L.; Brus, L. E.; Winge, D. R. Biosynthesis of cadmium sulphide quantum semiconductor crystallites. *Nature*, **1989**, *338*, 596-597.

[311] Braun, E.; Eichen, Y.; Sivan, U.; Yoseph, G. B. DNA Templated self assembly of a conductive wire connecting two electrodes. *Nature*, **1998**, *391*, 775-778.

[312] Shenton, W.; Douglas, T.; Young, M.; Stubbs, G.; Mann, S. Inorganic-organic nanotube composites from template mineralization of Tobacco Mosaic Virus. *Adv. Mater.*, **1999**, *11*, 253-256.

[313] Slocik, J. M.; Moore, J. T.; Wright, D. W. Monoclonal antibody recognition of Histidine-rich peptide encapsulated nanoclusters. *Nano Letts.*, **2002**, *2*, 169-173.

[314] Reese, R. N.; Winge, D. R. Sulfide stabilization of the cadmium-gamma-glutamyl peptide complex of *Schizosaccharomyces pombe*. *J. Biol. Chem.*, **1988**, *263*, 12832-12835.

[315] Storhoff, J. J.; Mirkin, C. A. Programmed materials synthesis with DNA. *Chem. Rev.*, **1999**, *99*, 1849-1862.

[316] Cui, H. N.; Zhang, H. J.; Xi, S. Q. Preparation and photoluminescence study of ultrafine cadmium sulfide particles. *J. Mater. Sci. Letts.*, **1998**, *17*, 913-915.

[317] Li, Z.; Du, Y. M. Biomimic synthesis of CdS nanoparticles with enhanced luminescence. *Mater. Letts.*, **2003**, *57*, 2480-2484.

[318] Willner, I.; Willner, B. Layered molecular optoelectronic assemblies. *J. Mater. Chem.*, **1998**, *8*, 2543-2556.

[319] Ballardini, B.; Balzani, V.; Credi, A.; Gandolfi, M. T.; Venturi, M. Artificial molecular-level machines: which energy to make them work? *Acc. Chem. Res.*, **2001**, *34*, 445-455.

[320] Raymo, F. M. Digital processing and communication with molecular switches. *Adv. Mater.*, **2002**, *14*, 393-396.

[321] Hopfield, J. J.; Onuchic, J. N.; Beratan, D. N. An electronic shift registers memory based on molecular electron transfer reactions. *J. Phys. Chem.*, **1989**, *93*, 6350-6356.

[322] Guisinger, N. P.; Yoder, N. L.; Hersam, M. C. Probing charge transport at the single-molecule level on silicon by using cryogenic ultra-high vacuum scanning tunneling microscopy. *Proc. Natl. Acad. Sci. USA*, **2005**, *102*, 8838-8843.

[323] Liu, Z.; Yasseri, A. A.; Lindsey, J. S.; Bocian, D. F. Molecular memories that survive silicon device processing and real-world operation. *Science*, **2003**, *302*, 1543-1545.

[324] Weizmann, Y.; Elnathan, R.; Lioubashevski, O.; Willner, I. Endonuclease based logic gates and sensors using magnetic force-amplified readout of DNA scission on cantilevers. *J. Am. Chem. Soc.*, **2005**, *127*, 12666-12672.

[325] Baron, R.; Lioubashevski, O.; Katz, E.; Niazov, T.; Willner, I. Logic gates and elementary computing by enzymes. *J. Phys. Chem. A*, **2006**, *110*, 8548-8553.

[326] Dubrovsky, T. B.; Hou, Z.; Stroeve, P.; Abbott, N. L. Self-assembled monolayers formed on electroless gold deposited on silica gel: A potential stationary phase for biological assays. *Anal. Chem.*, **1999**, *71*, 327-332.

[327] Sambrook, J.; Fritsch, E. F.; Maniatis, T. *Molecular Cloning*, 3rd ed.; Cold Spring Harbor Laboratory Press: New York; **1989**.

[328] Kim, S. U.; Kim, Y. J.; Choi, S. G.; Yea, C. H.; Singh, R. P.; Min, J.; Oh, B. K.; Choi, J. W. Direct immobilization of cupredoxin azurin modified by site-directed mutagenesis on gold surface. *Ultramicroscopy*, **2008**, *108*, 1390-1395.

[329] Yagati, A. K.; Kim, S. U.; Min, J.; Choi, J. W. Write once read many times (WORM) biomemory device consisting of cysteine modified ferredoxin. *Electrochem. Commun.*, **2009**, *11*, 854-858.

[330] Yagati, A. K.; Kim, S. U.; Min, J.; Choi, J. W. Multi-bit biomemory consisting of recombinant protein variants, azurin. *Biosens. Bioelectron.*, **2009**, *24*, 1503-1507.

[331] Ahn, J. M.; Kim, B. C.; Gu, M. B. Characterization of gltA: lux *CDABE* fusion in *Escherichia coli* as a toxicity biosensor. *Biotech. Bioproc. Eng.*, **2006**, *11*, 516-521.

[332] Choi, J. W.; Nam, Y. S.; Park, S. J.; Lee, W. H.; Kim, D.; Fujihira, M. Rectified photocurrent of molecular photodiode consisting of cytochrome c/GFP hetero thin films. *Biosens. Bioelectron.*, **2001**, *16*, 819-825.

[333] Choi, J. W.; Fujihira, M. Molecular-scale biophotodiode consisting of a green fluorescent protein/cytochrome c self-assembled heterolayer. *Appl. Phys. Letts.*, **2004**, *84*, 2187-2189.

[334] Maruccio, G.; Marzo, P.; Krahne, R.; Passaseo, A.; Cingolani, R.; Rinaldi, R. Protein conduction and negative differential resistance in large-scale nanojunction arrays. *Small*, **2007**, *3*, 1184-1188.

[335] Roth, K. M.; Lindsey, J. S.; Bocian, D. F.; Kuhr, W. G. Characterization of charge storage in redox active self assembled monolayers. *Langmuir*, **2002**, *18*, 4030-4040.

[336] Tomizaki, K.; Mihara, H. Phosphate mediated molecular memory driven by two different protein kinases as information input elements. *J. Am. Chem. Soc.*, **2007**, *129*, 8345-8352.

[337] Lee, W.; Park, K. S.; Kim, Y. W.; Lee, W. H.; Choi, J. W. Protein array consisting of sol-gel bioactive platform for detection of E. coli O157:H7. *Biosens. Bioelectron.*, **2005**, *20*, 2292-2299.

[338] Oh, B. K.; Kim, Y. K.; Park, K. W.; Lee, W. H.; Choi, J. W. Surface plasmon resonance immunosensor for the detection of Salmonella typhimurium. *Biosens. Bioelectron.*, **2004**, *19*, 1497-1504.

[339] Oh, B. K.; Lee, W.; Chun, B. S.; Bae, Y. M.; Lee, W. H.; Choi, J. W. Fabrication of protein chip based on surface plasmon resonance for detection of pathogens. *Biosens. Bioelectron.*, **2005**, *20*, 1847-1850.

[340] Ouyang, J.; Chu, C. H.; Szmanda, C. R.; Ma, L.; Yang, Y. Programmable polymer thin film and non-volatile memory device. *Nat. Mater.*, **2004**, *3*, 918-922.

[341] Lee, T.; Kim, S.U.; Min, J.; Choi, J.W. Biomolecular memory device composed of Cytochrome c on a self-assembled 11-mercaptoundecanoic acid layer. *Jpn. J. Appl. Phys.*, **2010**, *49*, 1-6.

[342] Kim, S. U.; Yagati, A. K.; Singh, R. P.; Min, J.; Choi, J. W. Charge storage investigation in self-assembled monolayer of redox-active recombinant azurin. *Curr. Appl. Phys.*, **2009**, *9*, e71-e75.

[343] Kim, S. U.; Yagati, A. K.; Min, J.; Choi, J. W. Nanoscale protein-based memory device composed of recombinant azurin. *Biomater.*, **2010**, *31*, 1293-1298.

[344] Lee, T.; Kim, S. U.; Min, J.; Choi, J. W. Multilevel biomemory device consisting of recombinant Azurin/Cytochrome c. *Adv. Mater.*, **2010**, *22*, 510-514.

CHAPTER 2

Metal Nanoparticles-Based Affinity Biosensors

Giovanna Marrazza[*]

Università di Firenze, Dipartimento di Chimica, Via della Lastruccia, 3; 50019 Sesto Fiorentino (Fi) Italy

Abstract: A new emerging field that combines nanoscale materials and biosensor technology is receiving increased attention. Nanostructures have been used to achieve direct wiring of enzymes to electrode surfaces, to promote electrochemical reactions, impose nano barcodes on biomaterials, and amplify the signal from biorecognition events. NP-based sensors have found wide spread applications in the environmental and medical applications for their sensitivity, specificity, rapidity, simplicity, and cost-effectiveness.

The aim of this chapter, without pretending to being exhaustive, is mainly to review recent important achievements about metal nanoparticles preparation, their bio modification and the new applications for protein detections by means of a set of selected recent publications.

Keywords: Biosensors; metal nanoparticles; biomodification; protein detections.

INTRODUCTION

According to the International Union of Pure and Applied Chemistry (IUPAC) a biosensor is a self-contained integrated device, which is capable of providing specific quantitative or semi-quantitative analytical information using a biological recognition element which is retained in direct spatial contact with an electrochemical transduction element.

Affinity biosensors are a subclass of biosensors. The sensing element is a highly specific receptor; it is generally biologic (bioreceptors) such as enzymes, antibodies and nucleic acids. In the last years, the use of artificial or semi-artificial receptors is increasing. This class includes PNA (Peptide Nucleic Acid) LNA (Locked Nucleic Acid), the MIPs (Molecular Imprinted Polymers), the oligopeptides, the aptamers, and recently affibodies.

In a recent review, the state of the art and the recent developments in immunosensor have been described [1]. Homogeneous immunosensor, heterogeneous immunosensor, integrated immunosensor and biochip format immunosensor based on optical, electrochemical, magnetic or mechanical detection/transduction systems are reviewed. Most of the developed immunosensors include a sensing layer supporting a particular immobilised antigen or antibody. The solid support used is generally in close contact with a transducer needed for the detection of the formed immune complex. The immunosensors are based either on competitive or sandwich assay, when applied to the detection of low and high molecular weight molecules, respectively (Fig. **1A, 1B**). Two approaches could be considered when dealing with competitive immunosensor. A first-one in which immobilised antibodies react with free antigens in competition with labelled antigens (Fig. **1a**). A second-one, using immobilised antigens and labelled antibodies, is generally preferred and prevents all the problems related to antibody immobilisation (*i.e.* loss of affinity, orientation) (Fig. **1b**).

In protein-sensing devices the immobilised compound determines the specificity of the device, and the immobilisation method frequently influences parameters such as lower detection limit, sensitivity, dynamic range, reusability or liability for unspecific binding. Thus, varieties of immobilisation approaches have been developed, which are applicable to different supports onto which, the compound has to be immobilised [2]. The immobilization procedure is dependent on the assay format and detection transducer.

*Address correspondence to Giovanna Marrazza: Università di Firenze, Dipartimento di Chimica, *Via* della Lastruccia, 3; 50019 Sesto Fiorentino (Fi) Italy; Tel. +39-055-5253320; E-mail: giovanna.marrazza@unifi.it; Web: www.unifi.it/dclabi

Songjun Li, Yi Ge and He Li (Eds)
All rights reserved - © 2012 Bentham Science Publishers

A) Competitive assay

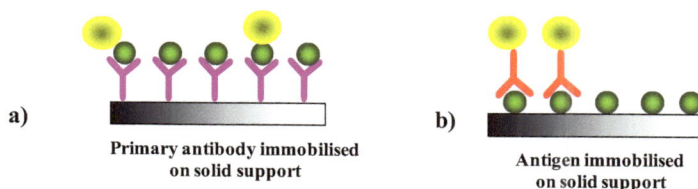

a)

**Primary antibody immobilised
on solid support**

b)

**Antigen immobilised
on solid support**

B) Sandwich assay

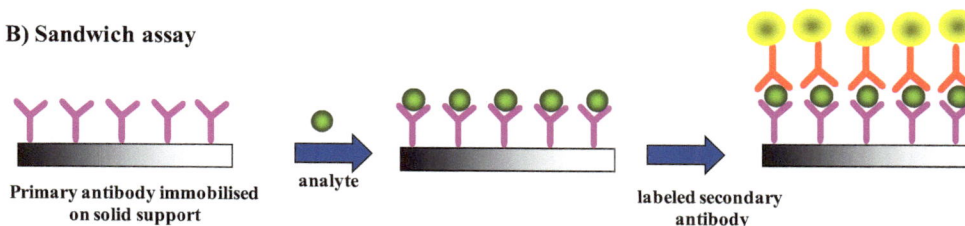

**Primary antibody immobilised
on solid support**

analyte

**labeled secondary
antibody**

Figure 1. A. Schematic representation of the competitive immunoassay: **1)** immobilised antibodies reacts with free antigens in competition with labeled antigens; **2)** immobilised antigens and labelled antibodies. **B.** Schematic representation of the sandwich immunoassay.

The coupling affinity biosensor with the metal nanoparticles (*i.e.* gold, silver, quantum dot *etc.*) provides good opportunities for building high-sensitivity bioassays.

The nanoparticles (NPs) have a diameter range 1-10 nm and would display electronic structures, reflecting the electronic band structure of the nanoparticles, owing to quantum-mechanical rules. The resulting physical properties are neither those of bulk metal nor those of molecular compounds, but they strongly depend on the particle size, inter-particle distance, nature of the protecting organic shell, and shape of the nanoparticles.

Gold nanoparticles have unique optical properties. The physical origin of this light absorption by gold nanoparticles is the coherent oscillation of the conduction electrons induced by the interacting electromagnetic field. Furthermore, they have a high surface area to volume ratio; the plasmon frequency is highly sensitive to the dielectric (refractive index) nature of its interface with the local medium, leading to colorimetric changes of the dispersions.

Quantum Dots (QDs) are metal nanoparticles of group II-VI compound like CdSe, ZnSe, CdTe, *etc.* As compared to organic fluorescent dyes, quantum dots are more photostable and the wavelength of the emitted light can be controlled by changing their size and composition of the materials. Furthermore, QDs have very broad excitation range but sharp emission, making it possible to excite different QDs with a single wavelength and yet result in variety of emission wavelengths.

Although the use of metal NPs in bioanalysis is a recent area of research, there are many publications on their medical applications for their unique biocompatibility, structural, electronic and catalytic properties. Among metal nanoparticles, silver nanoparticles (AgNPs) and gold nanoparticles (AuNPs) have several effective applications. AuNPs are important in imaging, as drug carriers, and for thermotherapy of biological targets [3-6]. AuNPs, nanoshells, nanorods, and nanowires have the extensive potential to be an integral part of our imaging toolbox and useful in the fight against cancer. AgNPs show improved antimicrobial activity.

In sensing and biosensing applications, the use of metal nanoparticle labels has proved to be particularly advantageous, due to the fact that optical or electrochemical analytical signals of the single biorecognition event (*i.e.* DNA hybridisation or immunoreaction) are significantly amplified. The versatile applications of NPs are strongly relating to the simplicity of synthesis, chemical and biological modifications. Particularly, the high affinity of thiols towards the surfaces of noble metals also facilitate the biofunctionalisation of these metallic nanostructure by utilizing the extensively developed and well-defined organic surface chemistry for biological modifications. Finally, NPs can be used as modifiers of the electrotransducer surfaces, creating nanostructurated surfaces in order to obtain better sensitivity, specificity and higher rates of recognition compared with current solutions.

In addition, metal nanoparticles have been widely applied to microanalytical systems (Lab on a chip) [7,8].

In the past few years, several excellent reviews have been published on the application of nanoparticles [9-15] and particularly on the use of gold nanoparticles [16-18] for the improvement of biosensing performance.

Here, without pretending to being exhaustive, the most recent applications of metal nanoparticles for electrochemical and optical immunosensors have been reported, highlighting some of their technical challenges and the new trends by means of a set of selected recent applications.

SYNTHESIS

Several physical and chemical processes for synthesis of metal nanoparticles were developed considering the nanoparticle applications in the nanobiotechnology area.

Chemical synthesis is usually preferred because NPs with uniform size, shape and surface functional groups by easy operation and control are obtained. NPs prepared by solution-based chemical reactions, are usually capped by organic shells called surface-capping agents or stabilizing agents. These agents contribute to colloidal stability and surface modification potential, thus preventing possible aggregation, and offering the possibility of a rich variety of functional groups and sites for biological modification.

Recently, there has been an increasing emphasis on the topic of green chemistry for the search of benign methods for the development nanoparticles and searching new natural compounds for biomedical applications (*i.e.* antibacterial, antioxidant, and antitumor activity). Biosynthetic processes have received much attention as a viable alternative for the development of metal nanoparticles where plant extract is used for the synthesis of nanoparticles without any chemical ingredients [19-25]. Leaf extracts of geranium, hibiscus, cinnamon, tamarind and coriander have also found suitable for the biosynthesis of silver and gold nanoparticles [26-30]. Room-temperature ionic liquids are attracting considerable interest in many fields of chemistry and industry, due to their potential as a green recyclable alternative to the traditional organic solvents. They are known to have the potential to enhance certain properties of metal nanoparticles, and also been used as stabilized agents to prepare inorganic nanoparticles [31].

In the following sections, the most common chemical methods for obtaining the metal NPs are reported.

GOLD NANOPARTICLES

One of the most simple and easily controlled processes is the citrate reduction technique of metal salt aqueous solutions. The method pioneered by J. Turkevich *et al.* in 1951 [32] and refined by G. Frens in 1973 [33], is the simplest one available. Generally, it is used to produce modestly monodisperse spherical gold nanoparticles suspended in water of around 10–20 nm in diameter. Larger particles can be produced, but this comes at the cost of mono-dispersion and shape.

In the citrate–gold process, a freshly prepared sodium citrate solution is introduced to a boiling solution of chloro auric acid (HAuCl$_4$). After a few minutes, the solution changes from colourless to a deep wine-red colour which suggests the formation of AuNPs. The Cetyl Trimethyl Ammonium Bromide (CTAB)-AuNPs are frequently used as seeds for synthesizing monodispersed gold nanorods with diverse aspect ratios. The CTAB solution is mixed with HAuCl$_4$ solution and ice-cold sodium borohydride (NaBH$_4$) solution is quickly added, resulting in the formation of a light-brown solution. Addition of the seed solution to a growth solution induces the growth of gold nanorods.

Additional methods have been developed for synthesizing AuNPs using different reductants and stabilizers.

The Brust-Schiffrin method for AuNP synthesis, published in 1994, has had a considerable impact on the overall field in less than a decade, because it allowed the facile synthesis of thermally stable and air-stable AuNPs of reduced dispersion and controlled size for the first time (ranging in diameter between 1.5 and 5.2 nm). Indeed, these AuNPs can be repeatedly isolated and redissolved in common organic solvents without irreversible aggregation or decomposition, and they can be easily handled and functionalised just as stable organic and molecular compounds [34].

A number of methods have been reported to synthesize the AuNPs in aqueous media in a recent review [35].

SILVER NANOPARTICLES

There are many different synthetic routes to silver nanoparticles.

The ion implantation has been shown to produce silver particles embedded in polyurethane, silicone, polyethylene, and polymethylmethacrylate. The particles grow in the substrate with the bombardment of ions. The existence of nanoparticles is proven with optical absorbance, though the exact nature of the particles created with this method is not known.

Typically, the wet chemical methods involve the reduction of a silver salt such as silver nitrate with a reducing agent like sodium borohydride in the presence of a colloidal stabilizer. Sodium borohydride has been used with polyvinyl alcohol, poly(vinylpyrrolidone), Bovine Serum Albumin (BSA), citrate and cellulose as stabilizing agents. It is important to note, not all nanoparticles are created equal for size and shape [35,36].

PLATINUM NANOPARTICLES

Platinum nanoparticles are usually in the form of a suspension or colloid of sub-micrometre-sized particles of platinum in a fluid, usually water.

Platinum nanoparticles are obtained by reduction of hexachloroplatinate. After dissolving hexachloroplatinate, the solution is rapidly stirred while a reducing agent is added. This causes platinum ions to be reduced to neutral platinum atoms. As more and more of these platinum atoms form, the solution becomes supersaturated and platinum gradually starts to precipitate in the form of sub-nanometre particles. The rest of the platinum atoms that form stick to the existing particles, and, if the solution is stirred vigorously enough, the particles will be fairly uniform in size.

To prevent the particles from aggregating, some sort of stabilizing agent or stabilizer that sticks to the nanoparticle surface is usually added [13].

QUANTUM DOTS

Typical dots are made of binary alloys such as cadmium selenide, cadmium sulphide, indium arsenide, and indium phosphide. There are colloidal methods to produce many different quantum dots. The synthesis is based on a three-component system composed of: precursors, organic surfactants, and solvents. When heating a reaction medium to a sufficiently high temperature, the precursors chemically transform into monomers. Once the monomers reach a high enough super-saturation level, the nanocrystal growth starts with a nucleation process. The temperature during the growth process is one of the critical factors in determining optimal conditions for the nanocrystal growth. Another critical factor that has to be stringently controlled during nanocrystal growth is the monomer concentration [13, 36].

BIOMODIFICATION

For further metal NPs applications in biosensor technology, the attaching the biomolecular recognition has to be readily achieved. The correct bio-receptor immobilisation on the metal NPs is crucial for the bioassay efficiency. A good strategy of immobilisation coupled to the adapted surface chemistry is the solution to ensure stability, correct orientation and retention of biological activity. There is no universal approach for the surface modification but more a set of methods which depend on the type of biomolecular recognition system and transducer method. The interaction between the biological active part and the NPs can be obtained with different approaches from fast and less stable processes to more time consuming and more robust ones.

The most used procedures for modifying the metal surface NPs are: electrostatic interactions, hydrophobic interactions used for specific recognition (antibody–antigen, biotin–avidin, *etc.*) and covalent coupling as shown in Fig. **2**.

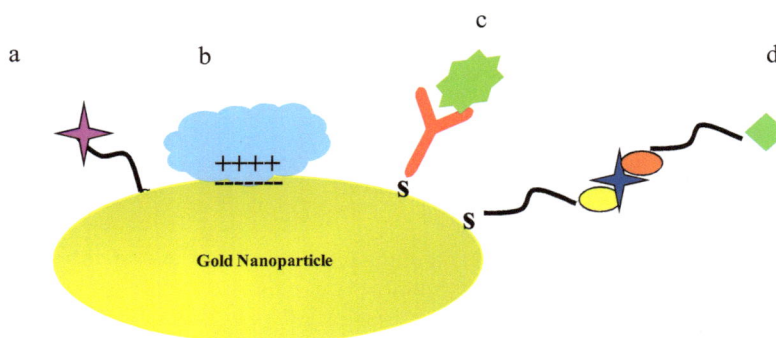

Figure 2. Schematic representation of modified gold nanoparticle probes. **a)** thiolated or disulfide modified ligand, **b)** electrostatic interaction, **c)** antibody-antigen association, **d)** streptavidin-biotin binding.

Physical adsorption on the solid surface is the most simple and fastest approach (no reagents or bioreceptor modifications are evolved). This method is based in weak interactions like Van der Waals, hydrogen bonding, hydrophobic or electrostatic interactions. Van der Waals interactions are based in dipole-dipole attractions. Biomolecules can create positive or negative dipoles in originally non polar areas due to intramolecular interactions that disturb the electron clouds. When the biomolecule are immobilised, their dipoles align to maximize the interaction with the electric dipoles of the molecules in the surface. Hydrogen bonding occurs when a hydrogen atom covalently bound to an electronegative element is attracted by another electronegative element creating a relatively strong interaction.

The hydrophobic interactions are related to the presence of amino acids as phenylalanine and leucine that are non polar and hence interact poorly with polar molecules like water. For this reason, most of the non polar residues are directed toward the interior of the molecule whereas such polar groups as aspartic acid and lysine are on the surface exposed to the solvent. When the surface is functionalised with a hydrophobic layer, it is energetically more favourable for the non polar residues to approach the surface creating a hydrophobic interaction.

Electrostatic interaction or physical adsorption is a simple process with the benefits of time saving and reduced complexity of ligand preparation. Its relative simplicity gives this approach certain advantages over the more complex covalent immobilisation methods. However, the immobilisation approaches result in a random orientation of the biomolecules since the orientation of the binding sites is not controlled. In addition, the biomolecules immobilisation can be disturbed by pH or temperature changes. This results in a strong non-specific interaction between the NPs and bioreceptors which leads to decreased detection selectivity; confirming the validity of the method, the non-specific signals are difficult to be minimised.

The covalent attachment on functionalised NPs, affinity immobilisation and self-assembling are, to date, the most successful approaches. This is mainly because each of these immobilisation strategies can lead to ordered sets of end-point attached and properly oriented binding sites. Moreover, such chemistries also allow controlling the conformational freedom of the bioreceptors and the corresponding inter-chain space through the modulation of the surface coverage. In comparison with the electrostatic interaction or physical adsorption immobilisation techniques discussed above, covalent binding is normally more complex, sometimes requiring intensive synthesis work on the ligands.

The most techniques reported for immobilizing ligands to gold and silver NPs surfaces are based on covalent bond formation between the sulphur containing ligands (*i.e.*, thiol, disulfide and thiolester) and the metal atoms on the particle surfaces. The organothiol ligands form generally well ordered and crystalline monolayers on the metal surfaces. For example, when exposed to the gold surface, normally in an ethanolic solution, the sulphur-gold bond is immediately formed, but it takes several hours to reach the crystalline close parking of the layer. The chain length is very important for the monolayer quality. Organothiols with more than 10 methylene groups give better protection to the sulphur-gold bond and avoid oxidation, thus creating stable and reproducible monolayers. Moreover, NPs can be stabilized with various molecules including alkanethiolates, glutathione, tiopronin, thiolated poly(ethyleneglycol) and so on, by their facile and robust interaction with thiol or disulfide groups.

A recent study on modified gold nanoparticles surface by the direct covalent linking of the antibody to the nanoparticles and assembling them onto the electrode surface is reported [37]. In the paper, the authors shown that direct linking of antibody to the AuNPs using covalent methods can be used for the development of immunosensors. This kind of approach provides a right orientation of the antibody with the free antigen binding sites for further immunoreaction without affecting the structure and function of the antibody. The AuNPs also helps in better immobilisation of the molecules onto the electrode preventing them from dissolving back into the bulk solution.

In the last years, synthetic bioreceptors such as aptamers and affibodies are being explored as tools affinity biosensor applications, as alternatives to traditionally used polyclonal or monoclonal antibodies.

Aptamers are nucleic acid ligands that have been designed through an *in vitro* selection process called SELEX (Systematic Evolution of Ligands by Exponential Enrichment). Because of the high diversity of molecular shapes of all possible nucleotide sequences, aptamers have been selected for a wide array of targets, including proteins, carbohydrates, lipids, or small molecules [38].

Affibody® are a class of small and robust affinity proteins that can be generated to interact with a variety of antigens, thus having the potential to provide useful tools for biotechnological research and diagnostic applications [39]. They are generated *via* selections from combinatorial constructed libraries displayed on phages. They are small (58 amino acids), three-helix bundle proteins that are lacking cysteines. They are built from the non-immunoglobulin scaffold named Z, which is derived from Staphylococcal protein A [40,41].

The synthetic receptors are usually coupled with nanomaterials by covalent linkage, non-covalent linkage and by streptavidin/biotin interaction. For instance, streptavidin-coated nanomaterials would easily adsorb biotinylated aptamers onto their surface. The hybridization of aptamers and nanoparticle tagged DNA has also been used for target detection, which is another popular format for combining aptamers with nanomaterials [42].

APPLICATIONS

In the following sections, the most recent NPs-based affinity biosensors for clinical targets using electrochemical and optical transducers are described.

ELECTROCHEMICAL DETECTION

Electrochemical affinity biosensors are attractive tools and have received considerable attention because they are easy and economical to mass production; they are robust, and achieve excellent detection limits with small analyte volumes. They have been widely used for several fields in electrochemical and electrical configurations. For biomedical diagnostics on biological samples, such as blood, electrochemical detection provides important route for non-transparent samples since the signal transduction can be accomplished by non-optical means.

The availability of a variety of new materials with unique properties at nanoscale dimension has attracted widespread attention in their utilisation for new electrochemical immunoassays.

The excellent electroactivity of gold and silver nanoparticles including different inorganic nanocrystals (*e.g.* ZnS, PbS, and CdS) and sensitive electrochemical techniques permit the detection of very low proteins concentrations (in the order of pM). NPs can be directly detected due to their own redox properties (*i.e.*, gold or heavy metals atoms constituents) or indirectly due to their electrocatalytic properties toward other species like silver reduction *etc.*

Affinity biosensors have emerged as a new technique for monitoring cancerous cells or their specific interaction with different analytes. This is because the conventional immunoassays require relatively long assay times and large, complicated detection devices. An excellent review has been published on the biosensor technology available today, areas which are currently being developed and researched for cancer markers diagnosis and a consideration of future prospects for the technology [43]. Some important cancer biomarkers, such as Alpha-Fetoprotein (AFP), Carcinoembryonic Antigen (CEA), carcinoma antigen 125 (CA125), Carbohydrate Antigen 19-9 (CA19-9), Human Chorionic Gonadotrophin (hCG) and prostate-specific antigen (PSA), have been detected by immunosensors for the diagnosis of liver, colon, and ovarian cancers, respectively.

Determination of the Carcinoma Antigen 125 (CA125) was carried out by using carbon electrodes modified with gold nanoparticles. The electrochemical immunosensor was designed by immobilizing anti-CA125 on a thionine and gold nanoparticles-modified carbon paste interface [44]. The same antibody was also immobilised on colloidal gold nanoparticles to form a bioconjugate stabilized with a cellulose acetate membrane on the surface of a GCE. A direct electrochemical immunoassay format was employed to detect CA125 antigen based on the current change before and after the antigen–antibody reaction. The current change was proportional to CA125 concentration ranging from 10 to 30 U/ml. The immunosensors were used to analyze CA125 in human serum specimens. Other approach was proposed for developing a competitive immunoassay format to detect CA125 antigen with HRP-labeled CA125 antibody as tracer, and o-phenylenediamine and hydrogen peroxide as enzyme substrates [45].

Another biosensor for CA125 detection was developed using anti-CA125 gold hollow microspheres and porous polythionine modified GCEs. The gold hollow microspheres greatly amplified the coverage of anti-CA125 molecules on the electrode surface. Electrochemical detection was accomplished by the amperometric changes occurring before and after the antigen–antibody interaction. The CA125 concentration was detected from 4.5 to 36.5 U/mL [46].

An immunoassay for human Chorionic Gonadotrophin (hCG) was proposed by using a conductive colloidal gold nanoparticle/titania sol–gel composite membrane deposited on a GCE *via* a vapour deposition method. Horseradish peroxidase-labeled hCG antibody (HRP-anti-hCG) was encapsulated into the composite architecture. The formation of the immunoconjugate between hCG and the immobilised HRP-anti-hCG introduced a barrier for the direct electrical communication between the immobilised enzyme and the electrode surface that can be monitorised by EIS. The hCG analyte could be determined in two linear ranges from 0.5 to 5.0 mIU/mL and 5.0 to 30 mIU/mL [47].

Idegami *et al.* [48] instead employed direct electrical detection of Au for the design of an immunosensor for the detection of hCG. Primary antibody was immobilised on screen-printed carbon strips and the captured antigen was sandwiched with Au-NP labelled secondary antibody. Gold nanoparticles were exposed to a preoxidation process and then reduced *via* DPV. Urine samples of pregnant and non-pregnant adults were tested and results compared with a standard ELISA test with good agreement, as evidenced by the fact that both tests were not able to detect any hCG in the urine sample of non-pregnant women.

Carbohydrate Antigen 19-9 (CA19-9) is one of the most important carbohydrate tumour markers, expressed in many malignancies as pancreatic, colorectal, gastric and hepatic carcinomas. An immunosensor for the rapid determination of CA19-9 in human serum has been developed by immobilisation of the antibody in colloidal gold nanoparticle-modified carbon paste electrodes and monitoring of the direct electrochemistry of HRP labelled to a CA19-9 antibody. The formation of the antigen–antibody complex blocked the electron transfer of HRP toward the electrode substrate that resulted in a significant current decrease; CA19-9 was detected from 2 to 30 U/ml [49].

Carcinoembryonic Antigen (CEA) is a well-known marker associated with progression of colorectal tumours. A current amplified immunosensor for its determination was fabricated by coating negatively charged polysulfanilic acid modified Glassy Carbon Electrodes (GCEs) with positively charged toluidine blue. This approach provided an interface containing amine groups to assemble gold nanoparticles for immobilisation of the carcinoembryonic antibody and HRP. Electrochemical Impedance Spectroscopy (EIS) and Cyclic Voltammetry (CV) were employed to characterize the electrochemical properties of the modified processes. The CVs reduction current of the immunosensor charged linearly in two concentration ranges of CEA from 0.5 to 5.0 and 5.0 to 120.0 ng/ml in presence of 0.3 mM H_2O_2 in analyte solution. [50].

Another immunosensor for CEA determination was proposed. A dual-amplification strategy was proposed *via* backfilling gold nanoparticles on (3-mercaptopropyl) trimethoxysilane sol–gel (MPTS) functionalized interface. MPTS acted as a building block for the electrode surface modification as well as a matrix for ligand functionalisation with first amplification. The second signal amplification strategy was based on the backfilling immobilisation of nanogold particles to the immunosensor surface. Using the non-competitive design, membrane potential change occurred before and after the antigen–antibody interaction. The dynamic concentration range obtained was from 4 to 86 ng/ml [51].

An example is the label-free immunoassay based on impedimetric measurements developed by Tang *et al.* [52] for CEA detection. Carcinoembryonic antibody was covalently attached to AuNPs and the composite immobilised on Au electrode by electrocopolymerisation with o-aminophenol. Electrochemical impedance spectroscopy studies demonstrated that the formation of antibody–antigen complexes increased the electron-transfer resistance of $[Fe(CN)6]^{3-/4-}$ redox probe at the poly-o-aminophenol/Carcinoembryonic antibody-AuNPs/Au electrode, thus monitoring of carcinoembryonic antigen concentration could be performed (detection limit 0.1 ng/mL).

A novel strategy for the fabrication of sensitive immunosensor to detect alfa-fetoprotein (AFP) in human serum has been proposed. The immunosensor was prepared by immobilizing AFP antigen onto the glassy carbon modified by gold nanoparticles and carbon nanotubes doped chitosan (AuNP/CNT/Ch) film. AuNP/CNT hybrids were produced by one-step synthesis based on the direct redox reaction. The electrochemical properties of AuNP/CNT/Ch films were characterized by impedance spectroscopy and cyclic voltammetry. Sample AFP, immobilised AFP, and alkaline phosphatase (ALP)-labeled antibody were incubated together for the determination based on a competitive immunoassay format. After the immunoassay reaction, the bound ALP label on the modified GC led to an amperometric response of 1-naphthyl phosphate which was changed with the different antigen concentrations in solution. Under the optimised experimental conditions, the resulting immunosensor could detect AFP with a detection limit of 0.6 ng/ml [53].

Sung at al. presented a novel electrochemical immunosensor which can be used for the detection of Vascular Endothelial Growth Factors (VEGFs) as a cancer-related biomarker. Gold nanoparticles were self-assembled onto an Indium Tin Oxide (ITO) electrode to prepare a modified sandwich type electrochemical immunoassay platform. VEGF antibodies were cleaved into two half-fragments and the fragments were immobilised onto the AuNP substrates by their thiol groups. The target molecules, VEGF, were then reacted with the AuNPs covered with the antibody fragments to form antigen–antibody complexes. Finally, the ferrocene-labeled VEGF antibodies were attached to VEGF. The redox current of ferrocene was measured by Differential Pulse Voltammetry (DPV) to quantify the VEGF concentration. Using the modified sandwich type immunoassay, 100 pg/ml of VEGF was detected [54].

In clinical diagnosis, detection of one biomarker cannot provide sufficient clinical information for all of the various cancer-related diseases, and the clinical information obtained from biomarkers is often related to the stage of tumori genesis, monitoring of treatment, and the state of the patient. Therefore, it is important to develop the microchip system with high multiplexing capabilities as well as an efficient detection method. Researches related to immunosensing microchips have achieved efficient multiplex detection of biomarkers.

A biobarcode assay (BCA) capable of achieving low detection limits and high specificity for both protein and DNA targets was developed by Goluch *et al.* [55]. The BCA utilizes gold nanoparticles (AuNPs) functionalised with oligonucleotides (the so-called biobarcodes, which serve as surrogate targets and amplifying agents) and a target-recognition element, which may be an antibody for protein detection or a unique oligonucleotide sequence for nucleic acid detection. The BCA also uses functionalised magnetic microparticles (MMPs), which are adorned with antibodies that bind to the target. In the presence of targets (protein or oligonucleotide molecules) in solution, the MMPs form a sandwich complex with the targets and gold nanoparticles, which can be localized and collected under an applied magnetic field. The barcode oligonucleotide molecules are then chemically released, identified, and quantified. The realisation of a BCA in a microfluidic format presents unique opportunities and challenges. A modified form of the BCA called the Surface Immobilised Biobarcode Assay (SI-BCA) was developed by the same research group [56]. The SI-BCA employs microchannel walls functionalized with antibodies that bind with the intended targets. Compared with the conventional BCA, it reduces the system complexity and results in shortened process time, which is attributed to significantly reduced diffusion times in the micro-scale channels. Raw serum samples, without any pre-treatment, were evaluated with this technique. Prostate specific antigen in the samples was detected at concentrations ranging from 40 pM to 40 fM. The entire assay, from sample injection to final data analysis was completed in 80 min.

There are several different strategies reported for electrochemical detection of analytes using NPs coupling with aptamers [57]. An example of using platinum nanoparticle (PtNP) as a catalyst label is reported [58]. The thrombin aptamer 1 was first immobilised on Au electrode and allowed to bind to thrombin. The PtNP tethered with thrombin aptamer 2 was then introduced to Form thrombin-aptamer-Pt NP sandwich complex. The PtNPs catalyzed H_2O_2 reduction to H_2O, resulting in cathodic currents which enabled thrombin detection with a detection limit as low as 1 nM. Similar methods using biocatalysts were also reported [59,60] but PtNPs showed higher sensitivity, probably due to the effectiveness of the PtNP catalyst.

An alternative application of Au-NPs was also reported by He *et al.* [61] for the development of label-free sensors for thrombin detection using aptamers. Anti-thrombin antibodies were immobilised on the microtitre plates to bind thrombin and the complex was sandwiched with aptamers conjugated to Au nanoparticles (biobarcode). After washing, a basic treatment allowed the collection of the barcode aptamers, which were further degraded and the amount of Adenine (proportional to the amount of bound thrombin) quantified by DPV. This assay takes advantage of the amplification potential of Au nanoparticles carrying numerous aptamer tags.

Enhancement of the analytical signal in detecting NPs can be achieved by catalytic deposition of another metal on the surface of the NP, followed by stripping and detection of this metal (Fig. **3**).

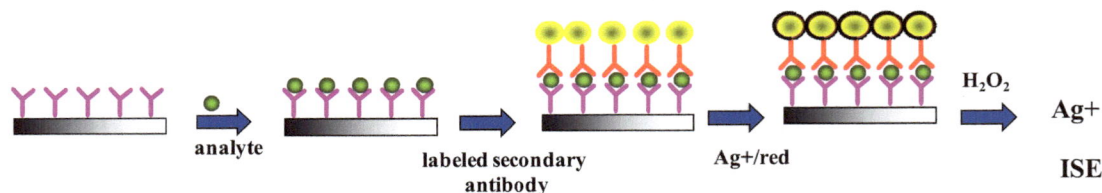

Figure 3. Potentiometric detection of sandwich immunoassay.

Other immunoassays were developed exploiting Au-NP labelling. Liao *et al.* [62] reported a highly sensitive assay based on the autocatalytic deposition of Au (III) onto AuNPs for the determination of rabbit IgG with square-wave stripping voltammetry (detection limit 1.6 fM).

Mao *et al.* [63] reported an example of this approach. They developed a novel electrochemical protocol for quantifying human IgG, based on precipitating copper on AuNP tags. The immunoassay was conducted by following the typical procedure for sandwich-type immunoreaction. Goat anti-human IgG was immobilised on the wells of microtiter plates. The human IgG analyte was first captured by the primary antibody and then sandwiched by secondary antibody labelled with gold nanoparticles. The copper enhancer solution was then added to deposit copper on the gold nanoparticle tags. After dissolved with HNO_3, the released copper ions were then quantified by ASV obtaining a LOD of 0.5 ng/ml.

Chu *et al.* [64] reported another catalytic approach, based on precipitating silver on AuNP labels using a silver-enhancement solution, and chemical reduction of silver ions to silver metal onto the surface of the AuNPs. After dissolving the silver metal in a nitric acid solution, silver ions were determined by ASV at a GCE. The method was evaluated using a non-competitive heterogeneous immunoassay of IgG as a model and achieving a LOD of 1 ng/mL, which is competitive with a colorimetric Enzyme-Linked Immunosorbent Assay (ELISA). The good performance of the method was attributed to the sensitive ASV determination of Ag(I) at a GCE and to the catalytic precipitation of a large number of silver ions on the colloidal gold-labelled Ab.

Yoomin and coworkers have described the development of a microchip-based multiplex electro-immunosensing system for simultaneous detection of cancer biomarkers using gold nanoparticles and silver enhancer. The microchip is composed of biocompatible poly (PDMS) and glass substrates. To fix the antibody-immobilised microbeads, they used pillar-type microfilters within a reaction chamber. An immunogold-silver staining (IGSS) method was used to amplify the electrical signal that corresponded to the immunecomplex. To demonstrate this approach, the authors simultaneously assayed three cancer biomarkers, Alpha-Fetoprotein (AFP), Carcinoembryonic Antigen (CEA), and Prostate-Specific Antigen (PSA) on the microchip. The electrical signal generated from the result of the immunoreaction was measured and monitored by a PC based system. The overall assay time was reduced from 3–8 h to about 55 min when compared to conventional immunoassays. The working range of the proposed microchip was from 10-3 to 10-1 μg/mL of the target antigen [65].

A novel silica coating magnetic nanoparticle-based Silver Enhancement Immunoassay (SEIA) for ricin toxin (RT) rapid electrical detection using interdigitated array microelectrodes (IDAMs) as electrodes was recently reported from Yan's group [66]. This novel system was developed by taking advantage of the separation and enrichment properties of magnetic nanoparticles (MNPs) and the catalytic properties of gold nanoparticles. In this system, MNPs labelled with anti-ricin A chain antibody 6A6 were used to capture ricin and GNPs labelled with anti-ricin B chain antibody 7G7 were used as detectors. To enhance the electrical signal, the catalytic properties of AuNPs were used to promote silver reduction. In the presence of ricin, a sandwich structure was formed which could be separated by a magnetic field. The sandwich complex was then transferred to IDAMs. The silver particles bridged the IDAM gaps and gave rise to an enhancing electrical signal that was detected by conductivity measurements.

OPTICAL DETECTION

Optical biosensors have been used widely over the past decade to analyze biomolecular interactions providing detailed information on the binding affinity and kinetics of interaction. Recently, nanoparticles, such as gold colloids and Quantum Dots (QDs), have been used for realizing new optical immunosensing. Recent reviews [42,67] reported the progress in aptamer and nanomaterial conjugates as molecular diagnostic and drug delivery agents in biomedical applications in optical sensing. In this paragraph, the progress of NP-based immunosensors, employing different receptors with various detection techniques including fluorometry, Surface Plasmon Resonance (SPR), Surface-Enhanced Raman Scattering (SERS), and colorimetry are summarized.

The fluorescent nanoparticles, such as quantum dots and luminophore-doped nanoparticles, have provided an attractive field of research for biosensor development because they have high photostability and narrow emission peaks, enabling greater multiplexing potential.

AuNPs are used as fluorescence quenchers for optical detection. Using thrombin as a detection model, Wang *et al.* investigated three different strategies (adsorption, covalent immobilisation, and hybridisation) for the AuNP surface modification with aptamers [68]. The thiolated aptamer immobilisation provided the best results with the highest constant affinity and the most sensitive detection limit (0.14 nM). For detection, after the thiolated aptamer was immobilised onto gold nanoparticles, dye-libelled complementary DNA was hybridised with the aptamer, which resulted in fluorescence quenching. With the addition of thrombin, the aptamer adopted a different conformation in order to bind with the target. Hence, the dye-labelled DNA was released from the AuNP surfaces, producing a detectable fluorescence signal.

A bioassay based on europium(III) and terbium (III)-doped fluorescent polystyrene and zirconia nanoparticles have been used for the detection of Prostate Specific Antigen (PSA) to a concentration <1 pg/ml [69].

The multiplexed detection of adenosine and cocaine using both QDs and gold nanoparticles (AuNPs) conjugated with aptamers was reported from Lu *et al.* [70]. In this work, aggregates of QDs with emission peak at 525 nm and AuNPs were formed by DNA aptamers specific for adenosine. In parallel, aggregates of QDs with emission peak at 585 nm and AuNPs were formed by a cocaine specific DNA aptamer. While fluorescence from the sensor composed of the mixture of both assemblies was initially quenched due to energy transfer from each QD to AuNPs, the signal was enhanced at 525 nm with the addition of adenosine, or at 585 nm in the presence of cocaine as analyte-aptamer interaction induced disassembly of QD-AuNP aggregates. Fluorescence increase at both 525 nm and 585 nm was observed when both analytes were present. A lateral flow device based on labelled AuNP and aptamer system was later developed from the same research group [71]. The lateral-flow device is composed of four components, a wicking pad, a glass fibre conjugation pad, a membrane, and an absorption pad. Biotin labelled AuNP aggregates containing cocaine aptamers were dropped onto the conjugation pad, and streptavidin was immobilised on a specific part of the membrane to capture biotinylated AuNPs. When the wicking pad of lateral flow device was dipped into a sample solution containing cocaine, the solution flowed along the conjugation pad, rehydrated and induced disassembly of AuNP aggregates. Since biotin was labelled on disassembled AuNPs, the AuNPs could be captured on the part of the membrane with streptavidin forming a red line. When the concentration of cocaine in the solution increased, the intensity of the red line formed on the membrane became higher and the estimated detection limit was 10 μM. Wang and co-workers demonstrated that labelled aptamer-AuNP colorimetric system could be used for detection of thrombin with extremely high sensitivity (with detection limit of 14 fM) using dot-blot arrays [72]. Furthermore, the same research group reported multiplex colorimetric sensing system based on smart AuNPs responsive to multiple stimuli with controlled cooperativity [73].

The Surface Plasmon Resonance (SPR) technique has been used mainly in protein–ligand and antibody–antigen interaction studies. SPR allows real time analysis of molecular association between an antibody and an antigen and subsequent kinetic characteristics derived from the changes in refractive index close to a metal surface on which the antigen or the antibody is attached. When a molecule from the solution binds to the immobilised molecule on the metal surface, the resonance angle changes, and the response is recorded in a Resonance Unit (RU) [74, 75].

The sensitivity of these label-free SPR-based immunosensors can be greatly enhanced by using for example AuNPs tags through the electronic coupling interaction between the Localized Surface Plasmon (LSP) of the NPs and the surface plasmon wave associated with the SPR gold film for example. SPR has been used to investigate metal NPs for applications in chemical sensing, imaging and targeting [76-78]. As an example, folic acid as a targeting moiety was conjugated to metal NPs and, its specific recognition by folic acid binding protein was demonstrated using SPR [79].

Choi *et al.* [80] reported the use of gold nanoparticles to detect prostate specific antigen using SPR. This sensor system showed a detection limit of 300 fM of the cancer marker.

Zou *el al.* have demonstrated the amplification effect of aptamer–Au NPs conjugates by using human immunoglobulin E (IgE) as model analyte by SPR [81]. Human IgE, captured by immobilised goat anti-human IgE on SPR gold film, is sensitively detected by SPR spectroscopy with a lowest detection limit of 1 ng/ml after anti-human IgE aptamer–Au NPs conjugates is used as amplification reagent. Meanwhile, the non-specific adsorption of aptamer–Au NPs conjugates on goat anti-human IgE is confirmed by SPR spectroscopy and then it is minimized by treating aptamer–Au NPs conjugates with 6-mercaptohexan-1-ol (MCH). These results confirm that aptamer–Au NPs conjugates is a powerful sandwich element and an excellent amplification reagent for SPR-based sandwich immunoassay

The direct assembly of AuNPs onto a SPR chip could enhance the signal in biomolecular interaction events are reported by Ko *et al.* [82]. Colloidal gold nanoparticles (AuNPs) were directly assembled onto a surface of SPR Au chip *via* 2-aminoethanethiol for the enhancement of sensitivity as a label-free detection system. A fusion protein was constructed by genetically fusing Gold Binding Polypeptides (GBP) to protein A (ProA) as a crosslinker for effective immobilisation of antibodies. The resulting GBP-ProA protein was directly self-immobilised onto both bare and AuNPs-assembled SPR chip surfaces *via* the GBP portion, followed by the oriented binding of human immunoglobulin G (hIgG) onto the ProA domain targeting the Fc region of antibodies and anti-hIgG in series. Furthermore, anti-*Salmonella* antibodies were immobilised onto both GBP-ProA layered chips for detection of *Salmonella typhimurium*. SPR analyses indicated the signal for successive binding of hIgG and anti-hIgG onto the GBP-ProA layered AuNPs-assembled chip was higher (about 92 and 30%, respectively) than that onto the identically treated bare chip. This signal enhancement in the AuNPs-assembled chip also caused a 10-fold increased sensitivity in detection of *S. typhimurium* compared to the bare one.

A novel Interference Localized Surface Plasmon Resonance (ILSPR) biosensor for the label-free detection of biomolecules in an arbitrary solution is reported by Tamiya *et al.* [83]. The experimental and simulation analysis of an original nanostructure design constructed with plasmonic gold nanoparticles and photonic thin-film multilayers of silicon dioxide (500 nm in thickness) and silicon on a substrate was presented. The nanostructure substrate showed a high sensitivity for various refractive index solutions and a prominent capacity for functionalizing alkanethiol molecules on the gold surface and demonstrates great potential in the development of a microfluidic-based biosensor for monitoring biotin-avidin interactions in real-time.

A reflection-based localized surface plasma resonance fiber-optic probe for chemical and biochemical sensing, called Fiber-Optic Localized Plasma Resonance (FO-LPR), has been proposed [84]. Bio-molecular recognition is detected the unique optical properties of self-assembled gold nanoparticles on the unclad portions of an optical fiber whose surfaces are modified with a receptor. To enhance the performance of the sensing platform, the sensing element is integrated with microfluidic chips to reduce sample and reagent volume, to shorten response time and analysis time, as well as to increase sensitivity. The main purpose of the study is to simulate the biochemical assays in the FO-LPR microfluidic chip and to investigate the effects parameters, such as inlet concentrations of analyte or the flow rate on the biochemical binding kinetics. The geometry of the grooved channel is also proposed to enhance the biochemical binding on the unclad optical fiber. The results reveal that the chaotic mixing generated by the grooves enhances the biochemical binding when the injected flow rate is high and, because of that, limits the performance of the molecular mixing. The enhancement of biochemical binding performance was significant, especially at the low injected concentration of analyte.

Chou *et al.* proposed a fiber-optic biosensor based on Localized Surface Plasmon Coupled Fluorescence (LSPCF). The fiber-optic biosensor is able to detect α-fetoprotein (AFP) concentration in human serum directly. LSPCF fiber-optic biosensor integrates a sandwich immunoassay with LSPs on the surface of gold NPs to improve specificity and sensitivity, simultaneously. In the experiments, a linear relationship between the fluorescence signal and the concentration of AFP in buffer solution is observed from the zero concentration to 100 ng/ml experimentally [85].

Raman scattering is an optical methodology utilizing the inelastic scattering of a photon. The optical signal can be significantly amplified when molecules are absorbed on metal surface, due to the extremely high electromagnetic fields produced on hot spots generated on the surface of metals; this phenomenon is called Surface Enhanced Raman Scattering (SERS).

SERS becomes more efficient on rough metal surfaces or on metal nanoparticle aggregates due to excitation of localized surface plasmon. The signal levels observed in SERS are several orders of magnitude higher than in conventional Raman scattering, providing the sensitivity required for bioanalytical and biomedical applications.

The detection of proteins or small molecules using SERS has also been reported using DNA aptamers. Wang and co-workers used aptamers to detect α-thrombin [86]. Thrombin has two binding sites specific for two different DNA aptamers with high affinity [87]. A substrate functionalised with aptamer 1 was first treated with thrombin and then incubated in AuNPs functionalised with aptamer 2 and SERS reporter. As a result, AuNPs were attached to the surface through aptamer-thrombin interaction. After Ag NP deposition, the NPs became larger, resulting in highly enhanced Raman scattering and thus highly sensitive signal. The detection limit was reported as 0.5 nM and the system had high selectivity over both β- and γ- thrombins. Jiang, Yu, and co-workers demonstrated a different methodology to detect adenosine using aptamers [88]. They used a dsDNA composed of both an adenosine aptamer strand and a partially complementary short DNA strand with a Raman reporter. In the presence of adenosine, the aptamer strand underwent structure switching, resulting in denaturation of dsDNA. Therefore the Raman probe labelled oligonucleotide was released in solution. The released strand later hybridized to the complementary DNA which was previously immobilised on a rough gold surface. This chain of processes resulted in an enhanced Raman signal.

For real-time and on-site use, however, it is more convenient if detection can be carried out without any equipment. Colorimetric detection provides an advantage in this regard, since it allows detection to be made by naked eye. The metallic nanoparticles, such as gold or silver nanoparticles, are ideal materials that allow colorimetric detection. AuNPs have very high extinction coefficient, making their colours distinguishable without any instrument at only a few nanomolar concentration [89]. Dispersed AuNPs smaller than 100 nm in solution originally have reddish colour. When AuNPs aggregate, their colour changes from red to blue, this is due to a shift of their surface plasmon resonance to a higher wavelength. The optical properties of AuNPs are strongly dependent on the interparticle separation distance, and aggregation that cause a massive shift in the extinction spectrum manifested as a colour change of suspensions from red to purple.

Jana and Ying [90] have presented a new method based on dot–blot assays for thrombin detection. They applied aptamer-conjugated silica-coated AuNPs for recognizing thrombin immobilised on nitrocellulose membrane. The brown/red colour would be observed in minutes with naked eye owing to the localization of the nanoparticles. The advantages of AuNPs with silica coating are its robust stability and enhanced optical cross-section. By depositing gold on the nanoparticle surfaces, the colour changes were highly amplified and the detection limit measured by naked eye was down to 25 nM.

Significantly, a direct colorimetric assay for cancer cells was developed by Tan and coworkers [91]. AuNPs modified with aptamers specific for CCRF-CEM cells (CCL-119 T-cell, human acute lymphoblastic leukemia) were incubated with target cells. Relying on the selective recognition property, aptamers directed the assembly of AuNPs on the membrane surfaces of target cells as a result. The assembled AuNPs acted like a larger scaled gold structure, which interacted efficiently with the light and exhibited significantly increased scattering and absorption coefficients compared to that of individual AuNP owing to the surface plasmon interaction. Thus, the target cells were labelled with distinct colour change while no spectra change of absorption could be observed even in complex control samples (fetal bovine serum). By measuring the absorbance changes of AuNPs in the presence of target cells and control cells, a high sensitivity of 90 cancer cells were demonstrated in addition to excellent selectivity.

Recently, a very simple, easily-operated and ultrasensitive multiplexed immunoassay system is proposed [92]. While paramagnetic microparticles conjugated with antibodies are used for fast antigens collection and separation, gold nanoparticles (Au-NPs) loaded with Horseradish Peroxidase (HRP)-labelled antibodies are employed for the naked-eye recognition or detection of the targets. The liver cancer-associated tumour markers, Carcinoembryonic Antigen (CEA) and Alfa-Fetoprotein (AFP) have been chosen as the model analytes for this work. Under optimal conditions, CEA and AFP antigens with naked eyes can be detected clearly. And the detection limit from spectrophotometric measurements is as low as 0.02 ng/ml. This proposed method will show significant clinical values for applications in cancer screening.

CONCLUSIONS

In this chapter, several kinds of metal nanoparticles applied in protein detections with electrochemical and optical transduction have been reviewed. Compared to current state-of-the art protein, detection significant advances have been made in terms of sensitivity, selectivity, and multiplexing capacity of important clinical targets using NPs.

Future works could be focused on: elimination of the non-specific adsorption of biomolecules on the transducer surface; simplifying the assay methodology; develop new detection systems under the areas of micro-/nanoelectronics; improving the integration and automation level and multianalyte detection. In addition, new synthetic receptors such as aptamers and affibodies will provide complementary excellent options to antibody-based assays. All these trends are important and should occur in parallel for affinity biosensors development. With the great research efforts, we can expect that in the future smaller and more sophisticated portable and low-cost affinity biosensors.

REFERENCES

[1]   Marquette, C. A.; Blum, L. J. State of the art and recent advances in immunoanalytical systems. *Biosens. Bioelectron.*, **2006**, *21*, 1424–1433.
[2]   Bilitewski, U. Protein-sensing assay formats and devices. *Anal. Chim. Acta.*, **2006**, *568*, 232–247.
[3]   Khlebtsov, N. G.; Dykman, L. A. Optical properties and biomedical applications of plasmonic nanoparticles. *J. Quant. Spectr. Rad. Trans.*, **2010**, *111*, 1–35.
[4]   Jain, P.K,; El-Sayed, I.H,; El-Sayed, M.A. Au nanoparticles target cancer. *Nano Today.*, **2007**, *2*, 18-29.
[5]   Hu, R,; Yong, K.T,; Roy, I,; Ding, H,; He, S,; Prasad, P. N. Metallic nanostructures as localized plasmon resonance enhanced scattering probes for multiplex dark-field targeted imaging of cancer cells. *J. Phys. Chem. C.*, **2009**,*113*, 2676–84.
[6]   De Jong, W.H.; Hagens, W.I.; Krystek, P.; Burger, M.C.; Sips, A.J.A.M.; Geertsma, R.E. Particle size-dependent organ distribution of gold nanoparticles after intravenous administration. *Biomat.*, **2008**, *29*, 1912–9.
[7]   Pumera, M.; Merkoci, A.; Alegret, S. New materials for electrochemical sensing VII. Microfluidic chip platforms. *Trends. Anal. Chem.*, **2006**, *25*, 3, 219-235.
[8]   Marrazza, G. Nanoparticle-based Microfluidific Biosensors in Challa Kumar: Microfluidic Devices in Nanotechnology. **2010**. John Wiley & Sons
[9]   De la Escosura-Muniz, A.; Ambrosi, A.; Merkoci, A. Electrochemical analysis with nanoparticle-based biosystems. *Trends. Anal. Chem.*, **2008**, *27*, 7, 568-584.
[10]   Gomez-Hens, A.; Fernandez-Romero, J.M.; Aguilar-Caballos, M.P. Nanostructures as analytical tools in bioassays. *Trends. Anal. Chem.*, **2008**, *27*, 5, 394-406.
[11]   Liu, G.; Lin, Y. Nanomaterial labels in electrochemical immunosensors and immunoassays. *Tal.*, **2007**, *74*, 308–317.
[12]   Ligler, F. S. Perspective on Optical Biosensors and Integrated Sensor Systems. *Anal. Chem.*, **2009**, *81*, 2, 519-526
[13]   De Dios, A. S.; Díaz-García, M. E. Multifunctional nanoparticles: Analytical prospects. *Anal. Chim. Acta.*, **2010**, *666*,1–22.
[14]   Wang, L.; Wei, Ma.; Liguang, Xu.; We, Chen.; Yingyue, Zhu.; Chuan lai, Xu.; Kotov, N. A. Nanoparticle-based environmental sensors. *Mater. Sci. Eng. R.*, **2010**, doi:10.1016/j.mser.2010.06.012.
[15]   Willner, I.; Baron, R., Willner B. Integrated nanoparticle-biomolecule systems for bisensi and bioelectronics. *Biosens. Bioelectron.*, **2007**, *22*, 1841-1852.
[16]   Pingarron, J. M., Yanez-Sedeno, P., Gonzalez-Cortes, A. Gold nanoparticle-based electrochemical biosensors., *Electrochim. Acta.*, **2008**, *53*, 5848–5866.
[17]   Wang, Z.; Ma, L. Gold nanoparticle probes. *Coordination Chem. Rev.*, **2009**, *253*, 1607–1618.
[18]   MDaniel, C. Didier Astruc. Gold Nanoparticles: Assembly, Supramolecular Chemistry, Quantum-Size-Related Properties, and Applications toward Biology, Catalysis, and Nanotechnology. *Chem. Rev.*, **2004**, *104*, 293-346.

[19]   Shashi P.D.; Manu, L.; Mika, S. Tansy fruit mediated greener synthesis of silver and gold nanoparticles. *Process. Biochem.,* **2010**, *45*, 1065–1071.

[20]   Parashar, V.; Parashar, R.; Sharma, B.; Pandey, A.C. Parthenium leaf extract mediated synthesis of silver nanoparticles: a novel approach towards weed utilization. *Digest J. Nanomater. Biostruct.,* **2009**, *4*, 45–50.

[21]   Philip, D. Biosynthesis of Au, Ag and Au–Ag nanoparticles using edible mushroom extract. *Spectrochim. Acta A.,* **2009**, *73*, 374–81.

[22]   Smitha, S. L.; Philip, D.; Gopchandran, K. G. Green synthesis of gold nanoparticles using Cinnamomum zeylanicum leaf broth. *Spectrochim. Acta A.,* **2009**, *74*, 735–9.

[23]   Song J. Y.; Jang, H.K.; Kim, B. S. Biological synthesis of gold nanoparticles using Magnolia kobus and Diopyros kaki leaf extracts. *Process. Biochem.,* **2009**, *44*, 1133–8.

[24]   Thakkar, K.N.; Mhatre, S. S.; Parikh, R.Y. Biological synthesis of metallic nanoparticles. *Nanomed. Nanotechnol. Biomed.,* **2010**, *6*, 257–62.

[25]   Shankar, S. S.; Rai, A.; Ahmad, A.; Sastry, M. Rapid synthesis of Au, Ag, and bimetallic Au core Ag shell nanoparticles using Neem (Azadirachta indica) leaf broth. *J. Colloid Interface Sci.,* **2004**, *275*, 496–502.

[26]   Shankar, S. S.; Ahmad, A.; Sastry, M. Geranium leaf assisted biosynthesis of silver nanoparticles. *Biotechnol. Prog.,* **2003**, *19*, 1627–31.

[27]   Philip, D. Green synthesis of gold and silver nanoparticles using Hibiscus rosa sinensis. *Physica. E.,* **2010**, *42*, 1417–24.

[28]   Sathishkumar, M.; Sneha, K.; Won, S.W.; Cho, C.W.; Kim, S.; Yun, Y.S. Cinnamon zeylanicum bark extract and powder mediated green synthesis of nano-crystalline silver particles and its bactericidal activity. *Colloids Surf B.,* **2009**, *73*, 332–8.

[29]   Narayanan, K. B.; Sakthivel, N. Coriander leaf mediated biosynthesis of gold nanoparticles. *Mater. Lett.,* **2008**, *62*, 4588–90.

[30]   Ankamwar, B.; Chaudhary, M.; Sastry, M. Gold nanotriangles biologically synthesized using tamarind leaf extract and potential application in vapor sensing. *Synth. React. Inorg. Met. Org. Nanomet. Chem.,* **2005**, *35*, 19–26.

[31]   Shaojun, Guoa.; Erkang, Wang. Synthesis and electrochemical applications of gold nanoparticles. *Anal. Chim. Acta.,* **2007**, *598*, 181–192.

[32]   Turkevich, J.; Stevenson, P. C.; Hillier, J. A study of the nucleation and growth processes in the synthesis of colloidal gold. *Discuss. Faraday. Soc.,* **1951**, *11*, 55-75.

[33]   Frens, G.; Controlled nucleation for the regulation of the particle size in monodisperse gold suspensions. *Nat.Phys.Sci.,* **1973**, *241*, 20–22.

[34]   Brust, M.; Walker, M. .; Bethell, D.; Schiffrin, D. J.; Whyman R. Synthesis of Thiol-derivatised Gold Nanoparticles in a Two-phase Liquid-Liquid System. *J. Chem. Soc. Chem. Commun.,* **1994**, 801-802.

[35]   Bhupinder, S.; Sekhon, Seema.; R, Kamboj. BPharm. Inorganic nanomedicine—Part 2. Nanomedicine: NBM **2010**;xx:1-7, doi:10.1016/j.nano.2010.04.003.

[36]   Kaittanis, C.; Santra, S.; Perez, J. M. Emerging nanotechnology-based strategies for the identification of microbial pathogenesis. *Adv. Drug Deliv. Rev.,* **2010**, *62*, 408–423.

[37]   Gautham, K.A.; Chanchal K.M. Gold nanoparticles based sandwich electrochemical immunosensor. *Biosens. Bioelectron.,* **2010**, *25*, 2016–2020.

[38]   Tombelli, S.; Minunni, M.; Mascini, M. Analytical applications of aptamers. *Biosens. Bioelectron.,* **2005**, *20*, 2424–2434.

[39]   Nygren, P.Å. Alternative binding proteins: affibody binding proteins developed from a small three-helix bundle scaffold. *The. FEBS. J.,* **2008**, *275*, 2668-76.

[40]   Orlova, A.; Tolmachev, V.; Pehrson, R.; Lindborg, M.; Tran, T.; Sandström, M.; Nilsson, F.Y.; Wennborg, A.; Abrahmsén, L.; Feldwisch, J. Synthetic affibody molecules: a novel class of affinity ligands for molecular imaging of HER2-expressing malignant tumors. *Cancer. Res.,* **2007**, *67*, 2178-86.

[41]   Gohring, J.T.; Dale, P.S.; Fan, X. Detection of HER2 breast cancer biomarker using the optofluidic ring resonator biosensor. *Sens. Actuators. B.,* **2010**, *146*, 226-30.

[42]   Guoqing, Wang.; Yunqing, Wang.; Lingxin, Chena.; Jaebum, Choo. Nanomaterial-assisted aptamers for optical sensing. *Biosens. Bioelectr.,* **2010**, 25, 1859–1868.

[43]   Tothill Ibtisam E. Biosensors for cancer markers diagnosis. *Sem. Cell Dev. Biol.,* **2009**, *20*, 55–62.

[44]   Tang, D.; Yuan, R.; Chai, Y. Electrochemical immuno-bioanalysis for carcinoma antigen 125 based on thionine and gold nanoparticles-modified carbon paste interface. *Anal. Chim. Acta.,* **2006**, *654*, 158.

[45]   Wu, L.; Chen, J.; Du, D.; Ju, H. Electrochemical immunoassay for CA125 based on cellulose acetate stabilized antigen/colloidal gold nanoparticles membrane. *Electrochim. Acta.,* **2006**, *51*, 1208.

[46]   Fu, X. H. Electrochemical Immunoassay for Carbohydrate Antigen-125 Based on Polythionine and Gold Hollow Microspheres Modified Glassy Carbon Electrodes. *Electroanal.,* **2007**, *19*, 1831.

[47]   Chen, J.; Tang, J.; Yan, F.; Ju, H. A gold nanoparticles/sol–gel composite architecture for encapsulation of immunoconjugate for reagentless electrochemical immunoassay. *Biomat.,* **2006**, *27*, 2313.

[48]   Idegami, K.; Chikae, M.; Kerman, K.; Nagatani, N.; Yuhi, T.; Endo, T.; Tamiya, E. Gold nanoparticle-based redox signal enhancement for sensitive detection of human chorionic gonadotropin hormone. *Electroanal.,* **2008**, *20*, 14-21.

[49]   Dan, D.; Xiaoxing, X.; Shengfu, W.; Aidong, Z. Reagentless amperometric carbohydrate antigen 19-9 immunosensor based on direct electrochemistry of immobilised horseradish peroxidise. *Tal.,* **2007**, *71*, 1257.

[50]   Li, X.; Yuan, R.; Chai, Y.; Zhang, L.; Zhuo, Y.; Zhang, Y. Amperometric immunosensor based on toluidine blue/nano-Au through electrostatic interaction for determination of carcinoembryonic antigen. *J. Biotechnol.,* **2006**, *123*, 356.

[51]   An, H.Z.; Yuan, R.T.; Tang, D.P.; Chai, Y.; Li, N. Dual-Amplification of Antigen-Antibody Interactions *via* Backfilling Gold Nanoparticles on (3-Mercaptopropyl) Trimethoxysilane Sol-Gel Functionalized Interface. *Electroanal.,* **2007**, *19*, 479.

[52]   Tang, H.; Chen, J. H.; Nie, L. H.; Kuang, Y. F.; Yao, S. Z. A label-free electrochemical immunoassay for carcinoembryonic antigen (CEA) based on gold nanoparticles (AuNPs) and nonconductive polymer film. *Biosens. Bioelectron.,* **2007**, *22*, 1061-1067.

[53]   Jiehua, Lin.; Chunyan, He.; Lijuan, Zhang.; Shusheng, Zhang. Sensitive amperometric immunosensor for alfa-fetoprotein based on carbon nanotube/gold nanoparticle doped chitosan film. *Anal. Biochem.,* **2009**, *384*, 130–135.

[54]   Gang-Il, Kima.; Kyung-Woo, Kimb.; Min-Kyu, Ohb.; Yun-Mo, Sunga. VEGF antibody fragments modified Au NPs/ITO electrode. *Biosens. Bioelectron.,* **2010**, *25*, 1717–1722.

[55]   Goluch, E.D.; Nam, J.-M.; Georganopoulou, D.G.; Chiesl, T.N.; Shaikh, K.A.; Ryu, K.S.; Barron, A.E.; Mirkin, C.A.; Liu, C. Abio-barcode assay for on-chip attomolar-sensitivity protein detection. *Lab. Chip.,* **2006**, *6*, 10, 1293–1299.

[56]   Goluch Edgar, D.; Savka I.Stoeva.; Jae-Seung, Lee.; Kashan, A. Shaikh.; Chad, A. Mirkin.; Chang, Liu. A microfluidic detection system based upon a surface immobilised biobarcode assay. *Biosens. Bioelecton.,* **2009**, *24*, 2397–2403.

[57]   Liang, C.H.; Wang, C.C.; Lin, Y.C.; Chen, C.H.; Wong, C.H.; Wu, C.Y. Iron oxide/gold core/shell nanoparticles for ultrasensitive detection of carbohydrate-protein interactions. *Anal. Chem.,* **2009**, *81*, 7750-6.

[58]   Polsky, R.; Gill, R.; Kaganovsky, L.; Willner, I. Nucleic Acid-Functionalized Pt Nanoparticles: Catalytic Labels for the Amplified Electrochemical Detection of Biomolecules. *Anal. Chem.,* **2006**, *78*, 2268.

[59]   Mir, M.; Vreeke, M.; Katakis, I. Different Strategies to Develop an Electrochemical Thrombin Aptasensor. *Electrochem. Commun.,* **2006**, *8*, 505.

[60]   Ikebukuro, K.; Kiyohara, C.; Sode, K. Novel Electrochemical Sensor Systemfor Protein Using the Aptamers in Sandwich Manner. *Biosens. Bioelectron.,* **2005**, *20*, 2168.

[61]   He, P.; Shen, L.; Caom Y.; Li, D. Ultrasensitive electrochemical detection of proteins by amplification of aptamer-nanoparticle biobarcode. *Anal. Chem.,* **2007**, *79*, 8024-8029.

[62]   Liao, K. T.; Huang, H. J. Femtomolar immunoassay based on coupling gold nanoparticle enlargement with square wave stripping voltammetry. *Anal. Chim. Acta.,* **2005**, *538*, 159-164.

[63]   Mao, X.; Jiang, J.; Luo, Y.; Shen, G.; Yu, R. Copper-enhanced gold nanoparticle tags for electrochemical stripping detection of human IgG. *Tal.,* **2007**, *73*, 420.

[64]   Chu, X.; Fu, X.; Chen, K.; Shen, G. L.; Yu, R. Q. An electrochemical stripping metallo-immunoassay based on silver-enhanced gold nanoparticle label. *Biosens. Bioelectron.,* **2005**, *20*, 1805-1812.

[65]   Yong-Jun, Ko.; Joon-Ho, Maeng.; Yoomin, Ahn.; Seung, Yong Hwang.; Nahm-Gyoo, Cho.; Seoung-Hwan, Lee. Microchip-based multiplex electro-immunosensing system for the detection of cancer biomarkers. *Electroph.,* **2008**, *29*, 3466–3476.

[66]   Jie, Zhuang. Tao. Cheng.; Lizeng, Gao.; Yongting, Luo.; Quan, Ren.; Di, Lu.; Fangqiong, Tang.; Xiangling, Ren.; Dongling, Yang.; Jing, Feng.; Jingdong, Zhu.; Xiyun, Yan. Silica coating magnetic nanoparticle-based silver enhancement immunoassay for rapid electrical detection of ricin toxin. *Toxicon.,* **2010**, *55*,145–152.

[67] Lee H. *et al.*, Molecular diagnostic and drug delivery agents based on aptamer-nanomaterial conjugates, *Adv. Drug Deliv. Rev.*, **2010**, doi:10.1016/j.addr.2010.03.003.

[68] Wang,W.; Chen, C.; Qian, M.; Zhao, X.S. Aptamer biosensor for protein detection using gold nanoparticles. *Anal. Biochem.*, **2008**, *373* (2), 213–219.

[69] Zhiqiang, Ye.; Mingqian, Tan.; Guilan, Wang.; Jingli, Yuan. Preparation, characterization and time-resolved fluorometric application of silica-coated terbium(III) fluorescent nanoparticles. *Anal. Chem.*, **2004**, *76*, 513–518.

[70] Liu, J.; Lee, J.H.; Lu, Y. Quantum Dot Encoding of Aptamer-Linked Nanostructures for One Pot Simultaneous Detection of Multiple Analytes. *Anal. Chem.*, **2007**, *79*, 4120.

[71] Wang, Y.; Liu, D. J.; Mazumdar, D.; Lu, Y. A. Simple and Sensitive "Dipstick" Test in Serum Based on Lateral Flow Separation of Aptamer-Linked Nanostructures. Angew. *Chem. Int. Ed.*, **2006**, *45*, 7955.

[72] Li, W. Ren.; Z. Liu.; S. Dong.; E. Wang. Ultrasensitive Colorimetric Detection of Protein by Aptamer-Au Nanoparticles Conjugates Based on a Dot-Blot Assay. *Chem. Commun.*, **2008**, *22*, 2520.

[73] Liu, J.; Lu, Y. Smart Nanomaterials Responsive to Multiple Chemical Stimuli with Controllable Cooperativity. *Adv. Mater.*, **2006**, *18,* 1667.

[74] Johnsson, B.; Lofas, S.; Lindquist, G. Immobilisation of proteins to a carboxymethyldextran-modified gold surface for biospecific interaction analysis in surface plasmon resonance sensors. *Anal. Biochem.*, **1991**, *198*, 2, 268–277.

[75] Johnsson, B.; Lofas, S.; Lindquist, G.; Edstrom, A.; Muller, Hillgren R.M.; A, Hansson. Comparison of methods for immobilisation to carboxymethyl dextran sensor surfaces by analysis of the specific activity of monoclonal antibodies. *J. Mol. Recognit.*, **1995**, *8*, 1–2, 125–131.

[76] Murphy, C.J.; Gole, A.M.; Hunyadi, S.E.; Stone, J.W.; Sisco, P.N.; Alkilany, A.; Kinard, B.E.; Hankins, P. Chemical sensing and imaging with metallic nanorods. *Chem. Commun.*, **2008**, *5*, 544–557.

[77] Huang, X.; Jain, P.K.; El-Sayed I.H.; El-Sayed, M.A. Gold nanoparticles: interesting optical properties and recent applications in cancer diagnostics and therapy. *Nanomed.*, **2007**, *2*, 5, 681–693.

[78] Wang, J.; Zhou, H.S. Aptamer-based au nanoparticles-enhanced surface plasmon resonance detection of small molecules. *Anal. Chem.*, **2008**, *80*, 18, 7174–7178.

[79] Sonvico, F.; Mornet, S.; Vasseur, S.; Dubernet, C.; Jaillard, D.; Degrouard, J.; Hoebeke, J.; Duguet, E.; Colombo, P.; Couvreur, P. Folate-conjugated iron oxide nanoparticles for solid tumor targeting as potential specific magnetic hyperthermia mediators: synthesis, physicochemical characterization, and *in vitro* experiments. *Bioconjug. Chem.*, **2005**, *16*, 5, 1181–118.

[80] Jianlong Wang, Ahsan Munir, Zhonghong Li H. Susan Zhou. Aptamer–Au NPs conjugates-enhanced SPR sensing for the ultrasensitive sandwich immunoassay. *Biosens. Bioelectron.*, **2009**, *25*, 124-129

[81] Sungho, Ko.; Tae Jung Park.; Hyo-Sop, Kim.; Jae-Ho, Kim.; Yong-Jin, Cho. Directed self-assembly of gold binding polypeptide-protein A fusion proteins for development of gold nanoparticle-based SPR immunosensors. Biosens. and Bioelectron., **2009**, 24, 8, 15, 2592-2597.

[82] Hiep, H. M.; Yoshikawa, H.; Saito, M.; Tamiya, E. Plasmon Resonance Biosensor Based on the Photonic Structure of Au Nanoparticles and SiO_2/Si Multilayers. *ACS. Nano.*, **2009**, *3*, 2, 446-452.

[83] Chun-Ping, Jen.; Ching-Te, Huang.; Yun-Hung, Lu.; Simulation of biochemical binding kinetics on the microfluidic biochip of fiber-optic localized plasma resonance (FO-LPR). *Microelect. Engin.*, **2009**, *86*, 1505–1510.

[84] Ying-Feng, Chang.; Ran-Chou, Chen.; Yi-Jang, Lee.; Shu-Chen, Chao.; Li-Chen, Su.; Ying-Chang, Li.; Chien, Chou. Localized surface plasmon coupled fluorescence fiber-optic biosensor for alpha-fetoprotein detection in human serum. *Biosens. Bioelectron.*, **2009**, *24,*1610–1614.

[85] Choi, J-W.; Kang, D-Y.; Jang, Y-H.; Kim, H-H.; Min, J.; Oh, B-K. Ultra sensitive surface plasmon resonance based immunosensor for prostate specific antigen using gold nanoparticale antibody complex. *Colloids Surf A: Physicochem. Eng. Aspects.*, **2008**, 313–314, 655–9.

[86] Wang, Y.; Wei, H.; Li, B.; Ren, W.; Guo, S.; Dong, S.; Wang, E. SERS Opens a New Way in Aptasensor for Protein Recognition with High Sensitivity and Selectivity. *Chem. Commun.*, **2007**, *48*, 5220.

[87] Tasset, D.M.; Kubik, M.F.; Steiner, W. Oligonucleotide Inhibitors of Human Thrombin That Bind Distinct Epitopes. *J. Mol. Biol.*, **1997**, *272*, 688.

[88] Chen, J.W.; Liu, X.P.; Feng, K.J.; Liang, Y.; Jiang, J.H.; Shen, G.L.; Yu, R.Q. Detection of Adenosine Using Surface-Enhanced Raman Scattering Based on Structure-Switching Signaling Aptamer. *Biosens. Bioelectron.*, **2008**, *24*, 66.

[89] Chen, J.; Saeki, F.; Wiley, B.J.; Cang, H.; Cobb M.J.; Li, Z.-Y.; Au. L.; Zhang, H.; Kimmey, M.B.; Li, X.; Xia, Y. Gold Nanocages: Bioconjugation and Their Potential Use as Optical Imaging Contrast Agents. *Nano Lett.*, 2005, *5*, 473.

[90]   Jana, N.R.; Ying, J.Y. Synthesis of Functionalized Au Nanoparticles for Protein Detection. *Adv. Mater.,* **2008**, *20* (3), 430–434.

[91]   Medley, C.D.; Smith, J.E.; Tang, Z.W.; Wu, Y.R.; Bamrungsap, S.; Tan, W.T. *Anal. Chem.,* **2008**, *80*, 4, 1067–1072.

[92]   Jing, Wang.; Ya, Cao.; Yuanyuan, Xu.; Genxi, Li. Colorimetric multiplexed immunoassay for sequential detection of tumor markers. *Biosens. Bioelectron.,* **2009**, 25, 532–536.

CHAPTER 3

Optical Sensors Based on Molecularly Imprinted Nanomaterials

Shanshan Wang, Xiaocui Zhu and Meiping Zhao[*]

Beijing National Laboratory for Molecular Sciences, MOE Key Laboratory of Bioorganic Chemistry and Molecular Engineering, College of Chemistry and Molecular Engineering, Peking University, Beijing, 100871, China

Abstract: This chapter focuses on recent developments in the construction of optical sensors based on intelligent molecularly imprinted nanomaterials. The first two parts review the general principles in the development of molecularly imprinted polymer (MIP)-based optical sensors. Four different ways to transform the binding events into measurable optical signals are discussed. In the third part, nanosized MIP materials are classified as nanoparticles (including core-shell nanoparticles), nanofibres/nanowires/ nanotubes and nanofilms. The principle, analytical properties and applications of recently reported optical sensors based on above three different nano-MIP formats are all reviewed in detail. Finally, some of the remaining unsolved issues to the nano-MIP-based optical sensors are briefly discussed for further development of the field.

Keywords: Optical biosensors; molecular imprinting; nanosized MIP materials.

1. INTRODUCTION

Sensors are analytical devices that generate quantifiable output signals upon the binding of the analyte to the recognition element [1]. They have shown distinct advantages in real-time detection of specific sample constituents in various fields, including clinical diagnostics, environmental analysis, food analysis and production monitoring. For biosensors, many biological receptors, such as antibodies, enzymes, aptamers and peptides, have all been used as the recognition elements, which are responsible for specifically recognizing and binding the target analyte in real samples [2-7]. However, these natural receptors have been suffering from high cost and poor chemical and physical stability.

Molecular imprinting technology is a powerful tool to generate tailor-made receptors for separation, catalytic reaction and detection [8, 9]. Compared with the biogenic antibodies, molecularly imprinted materials offer the advantages of ease of preparation, reusability and robustness for chemical and physical stresses. Molecularly Imprinted Polymers (MIPs) can be prepared in a variety of physical forms to suit the final application desired [10-12]. In recent years, remarkable progress has been made in fabrication of nanosized imprinted materials [13, 14], which provide compactness, significantly increased specific surface area and better accessibility to the imprinted cavity. These lead to fast equilibration with the analyte, which is especially beneficial for developing sensors.

Optical sensing uses light as the transduced signal and shows the merits of flexibility, high sensitivity, environmental stability, ease of miniaturization, inexpensiveness and non-destructive analyte analysis [10]. This review will focus on the recent achievements in the development of optical sensors based on MIP materials prepared in nanometer range, with an emphasis on optical transduction methods. For more information on molecular imprinting technology and other applications, the readers may refer to several other excellent general reviews that have appeared over the past few years [10, 11, 15-20].

2. GENERAL PRINCIPLES IN THE DEVELOPMENT OF MIP-BASED OPTICAL SENSORS

One of the most important aspects in the design of MIP-based sensors is transforming the binding events

*Address correspondence to Meiping Zhao:** Beijing National Laboratory for Molecular Sciences, MOE Key Laboratory of Bioorganic Chemistry and Molecular Engineering, College of Chemistry and Molecular Engineering, Peking University, Beijing, 100871, China; Tel: 86-10-62758153; Email: mpzhao@pku.edu.cn

Songjun Li, Yi Ge and He Li (Eds)
All rights reserved - © 2012 Bentham Science Publishers

into measurable signals. So far there are generally four types of signal transduction ways for the rebinding of MIPs to the target analytes [10, 21].

2.1. Application of Fluorescent Templates and Analogues

For a target analyte that has a special optical property, such as fluorescence, it can be directly used for detection. For example, a MIP fluorimetric sensor for monoamine naphthalene compounds was developed by Valero-Navarro *et al.* [22] using non-covalent molecular imprinting techniques and naphthalene as template. The system is based on the measurement of the native fluorescence signals of monoamine naphthalene compounds when they are adsorbed on-line on the MIP. It can be used for simultaneous determination of 1-naphthylamine and 2-naphthylamine at ng mL^{-1} level with a response time of 2 min.

A potential problem with above method is that residual template molecules in the polymer matrix may lead to a high background signal and result in decreased sensitivity. A remedy could be to imprint the polymer with a nonfluorescent analyte analogue. On the other hand, when the analyte does not display optical properties for the spectroscopic analysis, it can be determined using a labeled template or analogue derivative in a displacement or competitive assay [23-25]. As an example, a chloramphenicol MIP fluorescent sensor was developed based on monitoring the competition of chloramphenicol and its dansylated derivative in binding to the imprinted sites [25]. As an alternative approach, Haupt [26] used non-related fluorescent probes for the detection of the herbicide and synthetic auxin 2, 4-dichlorophenoxyacetic acid.

Benito-Peňa *et al.* [27] developed a fluorescence competitive assay for penicillin G analysis using pyrenemethylacetamidopenicillanic acid as the labeled competitor and successfully applied it to a pharmaceutical formulation analysis. An automated molecularly imprinted sorbent based assay for the rapid and sensitive analysis of penicillintype β-lactam antibiotics was proposed by Urraca *et al.* using penicillin G procaine salt as template and a stoichiometric quantity of a urea-based functional monomer [28]. Highly fluorescent competitors containing pyrene labels while keeping intact the 6-aminopenicillanic acid moiety for efficient recognition by the cross-linked polymers were tested as analyte analogues in the competitive assay. Pyrenemethy Lacetamido Penicillanic Acid (PAAP) was the tagged antibiotic providing for the highest selectivity when competing with penicillin G for the specific binding sites in the MIP. Upon desorption from the MIP, the emission signal generated by the PAAP was related to the antibiotic concentration in the sample.

Recently, González *et al.* [29] described a flow-injection optical sensor for digoxin by combination of sensor technology with MIP as the recognition phase. The MIP was packed into a flow cell and placed in a spectrofluorimeter to integrate the reaction and detection systems. The new fluorosensor showed high selectivity and sensitivity with a detection limit of 17 ng l^{-1}. The method was successfully applied for the determination of digoxin concentration of human serum samples. A fluorescent indicator-displacement molecular imprinting sensor array based on phenylboronic acid functionalized mesoporous silica was developed for discriminating saccharides [30].

2.2. Incorporation of Fluorescent Reporter within MIP Structures

A more widely applicable approach for generation of optical signals in MIP binding is to incorporate responsive chromophores or fluorophores into the polymer matrix [31-35]. When the analytes bind to the imprinted cavities, the microenvironment (*e.g.* polarity, pH) around the fluoro/luminophore is altered, resulting in quenching or enhancement of the fluorescence or energy transfer [36-40].

Turkewitsch *et al.* developed a MIP sensor for cyclic adenosine 3′, 5′-monophosphate (cAMP) by using a fluorescent functional monomer *trans*-4-[p-(N,N-dimethylamino) styryl]-N-vinylbenzylpyridinium chloride together with a conventional functional monomer [31]. Upon binding to the imprinted sites, the analyte interacts with the fluorescent groups and quenches their fluorescence, allowing the analyte to be quantified. One of the limitations of this strategy is the high background signal of the MIP. The fluorescence signal only changed by 20% on binding cAMP. A possible reason for this is that many of the fluorescent monomers were not incorporated into binding sites and were unresponsive to the bound analytes.

Takeuchi *et al.* [41] prepared an imprinted polymer for (-)-cinchonidine by the combined use of methacrylic acid and vinyl-substituted zinc(II) porphyrin as functional monomers. The MIPs showed significant fluorescence quenching during binding of (-)-cinchonidine in the low concentration range, which appeared to act as a fluorescence sensor selectively responded to the template molecule. In another approach, Tong *et al.* [42] used zinc(II)-protoporphyrin (ZnPP) as a functional monomer and developed a fluorescent sensor for histamine. The ZnPP has a Lewis acid binding site Zn and binds with the imidazolyl group of histamine through coordination, leading to decreased fluorescence intensity upon exposure to histamine.

Sánchez-Barragán *et al.* [43] proposed a novel concept for optosensing by introducing heavy-atom effect in the MIP sensing system. The polymer allows one to perform Room-Temperature Phosphorescence (RTP) transduction of the analyte. The noncovalent MIP was synthesized using tetraiodobisphenol A as one of the polymeric precursors and fluoranthene as template. Once recognized by the MIP, the iodide included in the polymeric structure induced efficient RTP emission from the analyte in the presence of an oxygen scavenger. The developed optosensing system has demonstrated a high specificity for fluoranthene against other polycyclic aromatic hydrocarbons. Detection limit for the target molecule was 35 ng/L (5-mL sample injections). The synthesized sensing material showed good stability and reusability. A molecularly imprinted fluorescent conjugated polymer material with an intrinsic capability for signal transduction was also synthesized for the detection of 2,4,6-trinitrotoluene (TNT) and related nitroaromatic compounds [44].

A potential fluorescent MIP sensor for (-)-ephedrine was developed by Nguyen and Ansell [45] with two novel polymerisable coumarins 6-styrylcoumarin-4-carboxylic acid and 6-vinylcoumarin-4-carboxylic acid as functional monomers and ethylene glycol dimethacrylate as a cross-linker. Both of the polymers exhibited a decrease of fluorescence in response to amines in acetonitrile, with some selectivity for the template over its enantiomer (+)-ephedrine and other structural analogues, though little response to ephedrine was observed in aqueous buffer.

A major drawback of above approaches is their negative fluorescence responses after rebinding of the templates. One strategy for developing more sensitive responsive MIPs is to design fluorescent monomers that turn on upon binding. For example, Takeuchi and co-workers [46] reported the use of the fluorescent monomer 2,6-bis(acrylamido)pyridine for the imprinting of cyclobarbital. An enhancement in fluorescence intensity upon binding with the template was observed, which could be caused by the increased rigidity of the monomer residues due to the formation of multiple hydrogen bonding upon complexation of the cyclobarbital with the polymeric recognition sites. Later, the same group designed a new imprinted polymer based on the fluorescent monomer 2-acrylamidoquinoline. The binding could be monitored by the enhanced emission of the monomer at 330 nm due to the inhibition of its photoinduced electron-transfer quenching mechanism [47].

Tan [48] *et al.* reported an ion imprinted mesoporous silica based fluorescence turn-on sensor array for discrimination of metal ions. A novel fluorescent functional monomer containing an 8-hydroxyquinoline moiety in combination with one-pot co-condensation method was employed to prepare fluorescent ion imprinted mesoporous silica for Zn^{2+} and Cd^{2+}. With the covalently anchored organic fluorophore in the inorganic mesoporous silica matrix, the binding of metal ions to the imprinting site was directly transformed into fluorescence signals.

In another approach, Subrahmanyam *et al.* [49] developed a fluorescent MIP sensor for creatine, an indicator of tissue degradation and kidney stress and also an abused drug by athletes. The polymer was synthesized based on a polymerisable thioacetale, formed by the reaction of o-phthalic dialdehyde and allylmercaptan. The MIP may form a fluorescent isoindole complex during reaction with primary amine. Wang and coworkers [34, 50] studied the imprinting of D-fructose using a fluorescent anthracene-boronic acid conjugate bearing a methacrylate moiety. Inclusion of boronic acid in the MIP enabled its high selectivity towards D-fructose when compared to other analogues such as D-glucose or D-mannose.

Graham *et al.* [51] prepared a MIP for the pesticide DDT *via* covalent imprinting strategy. An environmentally sensitive fluorescent probe, 7-nitrobenz-2-oxa-1,3-diazole (NBD), was incorporated into the polymer matrix to generate a sensor. The polymeric films were deposited on microscope slides and studied by fluorescent microscopy. Upon binding DDT, fluorescence intensity of NBD located adjacent to the binding sites showed an enhancement, allowing quantitative detection of DDT in water with a detection limit of 50 ppt and a response time of 60 s. The limitation of current design is the relatively minor changes in fluorescence intensity upon binding DDT.

2.3. Interfacing the MIP with the Transducer

The third promising approach is to interface the MIP with a transducer element in an appropriate way. To this end, the MIP can either be synthesized *in situ* at the transducer surface [52, 53] or preformed followed by immobilization on the surface of the transducer [54].

Very recently, Medina-Castillo *et al.* proposed an efficient strategy for constructing magnetic optical sensors [55]. They synthesized and characterized several types of magnetic MIP nanoparticles for the model analyte pyrene, followed by optical readout from the outside. The obtained optical magnetic MIP sensor shows high sensitivity (detection limit of 20 ng mL^{-1}), good imprinting effect (MIP/NIP ratio of 2.4), high selectivity and highly magnetic properties. The magnetic separator collected the Mag-MIPs at the tip of the optical fiber probe, and the bound analytes were detected very efficiently by measuring the intrinsic fluorescence of the target analyte.

2.4. Signals Generated by MIP Catalytic Reactions

The spectroscopically active species produced by a reaction catalyzed by the MIP may also be employed for sensing purposes [56]. A non-covalent MIP was fabricated using a thiol-specific fluoro-tagging agent (N-(1-pyrenyl)maleimide) labeled DL-homocysteine as template. Following *in situ* fluorescent derivatization, luminescent response of the MIP was found to correlate linearly with the concentration of DL-homocysteine. More importantly, the MIP material was found to be able to specifically enhance the rate of derivatization reaction between DL-homocysteine and N-(1-pyrenyl)maleimide. Thus the imprinting process enhanced the specificity of the derivatization reaction toward the template and may serve an efficient transduction way for the analyte.

3. OPTICAL SENSORS BASED ON MIP WITH DIFFERENT NANOSTRUCTURES

3.1. Imprinted Nanoparticles and Core-Shell Nanoparticles with Imprinted Shell

Imprinted nanoparticles can be prepared by many techniques such as precipitation and miniemulsion [57-61]. In precipitation polymerization, the functional monomers, cross-linkers and initiators are mixed homogenously in a dilute solution of porogenic solvent. As the polymerization proceeds, the expanding polymer becomes insoluble and aggregates into particles, which are stabilized against coagulation either sterically or by their rigid cross-linked surfaces. Ye *et al.* [62] reported the first synthesis of imprinted nanoparticles using the precipitation polymerization technique. They demonstrated that both hydrophobic and hydrophilic template molecules could be successfully imprinted using this method.

Miniemulsion polymerization is a method that uses the target molecule as the head of a surfactant which is directed to the inner surface of the micelle, thus leaving an imprinted recognition site at the surface of the nanoparticle. Pérez *et al.* presented a successful approach for the preparation of cholesterol-imprinted core-shell nanoparticles by emulsion polymerization [63, 64]. In their method, non-imprinted core particles of 30-40 nm in diameter were first prepared. These core particles were subsequently used in a second-stage polymerization where the imprinted shell was produced using a covalent imprinting approach. Cholesterol-specific recognition sites were prepared and then the template was removed from the shell of the nanoparticle by hydrolytic cleavage.

The main advantages of the core-shell approach are that the MIP layer can be very thin and uniform. Meanwhile, other functionalities such as magnetic and fluorescent properties can be conveniently built into the MIP as the core particle.

Quantum Dots (QDs) are semiconductor nanocrystals ranging from 2-10 nm in diameter. They provide narrow, symmetric and tunable emission spectra and show higher resistance to photobleaching than organic fluorescent dyes [65-68]. The QDs are now extensively used as fluorescent labels for chemical sensing with appropriate surface modified functional receptors.

MIPs embedded CdSe/ZnS QDs for templates uracil and caffeine were prepared by Lin *et al.* using 4-vinyl-pyridine and ethylene glycol dimethacrylic acid [69,70]. The obtained material provided positive fluorescent responses after rebinding with the templates. Stringer *et al.* developed a MIP sensor for highly explosive compounds by crosslinking fluorescent quantum dots to the MIP particles [71]. The system is capable of detecting 2,4-dinitrotoluene (DNT) as well as TNT compounds in solution with high sensitivity. The lower limits of detection for DNT and TNT are 30.1 and 40.7 μM, respectively. The detection is also rapid, with response time as low as 1 min for TNT.

Diltemiz *et al.* developed a CdS QD nanosensor for guanosine and its analogues by attaching a new metal-chelating monomer methacryloylamidohistidine-platinium to CdS QDs. [72] The guanosine can simultaneously chelate to Pt(II) metal ion and fit into the shape-selective cavity on the surface of the MIP. Another MIP-based fluorescence nanosensor was developed by Haibing Li *et al.* [73] by anchoring the MIP layer on the surface of silica nanospheres embedded CdSe QDs *via* a surface molecular imprinting process (shown in Fig. **1**). The uniform core-shell Lambda-Cyhalothrin (LC)-imprinted silica nanospheres (CdSe@SiO2@MIP) shows higher photostability, and allows a highly selective and sensitive determination of LC *via* FL intensity decreasing when removal of the original templates. The CdSe@SiO$_2$@MIP was applied to detect trace LC in water without the interference of other pyrethroids and ions. Under optimal conditions, the relative FL intensity of CdSe@SiO$_2$@MIP decreased linearly with the increasing LC in the concentration in the range of 0.1-1000 μM with a detection limit of 3.6 μg L^{-1}.

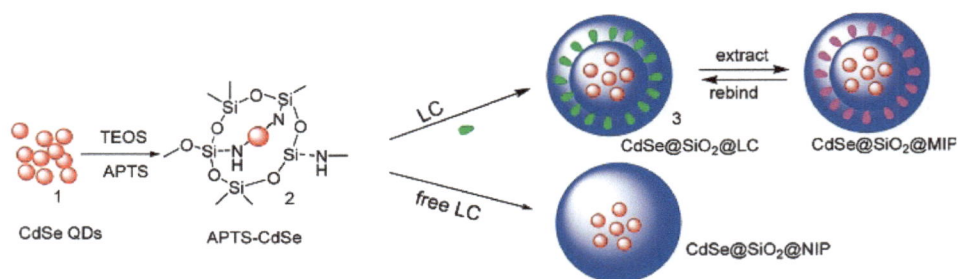

Figure 1. Schematic illustration for the molecular imprinting process of lambda-cyhalothrin -imprinted silica nanospheres embedded CdSe quantum dots [73].

3.2 Imprinted Nanofibers, Nanowires and Nanotubes

Molecular imprinting approach has also been extended to other types of nanomaterials, such as nanowires, nanotubes and nanofibres. Yang *et al.* successfully prepared molecularly imprinted nanowires by a nanomoulding process, with the molecular recognition sites at the surface of the nanowires [74]. In the approach, the pore walls of a nanoporous alumina membrane were first functionalized with the template. Then the nanopores were filled with the monomer mixture and polymerized, followed by chemical dissolution of the alumina membrane, leaving behind polypyrrole nanowires with glutamic acid binding sites situated at the surface.

Xie *et al.* [75] described the preparation of surface-imprinted nanowires/nanotubes with the templates TNT assembled onto aminopropyl group modified alumina pore walls by a strong charge-transfer complexing interaction between amino groups and electron deficient nitroaromatics. The obtained imprinted nanowires/nanotubes exhibit fast binding rate and high binding capacity for TNT molecules. The advantages of high affinity and selectivity of the MIP nanotube membrane for the chemical separation of a number of hormones has also been demonstrated [76].

As a different approach, electrospinning has also been extensively used to produce various types of polymeric nanofibers [77]. The technique is based on the ejection of a liquid seed material from a capillary tube under application of a high voltage. The ejected fibrous materials are instantaneously set and collected on an electrode plate as a mat. Ye and co-workers introduced the electrospinning technique for the preparation of MIP nanofibers and succeeded in obtaining mats imprinted with 2,4-dichlorophenoxyacetic acid (2,4-D), theophylline and 17-β-estradiol with fiber diameters of 150-300 nm [78-80]. The nanofibers can be imprinted by mixing the polymer fluid either with a template molecule [87] or with an MIP-nanoparticle [78, 80]. Initial reports have also demonstrated their potential applications as sensing elements using scintillation as the transduction method [81].

Menaker *et al.* synthesized surface imprinted microrods for avidin recognition using.precisely sized cylindrical pores (8 μm) of a track-etch Polycarbonate Membrane (PCM) filter served as sacrificial microreactor [82]. As shown in Fig. **2**, target proteins were fixed onto the pore walls by physical adsorption on the hydrophobic membranes. Conducting polymer rods of poly-3, 4-ethylenedioxythiophene doped with polystyrene sulphonate were electrochemically grown in the microreactors positioned on the surface of a gold electrode. After polymerizing the microrods into the pores, the PCM could be easily removed by dissolution in chloroform. The removal of the sacrificial material resulted in the formation of microrods confined to the surface of the gold electrode possessing the complementary imprint of the target protein on their surface.

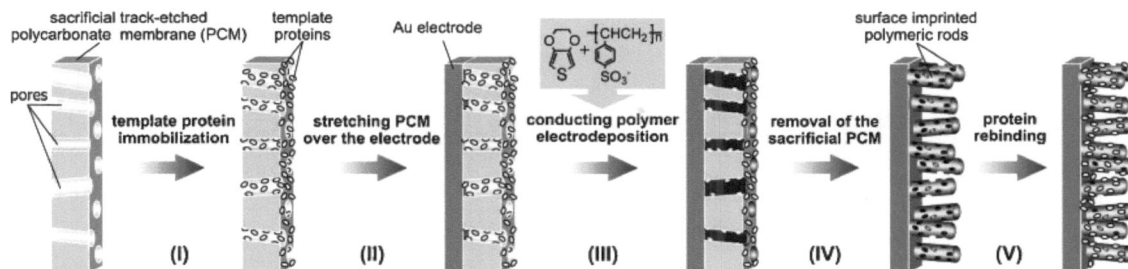

Figure 2. Schematics of the surface-imprinting strategy for fabrication of MIPs for protein assays. [82]

3.3 Imprinted Nanofilms

MIP films are essentially two-dimensional MIPs, prepared *in situ* by electropolymerization [83-86], photografting technique [87-89] and spin or drop coatings [90-92] in the presence of the analyte or by casting an imprinted polymer in the pores of an inert support membrane [93-95]. MIP films facilitate mass transfer, which is critical to rapid signal generation and analysis. They have been demonstrated for applications in combination with Surface Plasmon Resonance (SPR), [96, 97] Quartz Crystal Microbalance (QCM) [98-100], fluorescence [101-103], FTIR [104,105] and other techniques.

3.3.1 SPR and other Similar Techniques

SPR is a technique that offers rapid, label-free, real-time detection of recognition events or chemical reactions by measuring changes in the refractive index occurring on thin metal films [106,107]. By coating a selective element to the film, a specific chemical sensor can be obtained. To date, numerous SPR biosensors combined with MIP films have been developed. However, there is an inherent drawback of this method due to its low sensitivity to changes in refractive index generated by molecules of small sizes (approximately 1000 Da and smaller), which limits its application in detecting small organic molecules at very low concentrations. To address this issue, Izenberg *et al.* [108] tried to amplify the signal by a secondary molecular recognition event between a water-soluble processable star MIPs and the analyte firstly captured by the SPR sensor. On the other hand, some researchers try to utilize the effect of the electronic coupling between the localized plasmon of the gold nanoparticles (AuNPs) and the surface plasmon wave to enhance the SPR response. Matsui *et al.* [109] reported an SPR sensor with high sensitivtity to a low weight molecular-dopamine by preparing a MIP gel with embedded AuNPs on a gold substrate of a chip. The Au-MIP/MIP-coated sensor chip even showed a significant shift of SPR angle to a nanomolar sample, suggesting that cooperative use of an analyte binding macromolecule and an Au nanoparticle was effective for development of sensitive sensor chips.

Riskin et al. [110] fabricated an ultrasensitive SPR sensor for detection of hexahydro-1,3,5-trinitro-1,3,5-triazine (RDX) with a detection limit of 12 fM. They demonstrated that electropolymerization of the functionalized AuNPs in the presence of Kemp's acid (template) yields an imprinted composite with high binding affinities and selectivty to RDX. Tokareva et al. [111] further developed a sandwich-structural SPR sensor with a plasmon-band shift of 56 nm after the rebinding of the template (1 mM cholesterol) due to the changing of the interaction between 10 nm thick gold nanoislands and 2 nm diameter nanoparticles which sandwich the MIP. A novel label-free and high-throughput SPR detection sensor was reported by Huang et al. [112] using a microfluidic system with integrated MIP films designed for detection of multiple biomolecules (progesterone, cholesterol and testosterone).

Reflectometric interference spectroscopy is a method based upon the shift of the interference patterns due to the uptake of an analyte by the recognition layer. This shift corresponds to the optical thickness (the product of the physical thickness and the refractive index) change of the layer. The RIfS method is simple and robust with the wide available materials as transducer substrates. Recently, Belmont et al. [113] prepared a RIfS sensor for the dection of atrazine. The MIPs were deposited on glass transducers by two different methods: spin-coating followed by *in situ* polymerization of thin films of monomers containing a sacrificial polymeric porogen, or autoassembly of MIP nanoparticles with the aid of an associative linear polymer. Both MIP films have good reproduciblity and showed distinct difference with non-imprinted control films. In addtion, results were obtained upon measurements of atrazine solutions in toluene with atrazine concentrations as low as 1.7 ppm.

3.3.2 Fluorescence Detection

Jenkins et al. [114] developed a portable miniaturized lanthanide europium (III)-based MIP sensor for the detection of the hydrolysis product of the nerve agent soman in water. The imprinted polymer was coated onto the tip of a multimode optical fibre, delivering the excitation light and collecting back the emission from the formation of the hydrolysed soman-europium (III) polymer in presence of the analyte. The lanthanide-based MIP sensor achieved a low detection limit of 7 ppt with the response time less than 10 min.

A fluorescence-based sol-gel MIP chemical sensor for fluorene was created by using a fluorene analogue functionalized silane [115]. After chemical removal of template *via* cleavage of a carbamate linkage, an amine group was left that provided an attachment site for the environmentally sensitive fluorescent probe 7-nitrobenz-2-oxa-1,3-diazole (NBD). Fluorene binding alters the local polarity around the NBD probe and induces a change in fluorescence intensity. Sensing films deposited on glass slides were shown to have response times of <60 s and detection limits below 10 ppt. Binding experiments demonstrated that the materials had good selectivity for fluorene over close structural analogues including naphthalene, fluoranthene, and anthracene. However, the sensing design is limited by a lack of reversibility following fluorene binding.

Güney et al. developed an MIP membrane fluorescent sensor for Hg^{2+} detection [116]. The polymers were synthesized with 4-vinylpyridine as a functional monomer and Hg^{2+} as a template. 9-Vinylcarbazole was used as both a complex-forming agent and a fluorescence probe. The template was ion-bonded to pyridine and carbazole groups in the polymer membrane formed within a semicylindrical Teflon mold. After removal of the template by acid treatment, the MIP membranes could be used to directly determine Hg^{2+} concentration in aqueous solutions by the monitoring of the fluorescence intensity of the carbazole groups quenched upon complex formation with metal ions.

Tao et al. [117] proposed an innovative sensor strategy for selective and quantitative determination of proteins by introducing luminescent reporter molecules to the protein imprinted xerogels to transduce the binding event into an appropriate signal. The method achieved a limit of detection of about 2 pg/mL.

3.3.3 Other Sensing Approaches

QCM is a type of mass-sensitive transducer that allows for label-free detection of the target analytes. The group of Dickert successfully template a polymer with immunoglobulins (*i.e.*, natural antibodies) and use these MIP as stencils for designing actual plastic replicas of the initial antibody [118]. They first produced MIP nanoparticles with the respective antibody as template. Then they applied the imprinted nanoparticles as template in a surface imprinting process and thus generated a structured polymer surface directly on a 10 MHz QCM for further sensing detection.

Using Fourier Transform Infrared Spectroscopy (FTIR) for signal transduction may improve the discrimination between different analytes and thus extends the scope of MIP-based sensors to applications where cross reactivities or nonspecific interactions would cause problems. Jakush *et al.* [119] developed an infrared evanescent wave sensor for the herbicide 2,4-dichlorophenoxyacetic acid using MIP immobilized onto ZnSe attenuated total reflection elements. Thick films (5 μm) were immobilized onto the surfaces of the transducers, and upon exposure to the test solutions, selective enrichment of the analyte in the MIP layer was measured by observing MIR bands at 1595 and 1410 cm^{-1}, assigned to the anionic form of the carboxyl groups. The approach may be further extended to develop nano-scale MIP sensors.

Similarly, Raman measurement allows not only quantification but also identification of an analyte, and it has also been used in combination with MIP [120]. But selective detection of low concentrations of target molecules in very small sample volumes is still a challenge. Surface-Enhanced Raman Scattering (SERS), which can provide comparable sensitivity as SRP and specific vibrational spectra characteristic of the compound adsorbed on the MIP, might be a possible solution. The interrogation of MIP layers by SERS was first reported by Wulff and coworkers [121]. Recently, Kantarovich *et al.* [122] developed a MIP-based SERS biochip. The MIP solution droplets were printed using a pipet or a nano fountain pen on SERS-active surfaces. The results have shown that SERS can be used to monitor MIP-based sensors on microscale. Later, they [123] established a chemical nanosensors based on composite MIP particles and SERS with the detection concentration of (S)-propranolol being reduced to 0.1 μM. The molecularly imprinted core-shell nanocomposite particles were synthesized by precipitation and seeded emulsion polymerization. The layer of gold colloids located between the polymeric core and the MIP shell, resulted in several orders of magnitude signal enhancement in a single molecularly imprinted composite particle by contrast of the conventional Raman measurements on plain MIP particles and on bulk propranolol.

By combination of colloidal-crystal and molecular-imprinting techniques, Hu *et al.* prepared two imprinted photonic-hydrogel films with theophylline and (1R,2S)-(-)-ephedrine as template molecules, respectively [124]. The imprinted photonic polymer consists of a three-dimensional (3D), highly-ordered and interconnected macroporous array with a thin hydrogel wall, where nanocavities complementary to analytes are distributed. The polymer can quickly and directly report the molecular recognition event based on Bragg diffractive shifts due to the lattice change of their 3D ordered macroporous arrays upon rebinding to the target molecules. The method was highly sensitive and specific, which provided a rapid and handy approach for stimulant assay and drug analysis in athletic sports.

Zhao *et al.* report a new type of suspension array for the multiplex label-free detection of biomolecules [125]. The microcarriers of their suspension array are MIP beads with photonic crystal structure, which not only provide diffraction peaks for encoding but also convert the slight physicochemical response signals to the obvious changes of optical signals. This technique combines the advantages of suspension arrays, molecular imprinting, and photonic crystal sensors.

4. CONCLUSIONS

To fabricate MIP-based optical sensors, either the signal transduction element was directly built into the MIP structure or the MIP is interfaced on the surface of a transducer. Thanks to the unique property of MIP at nanoscale, optical sensors developed on the basis of nano-MIP exhibit excellent performance in comparison with their conventional sized counterparts, typically represented by the thorough removal of the template, fast response time and easiness of combination with other functionalities.

Despite the encouraging results of some recently published work, there remain several key issues to be addressed in the development of nano-MIP-based optical sensors. The first critical problem is the nonspecific binding of undesired components by the polymer matrix. To improve the selectivity of the MIP, functional monomers need to be designed more precisely. Another limitation of current MIP optosensing system is the relatively lower sensitivity. Possible solutions are to employ fluorescent labels with higher quantum yield or to incorporate efficient amplification steps.

Catalytic sensors are often superior to the sensors solely dependent on the binding event itself. Detection of the product from the analyte conversion may increase the sensitivity and avoid interference by nonspecific binding. However, only a few related reports are found in the literature.

Another important trend in this field goes toward miniaturization, high-throughput and multisensing. This relies on integration of the advanced microfabrication and signal processing techniques. With the significant improvement of the performance, nano-MIP-based sensors will soon find their wide applications in healthcare, environmental monitoring and other analytical areas.

REFERENCES

[1]   Potyrailo, R. A.; Mirsky, V. M. Combinatorial and high-throughput development of sensing materials: The first 10 years. *Chem. Rev.*, **2008**, *108*, 770-813.

[2]   Burton, D. R. Monoclonal antibodies from combinatorial libraries. *Acc. Chem. Res.*, **1993**, *26*, 405-411.

[3]   Osborne, S. E.; Ellington, A. D. Nucleic acid selection and the challenge of combinatorial chemistry. *Chem. Rev.*, **1997**, *97*, 349-370.

[4]   Clark, S. L.; Remcho, V. T. Aptamers as analytical reagents. *Electrophoresis*, **2002**, *23*, 1335-1340.

[5]   Srinivasan, N.; Kilburn, J. D. Combinatorial approaches to synthetic receptors. *Curr. Opin. Chem. Biol.*, **2004**, *8*, 305-310.

[6]   Schrader, T.; Hamilton, A. D. *Functional Synthetic Receptors*; Wiley-VCH: Germany, **2005**.

[7]   Klussmann, S. *The Aptamer Handbook: Functional Oligonucleotides and Their Applications*; Wiley-VCH: Germany, **2006**.

[8]   Wulff, G. Molecular imprinting in cross-linked materials with the aid of molecular templates-A way towards artificial antibodies. *Angew. Chem. Int. Ed.*, **1995**, *34*, 1812-1832.

[9]   Mayes, A.G.; Mosbach, K. Molecularly imprinted polymers: useful materials for analytical chemistry? *Trends Anal. Chem.*, **1997**, *16*, 321-332.

[10]   Henry, O. Y. F.; Cullen, D. C.; Piletsky, S. A. Optical interrogation of molecularly imprinted polymers and development of MIP sensors: a review. *Anal. Bioanal. Chem.*, **2005**, *382*, 947-956.

[11]   Haupt, K. Molecularly imprinted polymers: The next generation. *Anal. Chem.*, **2003**, *75*, 376A-383A.

[12]   Ge, Y.; Turner, A. P. F. Molecularly imprinted sorbent assays: Recent developments and applications. *Chem. Eur. J.*, **2009**, *15*, 8100-8107.

[13]   Flavin, K.; Marina Resmini, M. Imprinted nanomaterials: a new class of synthetic receptors. *Anal. Bioanal. Chem.*, **2009**, *393*, 437-444.

[14]   Tokonami, S.; Shiigi, H.; Nagaoka, T. Review: Micro- and nanosized molecularly imprinted polymers for high-throughput analytical applications, *Anal. Chim. Acta*, **2009**, *641*, 7-13.

[15]   Sellergren, B. Noncovalent molecular imprinting: antibody-like molecular recognition in polymeric network materials. *Trends Anal. Chem.*, **1997**, *16*, 310-320.

[16]   Kandimalla, V. B.; Ju, H. X. Molecular imprinting: a dynamic technique for diverse applications in analytical chemistry. *Anal. Bioanal. Chem.*, **2004**, *380*, 587-605.

[17]   Wei, S. T.; Jakusch, M.; Mizaikoff, B. Capturing molecules with templated materials-analysis and rational design of molecularly imprinted polymers. *Anal. Chim. Acta*, **2006**, *578*, 50-58.

[18]   Alexander, C.; Andersson, H. S.; Andersson, L. I.; Ansell, R. J.; Kirsch, N.; Nicholls, I. A.; O'Mahony, J.; Whitcombe, M. J. Molecular imprinting science and technology: a survey of the literature for the years up to and including 2003. *J. Mol. Recognit.*, **2006**, *19*, 106-180.

[19]   Turner, N. W.; Jeans, C. W.; Brain, K. R.; Allender, C. J.; Hlady, V.; Britt, D. W. From 3D to 2D: A review of the molecular imprinting of proteins. *Biotechnol. Prog.*, **2006**, *22*, 1474-1489.

[20]   Patek, M.; Drew, M. Chemical synthesis in nanosized cavities. *Curr. Opin. Chem. Biol.*, **2008**, *12*, 332-339.

[21] Haupt, K.; Mosbach, K. Molecularly Imprinted Polymers and Their Use in Biomimetic Sensors. *Chem. Rev.*, **2000**, *100*, 2495-2504.

[22] Valero-Navarro, A.; Salinas-Castillo, A.; Fernández-Sánchez, J. F.; Segura-Carretero, A.; Mallavia, R.; Fernández-Gutiérrez, A. The development of a MIP-optosensor for the detection of monoamine naphthalenes in drinking water. *Biosens. Bioelectron.*, **2009**, *24*, 2305-2311.

[23] Haupt, K.; Mayes, A. G.; Mosbach, K. Herbicide assay using an imprinted polymer-based system analogous to competitive fluoroimmunoassays. *Anal. Chem.*, **1998**, *70*, 3936-3939.

[24] Levi, R.; McNiven, S.; Piletsky, S. A.; Cheong, S. H.; Yano, K.; Karube, I. Optical detection of chloramphenicol using molecularly imprinted polymers. *Anal. Chem.*, **1997**, *69*, 2017-2021.

[25] Suarez-Rodriguez, J. L.; Diaz-Garcia, M. E. Fluorescent competitive flow-through assay for chloramphenicol using molecularly imprinted polymers. *Biosens. Bioelectron.*, **2001**, *16*, 955-961.

[26] Haupt K. Molecularly imprinted sorbent assays and the use of non-related probes. *React. Funct. Polym.*, **1999**, *41*, 125-131.

[27] Benito-Peňa, E.; Moreno-Bondi, M. C.; Aparicio, S.; Orellana, G.; Cederfur, J.; Kempe, M. Molecular engineering of fluorescent penicillins for molecularly imprinted polymer assays. *Anal. Chem.*, **2006**, *78*, 2019-2027.

[28] Urraca, J. L.; Moreno-Bondi, M. C.; Orellana, G.; Sellergren, B.; Hall, A. J. Molecularly imprinted polymers as antibody mimics in automated on-line fluorescent competitive assays. *Anal. Chem.*, **2007**, *79*, 4915-4923.

[29] Gonzalez, G. P.; Hernando, P. F.; Alegria, J. S. D. Determination of digoxin in serum samples using a flow-through fluorosensor based on a molecularly imprinted polymer. *Biosens. Bioelectron.*, **2008**, *23*, 1754-1758.

[30] Tan, J.; Wang, H. F.; Yan, X. P. Discrimination of saccharides with a fluorescent molecular imprinting sensor array based on phenylboronic acid functionalized mesoporous silica. *Anal. Chem.*, **2009**, *81*, 5273-5281.

[31] Turkewitsch, P.; Wandelt, B.; Darling, G. D.; Powell, W. S. Fluorescent functional recognition sites through molecular imprinting. A polymer-based fluorescent chemosensor for aqueous cAMP. *Anal. Chem.*, **1998**, *70*, 2025-2030.

[32] Liao, Y.; Wang, W.; Wang, B. H. Building fluorescent sensors by template polymerization: The preparation of a fluorescent sensor for L-tryptophan. *Bioorg. Chem.*, **1999**, *27*, 463-476.

[33] Matsui, J.; Tachibana, Y.; Takeuchi, T. Molecularly imprinted receptor having metalloporphyrin-based signaling binding site. *Anal. Commun.*, **1998**, *35*, 225-227.

[34] Wang, W.; Gao, S. H.; Wang, B. H. Building fluorescent sensors by template polymerization: The preparation of a fluorescent sensor for D-fructose. *Org. Lett.*, **1999**, *8*, 1209-1212.

[35] Jenkins, A. L.; Uy, O. M.; Murray, G. M. Polymer-based lanthanide luminescent sensor for detection of the hydrolysis product of the nerve agent soman in water. *Anal. Chem.*, **1999**, *71*, 373-378.

[36] Dickert, F. L.; Hayden, O. Molecular imprinting in chemical sensing. *Trends Anal. Chem.*, **1999**, *18*, 192-199.

[37] Dickert, F. L.; Tortschanoff, M.; Bulst, W. E.; Fischerauer, G. Molecularly imprinted sensor layers for the detection of polycyclic aromatic hydrocarbons in water. *Anal.Chem.*, **1999**, *71*, 4559-4563.

[38] Al-Kindy, S.; Badia, R.; Suarez-Rodriguez, J. L.; Diaz-Garcia, M. E. Molecularly imprinted polymers and optical sensing applications. *Crit. Rev. Anal. Chem.*, **2000**, *30*, 291-309.

[39] Ellwanger, A.; Berggren, C.; Bayoudh, S.; Crecenzi, C.; Karlsson, L.; Owens, P. K.; Ensing, K.; Cormack, P.; Sherrington, D.; Sellergren, B. Evaluation of methods aimed at complete removal of template from molecularly imprinted polymers. *Analyst*, **2001**, *126*, 784-792.

[40] Salinas-Castillo, A.; Sanchez-Barragan, I.; Costa-Fernandez, J. M.; Pereiro, R.; Ballesteros, A.; Gonzalez, J. M.; Segura-Carretero, A.; Fernandez-Gutierrez, A.; Sanz-Medel, A. Iodinated molecularly imprinted polymer for room temperature phosphorescence optosensing of fluoranthene. *Chem.Commun.*, **2005**, *25*, 3224-3226.

[41] Takeuchi, T.; Mukawa, T.; Matsui, J.; Higashi, M.; Shimizu, K. D. Molecularly imprinted polymers with metalloporphyrin-based molecular recognition sites coassembled with methacrylic acid. *Anal. Chem.*, **2001**, *73*, 3869-3874.

[42] Tong, A. J.; Dong, H.; Li, L. D. Molecular imprinting-based fluorescent chemosensor for histamine using zinc(II)-protoporphyrin as a functional monomer. *Anal. Chim. Acta.*, **2002**, *466*, 31-37.

[43] Sanchez-Barragan, I.; Costa-Fernandez, J. M.; Pereiro, R.; Sanz-Medel, A.; Salinas, A.; Segura, A.; Fernandez-Gutierrez, A.; Ballesteros, A.; Gonzalez, J. M. Molecularly imprinted polymers based on iodinated monomers for selective room-temperature phosphorescence optosensing of fluoranthene in Water. *Anal. Chem.*, **2005**, *77*, 7005-7011.

[44]   Li, J. H.; Kendig, C. E.; Nesterov, E. E. Chemosensory performance of molecularly imprinted fluorescent conjugated polymer materials. *J. Am. Chem. Soc.*, **2007**, *129*, 15911-15918.

[45]   Nguyen, T. H.; Ansell, R. J. Fluorescent imprinted polymer sensors for chiral amines. *Org. Biomol. Chem.*, **2009**, *7*, 1211-1220.

[46]   Kubo, H.; Nariai, H.; Takeuchi, T. Multiple hydrogen bonding-based fluorescent imprinted polymers for cyclobarbital prepared with 2,6-bis(acrylamido)pyridine. *Chem. Commun.*, **2003**, *22*, 2792-2793.

[47]   Kubo, H.; Yoshioka, N.; Takeuchi, T. Fluorescent imprinted polymers prepared with 2-acrylamidoquinoline as a signaling monomer. *Org. Lett.*, **2005**, *7*, 359-362.

[48]   Tan, J.; Wang, H. F.; Yan, X. P. A fluorescent sensor array based on ion imprinted mesoporous silica. *Biosens. Bioelectron.*, **2009**, *24*, 3316-3321.

[49]   Subrahmanyam, S.; Piletsky, S. A.; Piletska, E. V.; Chen, B. V.; Day, R.; Turner, A. P. F. "Bite-and-switch" approach to creatine recognition by use of molecularly imprinted polymers. *Adv. Mat.*, **2000**, *12*, 722-724.

[50]   Gao, S. H.; Wang, W.; Wang, B. H. Building fluorescent sensors for carbohydrates using template-directed polymerizations. *Bioorg. Chem.*, **2001**, *29*, 308-320.

[51]   Graham, A. L.; Carlson, C. A.; Edmiston, P. L. Development and characterization of molecularly imprinted sol-gel materials for the selective detection of DDT. *Anal. Chem.*, **2002**, *74*, 458-467.

[52]   Haupt, K.; Noworyta, K.; Kutner, W. Imprinted polymer-based enantioselective acoustic sensor using a quartz crystal microbalance. *Anal. Commun.*, **1999**, *36*, 391-393.

[53]   Kugimiya, A.; Takeuchi, T. Molecularly imprinted polymer-coated quartz crystal microbalance for detection of biological hormone. *Electroanalysis*, **1999**, *11*, 1158-1160.

[54]   Kriz, D.; Ramstrum, O.; Svensson, A.; Mosbach, K. Introducing biomimetic sensors based on molecularly imprinted polymers as recognition elements. *Anal. Chem.*, **1995**, *67*, 2142-2144.

[55]   Medina-Castillo, A. L.; Mistlberger, G.; Fernandez-Sanchez, J. F.; Segura-Carretero, A.; Klimant, I.; Fernandez-Gutierrez, A. Novel strategy to design magnetic, molecular imprinted polymers with well-controlled structure for the application in optical sensors. *Macromolecules*, **2010**, *43*, 55-61.

[56]   Chow, C. F.; Lam, M. H. W.; Leung, M. K. P. Fluorescent sensing of homocysteine by molecular imprinting. *Anal. Chim. Acta*, **2002**, *466*, 17-30.

[57]   Vaihinger, D.; Landfester, K.; Kräuter, I.; Brunner, H.; Tovar, G. E. M. Molecularly imprinted polymer nanospheres as synthetic affinity receptors obtained by miniemulsion polymerisation. *Macromol. Chem. Phys.*, **2002**, *203*, 1965-1973.

[58]   Ye, L.; Cormack, P. A. G.; Mosbach, K. Molecular imprinting on microgel spheres. *Anal. Chim. Acta*, **2001**, *435*, 187-196.

[59]   Zhu, Q. J.; Tang, J.; Dai, J.; Gu, X. H.; Chen, S.W. Synthesis and characteristics of imprinted 17-β-estradiol microparticle and nanoparticle with TFMAA as functional monomer. *J. Appl. Polym. Sci.*, **2007**, *104*, 1551-1558.

[60]   Yoshimatsu, K.; Reimhult, K.; Krozer, A.; Mosbach, K.; Sode, K.; Ye, L. Uniform molecularly imprinted microspheres and nanoparticles prepared by precipitation polymerization: The control of particle size suitable for different analytical applications. *Anal. Chim. Acta*, **2007**, *584*, 112-121.

[61]   Wei, S. T.; Molinelli, A.; Mizaikoff, B. Molecularly imprinted micro and nanospheres for the selective recognition of 17β-estradiol. *Biosens. Bioelectron.*, **2006**, *21*, 1943-1951.

[62]   Ye, L.; Cormack, P. A. G.; Mosbach, K. Molecularly imprinted monodisperse microspheres for competitive radioassay. *Anal. Commun.*, **1999**, *36*, 35-38.

[63]   Pérez, N.; Whitcombe, M. J.; Vulfson, E. N. Molecularly imprinted nanoparticles prepared by core-shell emulsion polymerization. *J. Appl. Polym. Sci.*, **2000**, *77*, 1851-1859.

[64]   Perez, N.; Whitcombe, M. J.; Vulfson, E. N. Surface imprinting of cholesterol on submicrometer core-shell emulsion particles. *Macromolecules*, **2001**, *34*, 830-836.

[65]   Basabe-Desmonts, L.; Reinhoudt, D.N.; Crego-Calama, M. Design of fluorescent materials for chemical sensing. *Chem. Soc. Rev.*, **2007**, *36*, 993-1017.

[66]   Jin, T.; Fujii, F.; Sakata, H.; Tamura, M.; Kinjo, M. Amphiphilic p-sulfonatocalix[4]arene-coated CdSe/ZnS quantum dots for the optical detection of the neurotransmitter acetylcholine. *Chem. Commun.*, **2005**, *34*, 4300-4302.

[67]   Jin, T.; Fujii, F.; Sakata, H.; Tamura, M.; Kinjo, M. Calixarene-coated water-soluble CdSe-ZnS semiconductor quantum dots that are highly fluorescent and stable in aqueous solution. *Chem. Commun.*, **2005**, *22*, 2829-2831.

[68]   Konishi, K.; Hiratani, T. Turn-on and selective luminescence sensing of copper ions by a water-soluble $Cd_{10}S_{16}$ molecular cluster. *Angew. Chem. Int. Ed.*, **2006**, *45*, 5191-5194.

[69]   Lin, C. I.; Joseph, A. K.; Chang, C. K.; Lee, Y. D. Molecularly imprinted polymeric film on semiconductor nanoparticles Analyte detection by quantum dot photoluminescence. *J. Chromatogr. A*, **2004**, *1027*, 259-262.

[70]   Lin, C. I.; Joseph, A. K.; Chang, C. K.; Lee, Y. D. Synthesis and photoluminescence study of molecularly imprinted polymers appended onto CdSe/ZnS core-shells. *Biosens. Bioelectron.*, **2004**, *20*, 127-131.

[71]   Stringer, R. C.; Gangopadhyay, S.; Grant, S. A. Detection of nitroaromatic explosives using a fluorescent-labeled imprinted polymer. *Anal. Chem.*, **2010**, *82*, 4015-4019.

[72]   Diltemiz, S. E.; Say, R.; Buyuktiryaki, S.; Hur, D.; Denizli, A.; Ersoz, A. Quantum dot nanocrystals having guanosine imprinted nanoshell for DNA recognition. *Talanta*, **2008**, *75*, 890-896.

[73]   Li, H. B.; Li, Y. L.; Cheng, J. Molecularly imprinted silica nanospheres embedded CdSe quantum dots for highly selective and sensitive optosensing of pyrethroids. *Chem. Mater.*, **2010**, *22*, 2451-2457.

[74]   Yang, H. H.; Zhang, S. Q.; Tan, F.; Zhuang, Z. X.; Wang, X. R. Surface molecularly imprinted nanowires for biorecognition. *J. Am. Chem. Soc.*, **2005**, *127*, 1378-1379.

[75]   Xie, C.G.; Zhang, Z. P.; Wang, D. P.; Guan, G. J.; Gao, D. M.; Liu, J. H. Surface molecular self-assembly strategy for TNT imprinting of polymer nanowire/nanotube arrays. *Anal. Chem.*, **2006**, *78*, 8339-8346.

[76]   Wang, H. J.; Zhou, W. H.; Yin, X. F.; Zhuang, Z. X.; Yang, H. H.; Wang, X. R. Template synthesized molecularly imprinted polymer nanotube membranes for chemical separations. *J. Am. Chem. Soc.*, **2006**, *128*, 15954-15955.

[77]   Agarwal, S.; Wendorff, J. H.; Greiner, A. Use of electrospinning technique for biomedical applications. *Polymer*, **2008**, *49*, 5603-5621.

[78]   Chronakis, I. S.; Jakob, A.; Hagström, B.; Ye, L. Encapsulation and selective recognition of molecularly imprinted theophylline and 17 ss-estradiol nanoparticles within electrospun polymer nanofibers. *Langmuir*, **2006**, *22*, 8960-8965.

[79]   Chronakis, I. S.; Milosevic, B.; Frenot, A.; Ye, L. Generation of molecular recognition sites in electrospun polymer nanofibers *via* molecular imprinting. *Macromolecules*, **2006**, *39*, 357-361.

[80]   Yoshimatsu, K.; Ye, L.; Lindberg, J.; Chronakis, I. S. Selective molecular adsorption using electrospun nanofiber affinity membranes. *Biosens. Bioelectron.*, **2008**, *23*, 1208-1215.

[81]   Yoshimatsu, K.; Ye, L.; Stenlund, P.; Chronakis, I. S. A simple method for preparation of molecularly imprinted nanofiber materials with signal transduction ability. *Chem. Commun.*, **2008**, *17*, 2022-2024.

[82]   Menaker, A.; Syritski, V.; Reut, J.; Öpik, A.; Horvath, V.; Gyurcsanyi, R. E. Electrosynthesized surface-imprinted conducting polymer microrods for selective protein recognition. *Adv. Mater.*, **2009**, *21*, 2271-2275.

[83]   Malitesta, C.; Losito, I.; Zambonin, P. G. Molecularly imprinted electrosynthesized polymers: New materials for biomimetic sensors. *Anal. Chem.*, **1999**, *71*, 1366-1370.

[84]   Ulyanova, Y. V.; Blackwell, A. E.; Minteer, S. D. Poly(methylene green) employed as molecularly imprinted polymer matrix for electrochemical sensing. *Analyst*, **2006**, *131*, 257-261.

[85]   Sreenivasan, K. Synthesis and evaluation of multiply templated molecularly imprinted polyaniline. *J. Mater. Sci.*, **2007**, *42*, 7575-7578.

[86]   Pietrzyk, A.; Kutner, W.; Chitta, R.; Zandler, M. E.; D'Souza, F.; Sannicolo, F.; Mussini, P. R. Melamine acoustic chemosensor based on molecularly imprinted polymer film. *Anal. Chem.*, **2009**, *81*, 10061-10070.

[87]   Huang, C.Y.; Syu, M.J.; Chang, Y.S.; Chang, C.H.; Chou, T.C.; Liu, B.D. A portable potentiostat for the bilirubin-specific sensor prepared from molecular imprinting. *Biosens. Bioelectron.*, **2007**, *22*, 1694-1699.

[88]   Delaney, T. L.; Zimin, D.; Rahm, M.; Weiss, D.; Wolfbeis, O. S.; Mirsky, V. M. Capacitive detection in ultrathin chemosensors prepared by molecularly imprinted grafting photopolymerization. *Anal. Chem.*, **2007**, *79*, 3220-3225.

[89]   Lakshmi, D.; Bossi, A.; Whitcombe, M. J.; Chianella, I.; Fowler, S. A.; Subrahmanyam, S.; Piletska, E. V.; Piletsky, S. A. Electrochemical sensor for catechol and dopamine based on a catalytic molecularly imprinted polymer-conducting polymer hybrid recognition element. *Anal. Chem.*, **2009**, *81*, 3576-3584.

[90]   Marx, S.; Zaltsman, A.; Turyan, I.; Mandler, D. Parathion sensor based on molecularly imprinted sol-gel films. *Anal. Chem.*, **2004**, *76*, 120-126.

[91]   Hsu, C.W.; Yang M.C. Electrochemical epinephrine sensor using artificial receptor synthesized by sol-gel process. *Sens. Actuators B: Chem.*, **2008**, *134*, 680-686.

[92]   Huang, C.Y.; Tsai, T. C.; Thomas, J. L.; Lee, M.-H.; Liu, B. D.; Lin, H. Y. Urinalysis with molecularly imprinted poly(ethylene-co-vinyl alcohol) potentiostat sensors. *Biosens. Bioelectron.*, **2009**, *24*, 2611-2617.

[93]   Hong, J.M.; Anderson, P. E.; Qian, J.; Martin, C. R. Selectively-permeable ultrathin film composite membranes based on molecularly-imprinted polymers. *Chem. Mater.*, **1998**, *10*, 1029-1033.

[94]   Prasad, K.; Prathish, K. P.; Gladis, J. M.; Naidu, G. R. K.; Rao, T. P. Molecularly imprinted polymer (biomimetic) based potentiometric sensor for atrazine. *Sens. Actuators B: Chem.*, **2007**, *123*, 65-70.

[95]   Liang, R. N.; Song, D. A.; Zhang, R. M.; Qin, W. Potentiometric sensing of neutral species based on a uniform-sized molecularly imprinted polymer as a receptor. *Angew. Chem. Int. Ed.*, **2010**, *49*, 2556-2559.

[96]   Lai, E. P. C.; Fafara, A.; VanderNoot, V. A.; Kono, M.; Polsky, B. Surface plasmon resonance sensors using molecularly imprinted polymers for sorbent assay of theophylline, caffeine, and xanthine. *Can. J. Chem.*, **1998**, *76*, 265-273.

[97]   Taniwaki, K.; Hyakutake, A.; Aoki, T.; Yoshikawa, M.; Guiver, M. D.; Robertson, G. P. Evaluation of the recognitionability of molecularly imprinted materials by Surface Plasmon Resonance (SPR) spectroscopy. *Anal. Chim. Acta*, **2003**, *489*, 191-198.

[98]   Shoji, R.; Takeuchi, T.; Kubo, I. Atrazine sensor based on molecularly imprinted polymer-modified gold electrode. *Anal. Chem.*, **2003**, *75*, 4882-4886.

[99]   Ersoz, A.; Denizli, A.; Ozcan, A.; Say, R. Molecularly imprinted ligand-exchange recognition assay of glucose by quartz crystal microbalance. *Biosens. Bioelectron.*, **2005**, *20*, 2197-2202.

[100]   Kikuchi, M.; Tsuru, N.; Shiratori, S. Recognition of terpenes using molecular imprinted polymer coated quartz crystal microbalance in air phase. *Sci. Technol. Adv. Mater.*, **2006**, *7*, 156-161.

[101]   Voicu, R.; Faid, K.; Farah, A. A.; Bensebaa, F.; Barjovanu, R.; Py, C.; Tao, Y. Nanotemplating for two-dimensional molecular imprinting. *Langmuir*, **2007**, *23*, 5452-5458.

[102]   Conrad II, P. G.; Nishimura, P. T.; Aherne, D.; Schwartz, B. J.; Wu, D. M.; Fang, N.; Zhang, X.; Roberts, M. J.; Shea, K. J. Functional molecularly imprinted polymer microstructures fabricated using microstereolithography. *Adv. Mater.*, **2003**, *15*, 1541-1544.

[103]   Vandevelde, F.; Leïchle, T.; Ayela, C.; Bergaud, C.; Nicu, L.; Haupt, K. Direct patterning of molecularly imprinted microdot arrays for sensors and biochips. *Langmuir*, **2007**, *23*, 6490-6493.

[104]   Jakusch, M.; Janotta, M.; Mizaikoff, B.; Mosbach, K.; Haupt, K. Molecularly imprinted polymers and infrared evanescent wave spectroscopy. A chemical sensors approach. *Anal. Chem.*, **1999**, *71*, 4786-4791.

[105]   Henry, O. Y. F.; Piletsky, S. A.; Cullen, D. C. Fabrication of molecularly imprinted polymer microarray on a chip by mid-infrared laser pulse initiated polymerization. *Biosens. Bioelectron.*, **2008**, *23*,1769-1775.

[106]   Phillips, K. S.; Cheng, Q. Recent advances in surface plasmon resonance based techniques for bioanalysis. *Anal. Bioanal. Chem.*, **2007**, *387*, 1831-1840.

[107]   Gordon, R.; Sinton, D.; Kavanagh, K. L.; Brolo, A. G. A new generation of sensors based on extraordinary optical transmission. *Acc. Chem. Res.*, **2008**, *41*, 1049-1057.

[108]   Izenberg, N. R.; Murrray, G. M.; Pilato, R. S.; Baird, L. M.; Levin, S. M.; Van Houten, K. A. Astrobiological molecularly imprinted polymer sensors. *Planet. Space Sci.*, **2009**, 57, 846-853.

[109]   Matsui, J.; Akamatsu, K.; Hara, N.; Miyoshi, D.; Nawafune, H.; Tamaki, K.; Sugimoto, N. SPR sensor chip for detection of small molecules using molecularly imprinted polymer with embedded gold nanoparticles. *Anal. Chem.*, **2005**, *77*, 4282-4285.

[110]   Riskin, M.; Tel-Vered, R.; Willner, I. Imprinted Au-nanoparticle composites for the ultrasensitive surface plasmon resonance detection of hexahydro-1,3,5-trinitro-1,3,5-triazine (RDX). *Adv. Mater.*, **2010**, *22*, 1387-1391.

[111]   Tokareva, I.; Tokarev, I.; Minko, S.; Hutter, E.; Fendler, J. H. Ultrathin molecularly imprinted polymer sensors employing enhanced transmission surface plasmon resonance spectroscopy. *Chem. Commun.*, **2006**, *77*, 3343-3345.

[112]   Huang, S. C.; Lee, G. B.; Chien, F. C.; Chen, S. J.; Chen, W. J.; Yang, M. C. A microfluidic system with integrated molecular imprinting polymer films for surface plasmon resonance detection. *J. Micromech. Microeng.*, **2006**, *16*,1251-1257.

[113]   Belmont, A.-S.; Jaeger, S.; Knoppc, D.; Niessner, R.; Gauglitz, G.; Haupt, K. Molecularly imprinted polymer films for reflectometric interference spectroscopic sensors. *Biosens. Bioelectron.*, **2007**, *22*, 3267-3272.

[114]   Jenkins, A. L.; Uy, O. M.; Murray, G. M. Polymer-based lanthanide luminescent sensor for detection of the hydrolysis product of the nerve agent soman in water. *Anal. Chem.*, **1999**, *71*, 373-378.

[115]   Carlson, C. A.; Lloyd, J. A.; Dean, S. L.; Walker, N. R.; Edmiston, P. L. Sensor for fluorene based on the incorporation of an environmentally sensitive fluorophore proximal to a molecularly imprinted binding site. *Anal. Chem.*, **2006**, *78*, 3537-3542.

[116]   Güney, O.; Cebeci, F. C. Molecularly imprinted fluorescent polymers as chemosensors for the detection of mercury ions in aqueous media. *J. Appl. Polym. Sci.*, **2010**, *117*, 2373-2379.

[117] Tao, Z. Y.; Tehan, E. C.; Bukowski, R. M.; Tang, Y.; Shughart, E. L.; Holthoff, W. G.; Cartwright, A. N.; Titus, A. H.; Bright, F. V. Templated xerogels as platforms for biomolecule-less biomolecule sensors. *Anal. Chimi. Acta*, **2006**, *564*, 59-65.

[118] Schirhagl, R.; Lieberzeit, P. A.; Dickert, F. L. Chemosensors for viruses based on artificial immunoglobulin copies. *Adv. Mater.*, **2010**, *22*, 2078-2081.

[119] Jakush, M.; Janotta, M.; Mizaikoff, B. Molecularly Imprinted Polymers and Infrared Evanescent Wave Spectroscopy. A Chemical Sensors Approach. *Anal. Chem.*, **1999**, *71*, 4786-4791.

[120] Popp, J.; Kiefer, W. In: *Encyclopedia of Analytical Chemistry*, Mayer, R.A., Ed.; John Wiley and Sons: New York, **2000**; Vol. *21*, pp. 13104

[121] Kostrewa, S.; Emgenbroich, M.; Klockow, D.; Wulff, G. Surface-enhanced raman scattering on molecularly imprinted polymers in water. *Macromol. Chem. Phys.*, **2003**, *204*, 481-487.

[122] Kantarovich, K.; Tsarfati, I.; Gheber, L. A.; Haupt, K.; Bar, I. Writing droplets of molecularly imprinted polymers by nano fountain pen and detecting their molecular interactions by surface-enhanced raman scattering. *Anal. Chem.*, **2009**, *81,* 5686-5690.

[123] Bompart, M.; Wilde, Y. D.; Haupt, K. Chemical nanosensors based on composite molecularly imprinted polymer particles and surface-enhanced raman scattering. *Adv. Mater.*, **2010**, *22*, 1-6.

[124] Hu, X. B.; Li, G. T.; Li, M. H.; Huang, J.; Li, Y.; Gao, Y. B.; Zhang, Y. H. Ultrasensitive Specific Stimulant Assay Based on Molecularly Imprinted Photonic Hydrogels. *Adv. Funct. Mater.*, **2008**, *18*, 575-583.

[125] Zhao, Y. J.; Zhao, X. W.; Hu, J.; Li, J.; Xu, W. Y.; Gu, Z. Z. Multiplex label-free detection of biomolecules with an imprinted suspension array. *Angew. Chem. Int. Ed.*, **2009**, *48*, 7350-7352.

CHAPTER 4

Thermo Sensitive Polymers for Prolong Delivery of Contraceptive Hormones in Women

Priyanka Singh[1*] and Sibao Chen[2]

[1]*1919 Elm Street, Fargo, ND 58102 and* [2]*Purdue Pharmaceuticals L.P. 4701 Purdue Drive, Wilson, NC 27893*

Abstract: This review discusses the various available controlled release products for contraception in women as well as elaborates about the thermosensitive polymers, their characterization and application for controlled delivery of contraceptive hormones. The thermosensitive polymers are free flowing solutions in water at room temperature and turn into gel at body temperature and deliver the incorporated hormones at controlled rate for longer duration after a single subcutaneous injection. These polymers are biodegradable, biocompatible, and hold a great promise for prolonged delivery of contraceptive hormones.

Keywords: Thermosensitive polymers; controlled release; contraceptive hormones.

INTRODUCTION

The Population Council began researching subdermal contraceptive implants in 1966 [1]. The idea of using subdermal capsules using silicon polymers for contraceptive hormone evolved from the fact that these polymers could form a reservoir for the prolonged release of a variety of lipophilic drugs. Currently available subdermal implant, levonorgestrel, is contained in six flexible, closed capsules made of silicon polymer. Each capsule contains 36 mg of the drug, levonorgestrel. Each capsule is 34 mm long and 2.4 mm in diameter [2]. Levonorgestrel implants are also available in rods (two rods, each contains 75mg of the levonorgestrel). Although silicon based subdermal implants can control the release of incorporated hormone, however, the removal of the implants after drug release can be challenging. One of the studies evaluated 1,253 removal procedures at 15 clinical settings. The removal usually took 30 min [3]. However, about 19% of removals lasted for more than 1h. A few patients had to return for a second removal procedure. About one-quarter of the women reported substantial pain. In addition to the above difficulties, pruritis (generally transient), infection at implant site, removal difficulties as well as damage to capsules were reported.

A retrospective analysis of 3,416 subdermal implant removals was performed, including women from 11 countries who participated in clinical trials [4]. Some of these women experienced difficult or complicated removals. Most complicated removals were due to the implants being broken during the removal procedure; embedding or displacement of the implants also led to difficult removals. Subdermal implants have become the target of litigation. Law suits claimed that implants caused a variety of problems in users. Subdermal implant litigation patterns have paralleled those relating to silicone breast implants [5, 6].

Injectables using microspheres or microcapsules containing one or more contraceptive hormones have been investigated [7]. A sterile solution suspends the time-released spheres. The microspheres contain a polymer commonly used in a biodegradable suture, poly-dl-lactide-co-glycolide. However, burst release from microspheres is a problem. Also, microspheres pose a significant manufacturing challenges requiring 5 to 6 major processing steps. In addition, microspheres may cause an acute tissue reaction (*e.g.* nodule) and, possible, transient irritation resulting in the presence of particles. In contrast, smart polymer based injectable solution is simple to prepare and forms an implant upon injection. Smart polymers are widely explored as potential drug-delivery systems [8-10]. Biodegradable, biocompatible, thermosensitive smart

---

*Address correspondence to Priyanka Singh: University of North Dakota, College of Medicine and Health Sciences, 1919 Elm Street, Fargo, ND 58102; E-mail: psingh@medicine.nodak.edu

Songjun Li, Yi Ge and He Li (Eds)
All rights reserved - © 2012 Bentham Science Publishers

polymer based drug delivery system offers several advantages, including controlled release of drugs, low burst release, low batch-to-batch variation in comparison to implants or microspheres, high drug loading, and ease of preparation. This chapter explores various steroidal hormones available as implants, thermosensitive polymers, their characterization and applications in drug delivery for contraceptives.

HORMONES FOR CONTRACEPTION

Control of fertility constitutes a global health issue, as overpopulation and unintended pregnancy have both major personal and societal impact. The contraceptive revolution in the 1960s led to the development of hormonal-based oral contraceptives for women. It has had a major impact on societal dynamics in several cultures and laid the foundations for women's liberation [11].

Oral contraception, or the pill, is used today by over 80 million women in the world, making it the third most popular method of family planning after female sterilization (210 million users) and intrauterine devices (156 million). The main drawback of combined oral contraceptives is that they must be taken daily. This drawback has over the past half-century fuelled a quest for alternatives that could be taken less frequently. Thereafter many new drug delivery systems were developed. Injectable progestogens (depot medroxyprogesterone acetate and norethindrone enanthate) were approved in some countries in the early 1980s. Combined injectables (containing both estrogen and progestogen which are administered monthly) are now widely used in Central and South America and have recently been approved in the USA. Progestogen-only contraceptive implants became widely available in the 1990. The addition of a progestogen to the intrauterine device has produced an IUD that is licensed for 5 years. At the end of this long list of new delivery systems come the contraceptive vaginal ring (worn in the vagina for 21 days and removed for 7 days) and a contraceptive transdermal patch. Also, transdermal gel and transnasal spray have been explored for delivering contraceptive hormones [12].

Contraceptive implants were originally conceived for long term continuous release of a steroid hormone in order to avoid over and under dosing periods and to avoidthe user from daily administration [13]. Long-term use is one of their most appealing features for many users because this is linked to a sense of comfort and reliability [14]. The first scientific publication on a contraceptive implant for women releasing a progestogen appeared in 1969.

Table 1. Contraceptive implants, available or being developed.

Implant	Distinctive Components	Registration	Lifespan (years)	Chief Mechanism of Action
Norplant	6 silicone capsules levonorgestrel	~ 60 countries	7[a]	Inhibits ovulation and makes cervical mucus impenetrable by sperm
Jadelle	2 silicone rods levonorgestrel	In some EU countries, USA, Thailand, and Indonesia	3	Inhibits ovulation and makes cervical mucus impenetrable by sperm
Implanon	1 polymer (resin) rod etonogestrel	Australia, Indonesia and many EU countries	3	Suppresses ovulation and endometrial development
Nestorone	1 silicone rod nestorone	Brazil	2[b]	Suppresses ovulation
Elcometrine	1 silicone capsule nestorone	Brazil	0.5[c]	Supresses ovulation

a. Approved for 5 years.

b. Intended life span

c. Approved

In 1983 the first usable implant, Norplant®, which releases the drug, levonorgestrel, through six capsules, was approved by the Finnish national drug regulatory authority. Since then, several more implants have been approved and others are under development (Table **1**). The contraceptive implants have been approved in more than 60 countries and used by ~11 million women worldwide.

THERMOSENSITIVE POLYMERS

Aqueous solutions of some polymers undergo sol-to-gel transition in response to temperature changes. The drugs can be mixed in a sol state and injected using a syringe into subcutaneous layers to form a depot system. The reverse thermo-responsive phenomenon is usually known as Reverse Thermal Gelation (RTG) and it constitutes one of the most promising strategies for the development of injectable systems for biomedical applications. Water solutions of these polymers display low viscosity at ambient temperature, and exhibit a sharp viscosity increase following a small temperature rise, producing a semi-solid gel at body temperature. There are numerous RTG displaying polymers such as Poly (N-isopropylacrylamide) (PNIPAAM), Poly (ethylene oxide)-poly (propylene oxide)-poly (ethylene oxide) triblocks (PEO-PPO-PEO), ethyl (hydroxyethyl) cellulose (EHEC) and poly (ethylene glycol)-poly (lactic acid)-poly (ethylene glycol) triblocks (PEG-PLA-PEG). Aqueous solutions of these polymers have a Lower Critical Solution Temperature (LCST), resulting in the viscosity increase, upon heating above this temperature [15].

Numerous polymers show abrupt changes in solubility as a function of environmental temperature. The poly-NIPAAM exhibits a rather sharp LCST of ~32°C. However, poly-NIPAAM exhibits toxicity therefore it is not suitable for biomedical applications. It activates platelets on contact with blood along with non-degradability, makes it difficult to get FDA approval. Therefore a vast majority of the drug delivery system which employs LCST, use block copolymers of PEO and PPO because of FDA approval. Triblock PEO-PPO-PEO copolymers (Pluronics or Poloxamers) are available in a variety of compositions and are of particular interest, as their gelation phenomena have been well studied. They have shown gelation at body temperature at a concentration over 15% w/w. However, these concentrations of a surfactant lead to notable cytotoxicity. Furthermore, elevated levels of plasma cholesterol and triglycerides resulting from the chronic administration of poloxamer containing drug formulations to patients may potentially hinder therapeutic outcome.

Block copolymers consisting of a hydrophobic polyester segment and a hydrophilic PEG segment have attracted large attention due to their biodegradability, biocompatibility, and tailor-made properties. Various kinds of block copolymers have been developed to date and can be classified according to their block structure as AB diblock, ABA or BAB triblock, multi-block, branched block, star-shaped block, and graft block copolymers (Fig. **1**), in which A is a hydrophobic block made up of biodegradable polyesters (PLA, PGA, or PLGA) and B is a hydrophilic PEG block. A wide variety of drug formulations, such as micro/nano-particles, micelles, hydrogels, and injectable drug delivery systems have been developed using PLGA-PEG block copolymers [16-19].

The use of block copolymers in drug delivery was first proposed in the early 1980s. The graft copolymers of PEG–g–PLGA and PLGA–g–PEG with sol-to-gel transitions at ~30°C were developed [20-21]. PEG–g–PLGA copolymers have hydrophilic backbones and form gels with short durability, whereas PLGA–g–PEG copolymers have hydrophobic backbones and form much more durable gels. By mixing the two copolymers, the durability of a gel can be controlled from one week to three months. Water-soluble PLA-PEG copolymers with a relatively low molecule weight PLA block have been found to self-disperse in water to form polymeric micelles and can be used to solubilize hydrophobic drugs (Riley *et al.*, 2001) [22].

RTG of biodegradable triblock copolymers has been reported. These polymers were triblock copolymers consisting of A-blocks and B-blocks arranged as BAB or ABA, where A is PLGA or PLA and B is PEG or PEO. The polymers are soluble in water, forming a free-flowing solution that spontaneously gels at body temperature to create a water-insoluble gel. The copolymerization of PEG and lactide or lactide/glycolide is now regarded as a suitable method to achieve new polymeric materials with novel physical, chemical, and biological properties adaptable to specific uses [23]. The control of drug release can be achieved by the adjustment of the triblock copolymer compositions [24,25]. These biodegradable, thermally reversible drug delivery systems are based on custom designed polymers/processes that control drug release and offer substantial advantages in administration and manufacture. They hold particular promise to provide solutions to the parenteral delivery problems of both polypeptide and steroidal hormones as well as other poorly soluble drugs. Thus, the biodegradable thermal gels hold high potential as injectable, long-term drug delivery systems.

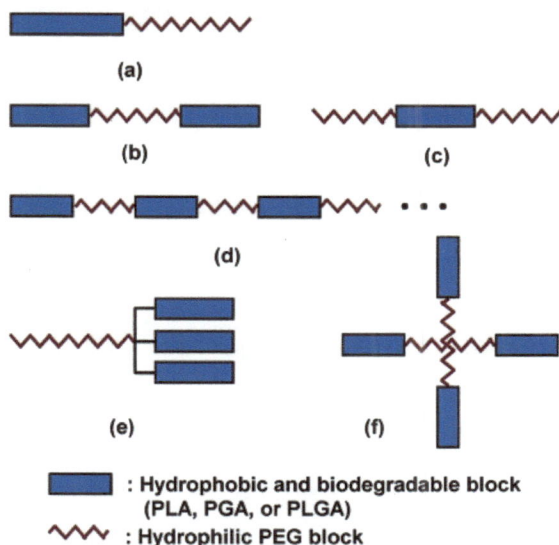

Figure1. Schematic presentation of block copolymer structures: (a) A-B diblock, (b) A-B-A, (C) B-A-B, (d) alternating multiblock, (e) multi-armed structure, and (f) star-sharp block.

A hydrogel consisting blocks of poly PEO-PLA-PEO has been reported [26]. Unfortunately, this system can only be loaded with bioactive molecules in an aqueous phase at an elevated temperature (around 45°C), where it forms a sol. This loading procedure limits the nature of the drugs that can be incorporated in the drug delivery system to those that are not prone to hydrolysis such as steroidal contraceptive hormones. Another ABA type triblock copolymers (PLA-PEO-PLA, PLGA-PEO-PLGA, PEO-PLGA-PEO) were developed, and their thermal properties and delivery characteristics for therapeutics have been investigated [27-31].

Zentner *et al.* [32]. developed an ABA type triblock copolymer (PLGA-PEG-PLGA, ReGel®), ReGel® is processed as a simple aqueous solution and sterilized by filtration, and injections are readily made using small needles (25-gauge). ReGel® can be degraded to the G. R. A. S. materials lactic acid, glycolic acid, and PEG, and it provides the formulation chemist with a flexible approach to protein and small molecule delivery that is simple to process and administer. Moreover, ReGel®'s inherent ability to solubilize and stabilize poorly soluble and sensitive drugs, including proteins is a substantial benefit, and the exclusive use of water as the solvent of choice in all purification and formulation steps involving ReGel® results in a drug delivery system that is safe for handlers and patients, and provides a delivery vehicle that is very gentle for proteins.

The sol-to-gel transition temperature of PEG–PLGA–PEG triblock copolymers in aqueous solution could be controlled over a temperature range of 15°C to 45°C by changing the molecular compositions such as PLGA length, PEG length and ratio of LA to GA in the middle block. The more hydrophobic the polymer, the stronger the shear stress required to make the gel flow. Therefore, the upper transition temperature could be controlled over a temperature range of 45°C to 60°C. The upper gel-to-sol transition is driven by a change in a three-dimensional micellar structure, which is disrupted in PEG–PLGA–PEG triblock copolymers. The PEG–PLGA–PEG triblock copolymer aqueous solutions (33 wt %) formed gel upon subcutaneous injection *in vivo* in rats. Such characteristics are desirable to use PEG–PLGA–PEG triblock copolymers in the formulation of a drug for a long-term injectable drug delivery system [33-36].

Table **2** presents various analytical techniques used to characterize thermosensitive polymers for their drug delivery applications.

Table 2: Analytical techniques used to characterize thermosensitive polymers for their drug delivery applications

Polymer Property	Analytical Technique/Method
Structural characterization [37-42]	Nuclear Magnetic Resonance (NMR) Fourier Transform Infrared Spectroscopy (FT-IR)
Molecular weight and molecular weight distribution [37,43,44]	Gel Permeation Chromatography (GPC)
Micellization and Size [40,45-47]	NMR (^{1}H and ^{13}C-NMR), Transmission Electron Microscopy (TEM), Dynamic Light Scattering (DLS)
Critical micelle concentration [40, 48]	Dye solubilization and Fluorescence Spectroscopy
Critical micelle temperature [45]	Static light scattering (SLS)
Phase inversion [43,49]	Tube inversion method, Dynamic mechanical analysis, Differential Scanning calorimetry (DSC)
	Ultraviolet Spectroscopy (UV)
Rheology/Viscosity [50]	Viscometry
Hydrogel microstructure [51]	Confocal Microscopy and Scanning Electron Microscopy (SEM)
Hydrogel swelling, *In vitro* degradation [52]	GPC, FTIR, GPC
In vitro and *in vivo* biocompatibility [38, 53]	Cell viability tests and Histological evaluation of implanted site using microscopy

APPLICATION OF THERMOSENSITIVE POLYMERS FOR CONTROLLED DELIVERY OF HORMONES

Hormones are widely being used as a controlled release dosage forms and devices. Controlled release dosage forms, especially non-biodegradable implants have been associated with high cost, inconvenience, and discomfort due to minor surgery which is required to insert or remove the device by professionals. Controlled release and biodegradable depot formulations have demonstrated promise to overcome many of the above problems. A recent application of thermosensitive polymers for delivery of contraceptive hormones have been given in Table **3**.

Table 3: Delivery applications of thermosensitive polymers for contraceptive hormones.

Polymer Type	Hormone Delivered	Remarks
PLA and PLGA-PEG-PLGA (Solution)	Testosterone	Testosterone was formulated and studied for prolonged release injectable implant delivery systems [54].
PLGA-PEG-PLGA	Levonorgestrel (LNG) and testosterone (TSN)	Controlled release of LNG was achieved with significant increase in bioavailability after single subcutaneous injection in rabbits [55].
Poly(organophosphazenes) (Hydrogel)	2-methoxyestradiol	Methoxyestradiol solubility increased leading to substantial inhibition of tumor growth and angiogenesis [56].
PLGA-PEG-PLGA with LA/GA (Hydrogel)	Levonorgestrel (LNG)	Prolonged release of LNG for female contraception was achieved [57].

CONCLUSION

Non-biodegradable implants pose a challenge during administration as well as removal from the body. It may also lead to pruritis and infection at the site of implant. Biodegradable and biocompatible thermosensitive triblock copolymers hold great promise for controlled delivery of contraceptive hormones with added advantages such as low burst release, low batch-to-batch variation in comparison to implants and microspheres, high drug loading and ease of preparation.

REFERENCES

[1]   Segal S.J. The development of Norplant implants. *Stud Fam Plann*, **1983**, *14*, 159-163.
[2]   Population Council. *Summary of Clinical Findings on NORPLANT Subdermal Contraceptive Implants*. New York: Population Council, **1989.**
[3]   Frank, M.L.; Poindexter, A.N.; Cornin, L.M.; Cox, C.A.; Bateman, L. One-year experience with subdermal contraceptive implants in the United States. *Contraception,* **1993**, *48*, 229-243.
[4]   Dunson, T.R.; Amatya, R.N.; Krueger, S.L. Complications and risk factors associated with the removal of Norplant implants. *Obstet Gynecol,* **1995**, *85*, 543-548.
[5]   Frank, M.L.; DiMaria, C. Levonorgestrel subdermal implants: contraception on trial. *Drug Safety*,**1997**, *17*, 360-368.
[6]   Connell, E.B. The exploitation of autoimmune disease: breast implant litigation and its dire implications for women's health. *J Women's Health*, **1998**, *7*, 329-338.
[7]   Singh, M.; Saxena, B.B.; Singh, R. Kaplan, J.; Ledger, W.J. Contraceptive efficacy of norethindrone encapsulated in injectable biodegradable biodegradable poly-dl-lactide-co-glycolide microspheres (NET-90: Phase III clinical study). *Adv Contracept,* **1997**, *13*, 1-11.
[8]   Klein, J. Smart polymer solutions. *Nature* **2000**, *405*, 745-747.
[9]   Kost, J.; Lapidot, S.A. Smart Polymers for Controlled drug Delivery. In: Wise, D.L.; (ed.), *Handbook of Pharmaceutical Controlled Release Technology*. New York, Marcel Dekker, **2000,** pp 65-87.
[10]   Kikuchi, A. and Okano, T. Pulsatile drug release control using hydrogels. *Adv. Drug Deliv, Rev.* **2002**, *54*, 53-77.
[11]   Pincus, G. The control of fertility, Academic Press, New York, **1965**.
[12]   Glasier, A. Historical perspective, Contraception-past and future. Nature Cell Biology & Nature Medicine, Fertility supplement. S3-S6, **2002.**
[13]   Segal, S.J.; Croxatto, H.B. Single administration of hormones for long term control or reproductive function. XXIII Meeting of American Fertility Society, Washington D.C.; **1967.**
[14]   Croxatto, H. B. Norplant, Levonorgestrel-releasing contraceptive implant. *Ann Med.;* **1993**, *25*,155-160.
[15]   Cohn, D.; Sosnik, A.; Levy, A. Improved reverse thermo-responsive polymeric systems. Biomaterials, **2003**, *24*, 3707-3714.
[16]   Jeong, B.; Bae, Y.H.; Lee, D.S.; Kim, S.W. Biodegradable block copolymers as injectable drug-delivery systems. *Nature*, **1997**, *388*, 860-862.
[17]   Modock, M.; Kissel, T.; Li, Y.X.; Koll, H.; Winter, G. Erythropoietin loaded microspheres prepared from biodegradable LPLG-PEO-LPLG triblock copolymers: protein stabilization and *in vitro* release properties. *J. Control. Release*, **1998**, *56*, 105-115.
[18]   Hoffman, A.S. Hydrogels for biomedical applications. *Adv. Drug Del. Rev.,* **2002**, *43*, 1-12.
[19]   Liggins, R.T.; Burt, H.M. Polyether-polyester diblock copolymers for the preparation of paclitaxol loaded polymeric micelle formations. *Adv. Drug Deliv. Rev.,* **2002,** *54*, 191-202.
[20]   Pratten, M.K.; Lloyd, J.B.; Horpel, G.; Ringsdorf, H. Micelle-forming block copolymers: Pinocytosis by macrophages and interaction with model membranes. Makromol. *Chem. Phys.,* **1985**, *186*, 725-733.
[21]   Jeong, B.; Bae, Y.H.; Kim, S.W. Thermoreversible gelation of PEG-PLGA-PEG triblock aqueous solutions. *J. Biomed. Mater. Res.,* **2000**, *50*, 171-177.
[22]   Riley, T.; Stolnik, S.; Heald, C.R.; Xiong, C.D.; Garnett, M.C.; Illum, L.; Davis, S.S. Physicochemical evaluation of nanoparticles assembled from PLA-PEG block copolymers as drug delivery vehicles. *Langmuir,* **2001**, *17*, 3168-3174.
[23]   Liu, L.; Li C.X.; Li, X.C.; Yuan, Z.; An Y.L.; He, B.L. Biodegradable polylactide/poly(ethylene glycol)/polylactide triblock copolymer micelles as anticancer drug carriers. *J. Appl. Polym. Sci.,* **2001**, *80*, 1976-1982.

[24]   Matsumoto, J.; Nakada, Y.; Sakurai, K.; Nakamura, T.; Takahashi, Y. Preparation of nanoparticles consisted of poly (L-lactide)–poly (ethylene glycol)–poly (L-lactide) and their evaluation *in vitro*. *Int. J. Pharm.*, **1999**, *185*, 93-101.

[25]   Metters, A.T.; Bowman, C.N.; Anseth, K.S. A statistical kinetic model for the bulk degradation of PLA-PEG-PLA hydrogel Networks, *J. Phys. Chem. B*, **2000**, *104*, 7043-7049.

[26]   Jeong, B.; Bae, Y.H.; Lee, D.S.; Kim, S.W. Biodegradable block copolymers as injectable drug-delivery systems. *Nature*, **1997**, *388*, 860-862.

[27]   Li, Y.X.; Volland, C. ; Kissel, T. In-vitro degradation and BSA release of the ABA triblock copolymers consisting of PLLA, or PLGA A-blocks attached to central PEO B- Blocks. *J. Control.Release*, **1994**, *32*, 121-128.

[28]   Molina, I.; Li, S.; Martinez, M.B.; Vert, M. Protein release from physically crosslinked hydrogels of the PLA/PEO/PLA triblock copolymer-type. *Biomaterials*, **2001**, *22*, 363-369.

[29]   Maglio, G.; Migliozzi, A.; Palumbo, R. Thermal properties of di- and triblock copolymers of PLLA with PEO or PCL. *Polymer*, **2003**, *44*, 369-375.

[30]   Witt, C.; Mader, K.; Kissel, T. The degradation, swelling and erosion properties of biodegradable implants prepared by extrusion or compression molding of PLGA and ABA triblock copolymers. *Biomaterials*, **2000**, *21*, 931-938.

[31]   Kwon, K.W.; Park, M.J.; Bae, Y.H.; Kim, H.D.; Char, K. Gelation behavior of PEO-PLGAPEO triblock copolymers in water. *Polymer*, **2002**, *43*, 3353-3358.

[32]   Zentner, G.M.; Rathi, R.; Shih, C.; McRea, J.C.; Seo, M.H.; Oh, H.; Rhee, B.G.; Mestecky, J.; Moldoveanu, Z.; Morgan, M.; Weitman, S. Biodegradable block copolymers for drug delivery of proteins and water-insoluble drugs. *J. Control. Release*, **2001**, *72*, 203-215.

[33]   Jeong, B.; Bae, Y.H.; Kim, S.W. Thermoreversible gelation of PEG-PLGA-PEG triblock copolymer aqueous solutions. *Macromolecules*, **1999**, *32*, 7064-7069.

[34]   Jeong, B.; Bae, Y.H.; Kim, S.W. Biodegradable thermosensitive micelles of PEG-PLGA-PEG triblock copolymers. *Colloids and Surfaces B: Biointerfaces*, **1999**, *16*, 185-193.

[35]   Jeong, B.; Bae, Y.H.; Kim, S.W. Thermoreversible gelation of PEG-PLGA-PEG triblock aqueous solutions. *J. Biomed. Mater. Res.*, **2000a**, *50*, 171-177.

[36]   Jeong, B.; Bae, Y.H.; Kim, S.W. Drug release from biodegradable injectable thermosensitive hydrogel of PEG-PLGA-PEG triblock copolymers. *J. Control. Release*, **2000**, *63*, 155-163.

[37]   Singh, S.; Webster, D.C.; Singh, J. Thermosensitive polymers: synthesis, characterization, and delivery of proteins. *Int J Pharm*, **2007**, *341*, 68-77.

[38]   Tang, Y.; Singh, J. Biodegradable and biocompatible thermosensitive polymer based injectable implant for controlled release of protein. *Int J Pharm*, **2009**, *365*, 34-43.

[39]   Cho, H.; Chung, D.; Jeongho, A. Poly(D,L-lactide-ran-epsilon-caprolactone)-poly(ethylene glycol)-poly(D,L-lactide-ran-epsilon-caprolactone) as parenteral drug-delivery systems. *Biomaterials*, **2004**, *25*, 3733-3742.

[40]   Jeong, B.; Han Bae, Y.; Wan Kim, S. Biodegradable thermosensitive micelles of PEG-PLGAPEG triblock copolymers. *Colloids and Surfaces B: Biointerfaces*, **1999**, *16*, 185-193.

[41]   Cui, Z.; Lee, B.H.; Vernon, B.L. New hydrolysis-dependent thermosensitive polymer for an injectable degradable system. *Biomacromolecules*, **2007**, *8*, 1280-1286.

[42]   Du, J.; Peng, Y.; Zhang, T.; Ding, X.; Zheng, Z. Study on pH-sensitive and thermosensitive polymer networks containing polyacetal segments. *J Appl Polym Sci Symp*, **2002**, *83*, 3002-3006.

[43]   Jeong, B.; Kibbey, M.R.; Birnbaum, J.C.; Won, Y.; Gutowska, A. Thermogelling biodegradable polymers with hydrophilic backbones: PEG-g-PLGA. *Macromolecules*, **2000**, *33*, 8317-8322.

[44]   Chen, S.; Pieper, R.; Webster, D.C. Singh, J. Triblock copolymers: synthesis, characterization, and delivery of a model protein. *Int J Pharm*, **2005**, *288*, 207-218.

[45]   Soga, O.; van Nostrum, C.F.; Ramzi, A.; Visser, T.; Soulimani, F.; Frederik, P.M.; Bomans, P.H.; Hennink, W.E. Physicochemical characterization of degradable thermosensitive polymeric micelles. *Langmuir*, **2004**, *20*, 9388-9395.

[46]   Soga, O.; van Nostrum, C.F.; Fens, M.; Rijcken, C.J.F.; Schiffelers, R.M.; Storm, G.; Hennink, W.E. Thermosensitive and biodegradable polymeric micelles for paclitaxel delivery. *J Control Release*, **2005**, *103*, 341-353.

[47]   Park, K.M.; Bae, J.W.; Joung, Y.K.; Shin, J.W.; Park, K.D. Nanoaggregate of thermosensitive chitosan-Pluronic for sustained release of hydrophobic drug. *Colloids Surf B Biointerfaces*, **2008**, *63*, 1-6.

[48]   Basu Ray, G.; Chakraborty, I.; Moulik, S.P. Pyrene absorption can be a convenient method for probing critical micellar concentration (cmc) and indexing micellar polarity. *J Colloid Interface Sci*, **2006**, *294*, 248-254.

[49]   Jeong, B.; Bae, Y.H.; Kim, S.W. Thermoreversible gelation of PEG–PLGA–PEG triblock copolymer aqueous solutions. *Macromolecules*, **1999**, *32*, 7064-7069.

[50]   Vermonden, T.; Besseling, N.A.M.; van Steenbergen, M.J.; Hennink, W.E. Rheological studies of thermosensitive triblock copolymer hydrogels. *Langmuir*, **2006**, *22*, 10180-10184.

[51]   Vermonden, T.; Jena, S.S.; Barriet, D.; Censi, R.; van der Gucht, J.; Hennink, W.E.; Siegel, R.A. Macromolecular diffusion in self-assembling biodegradable thermosensitive hydrogels. *Macromolecules*, **2010**, *43*, 782-789.

[52]   Hu, D.S.; Liu, H. Structural analysis and degradation behavior in polyethylene glycol/poly(Llactide) copolymers. *J Appl Polym Sci Symp*, **1994**, *51*, 473-482.

[53]   Chen, S.; Singh, J. Controlled release of growth hormone from thermosensitive triblock copolymer systems: *In vitro* and *in vivo* evaluation. *Int J Pharm*, **2008**, *352*, 58-65.

[54]   Chen, S.; Singh, J. Controlled delivery of testosterone from smart polymer solution based systems: *in vitro* evaluation. *Int J Pharm*, **2005**, *295*, 183-190.

[55]   Chen, S.; Pederson, D.; Oak, M.; Singh, J. *In vivo* absorption of steroidal hormones from smart polymer based delivery systems. *J Pharm Sci*, **2010**, *99*, 3381-3388.

[56]   Cho, J.; Hong, K.; Park, J.W.; Yang, H.; Song, S. Injectable delivery system of 2-methoxyestradiol for breast cancer therapy using biodegradable thermosensitive poly(organophosphazene) hydrogel. *J Drug Target.*, **2010**, doi:10.3109/1061186X.2010.499461.

[57]   Chen, S.; Singh, J. *In vitro* release of levonorgestrel from phase sensitive and thermosensitive smart polymer delivery systems. *Pharm Dev Technol*, **2005**, *10*, 319-325.

CHAPTER 5

Prospects of Nanosensors in Environmental and Biomedical Fields

Salaimutharasan Gnanamani[1*], Siva Chidhambaram[1], and Mani Prabaharan[2]

[1]*Department of Nanotechnology, Faculty of Engineering and Technology, SRM University, Kattankulathur-603 203, India and* [2]*Department of Chemistry, Faculty of Engineering and Technology, SRM University, Kattankulathur-603 203, India*

Abstract: A nanosensor is a sensor that is built on the nanoscale, whose purpose is mainly to obtain data on the atomic scale and transfer it into data that can be easily analyzed. Nanosensors have a wide application in the fields of environmental protection, biotechnology, medical diagnostics, drug screening, food safety and security. This review is an attempt to give an overview on different types of nanosensors based on carbon nanotubes, metal and metal oxide naoparticles and their application in environmental and biomedical fields. Due to the increased gas sensing properties, metal oxide based nanosensors were found to be potential candidates for NO_x, ethanol, ammonia and ozone sensing applications.

Keywords: Nanosensors; environment; biomedicare; metal oxide nanowires.

1. INTRODUCTION

Nanotechnology is enabling the development of devices in a scale ranging from one to a few hundred nanometers. At this scale, novel nanomaterials show new properties and behaviors not observed at the microscopic level. The aim of nanotechnology is on creating nano-devices with new functionalities stemming from these unique characteristics, not on just developing miniaturized classical machines. One of the early applications of nanotechnology is in the field of nanosensors [1]. A nanosensor is not necessarily a device merely reduced in size to a few nanometers, but a device that makes use of the unique properties of nanomaterials to detect and measure new types of events in the nanoscale. For example, nanosensors can detect chemical compounds in concentrations as low as one part per billion [2, 3], or the presence of different infectious agents such as virus or harmful bacteria [4]. Nanoparticles, nanotubes, nanorods, embedded nanostructures, porous silicon, and self-assembled materials are some of the nanostructures that are used for the development of nanosensors [5]. Most related nanostructures for environmental and biomedical applications are nanotubes and self-assembled materials [6].

In this chapter, we discuss various types of nanosensors and their application in environmental and biomedical fields. The types of nanosensors discussed in this chapter are physical sensor, chemical and biosensor, deployable nanosensors, localized and propagating surface plasmon resonance sensors, sensors based on Carbon Nanotubes (CNTs), sensors based on bulk nanostructured materials, sensors based on porous silicon, sensors based on self-assembled nanostructures and metal oxide nanosensors. Special emphasis has been given to the gas sensing properties and application of metal oxide nanostructures.

2. PRODUCTION METHODS OF NANOSENSORS

Nanosensors can be prepared by using different methods. The three most commonly known methods are top-down lithography, bottom-up assembly, and molecular self-assembly [7]. Researchers have also found a way to manufacture a nanosensor using semiconducting nanowires, which is said be an "easy-to-make" method of producing a type of nanosensor. Other methods of creating the sensors include the use of Carbon Nano Tubes (CNTs), as well as one method using a material found in blue crabs.

Address correspondence to Salaimutharasan Gnanamani: Department of Nanotechnology, Faculty of Engineering and Technology, SRM University, Kattankulathur-603 203, India; E-mails: salaimutharasan@gmail.com

Songjun Li, Yi Ge and He Li (Eds)
All rights reserved - © 2012 Bentham Science Publishers

The top-down lithography method is quite simple in concept. It is the method of starting out with a larger block of material and carving out the desired form of hat. The pieces that are carved out are used as the components to use in specific microelectronic systems such as sensors. In this case the components that are carved out are of the nanosized scale. This is the method that is used in the creation of many integrated circuits. In the case of nanosensors, it is common to use a silicon wafer as the base for this method. A layer of photoresist is then added to the wafer, then using lithography to shine a light on parts of the wafer to carve away parts of the wafer to create the component you desire. This piece of material can then be doped and modified using other materials to be used for things such as nanosensors [8].

The method of bottom-up assembly is a bit more difficult to accomplish, however, simple in concept. This method uses atomic sized components as the basis of the sensor. These components are moved one by one into position to create the sensor. This is an extremely difficult method to use especially in mass production because at this point in time it has only been achieved in a laboratory using atomic force microscopes.

Third method of producing nanosensors is based on the self-assembly or growing of particular nanostructures (Fig. **1**). There are two methods to the concept of molecular self-assembly, also known as "growing" nanostructures [9, 10]. The first of these methods uses a piece of previously created or even naturally formed nanostructure as the base and immersing it in free atoms of its own kind. Over time, the structure would begin to take a shape with an irregular surface that would then cause the structure to become more prone to attracting more molecules, continuing the pattern of capturing more of the free atoms and forming more of itself, creating a larger component of the nanosensor. The second method of self-assembly is more difficult. It begins with a complete set of components that automatically assemble themselves into the finished product, in this case the nanosensor. This has only been accomplished in the manufacturing of micro-sized computer chips, and has yet to be accomplished at the nanoscale. However, if this were to be perfected at the nanoscale, the sensors would be able to be made accurately, at a quicker rate and for a cheaper cost. This is because they would assemble themselves without having to manually assemble each individual sensor.

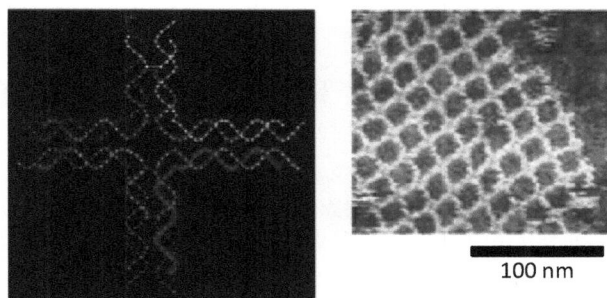

100 nm

Figure 1. An example of a DNA molecule used as a starter for larger self-assembly and an AFM image of a self-assembled DNA nanogrid.

3. MAJOR TYPES OF SENSORS

3.1. Physical Sensors

Researchers at the Georgia Institute of Technology developed the world's smallest "balance" by taking advantage of the unique electrical and mechanical properties of carbon nanotubes[11]. They mounted a single particle on the end of a Carbon Nanotube (CNT) and applied an electrical charge on it. The mass of the particle was calculated from changes in the resonance vibrational frequency with and without the particle on CNT. This approach may be used to determine the mass of individual biomolecules.

3.2. Chemical and Biosensors

Various gas sensors based on nanotubes have been reported in the past few years. Modi *et al.* have developed a miniaturized gas ionization detector based on CNTs [12]. The sensor can be used in gas chromatography. Titania nanotubes have been incorporated in a wireless sensor network to detect hydrogen concentrations in the atmosphere [13]. Kong *et al.* have developed a chemical sensor based on nanotube molecular wires for the detection of gaseous molecules such as NO_2 and NH_3 in the environment [14]. Datskos and Thundat fabricated nanocantilevers using a focused ion beam technique. They developed an electron transfer transduction approach to measure cantilever motion[15]. The device developed in this work might have enough sensitivity to detect single chemical and biological molecule.

Nanotechnology enables the very selective and sensitive detection of a broad range of biomolecules. By using the sequential electrochemical reduction of the metal ions onto an alumina template, one can now create cylindrical rods made up of metal sections having the length of 50 nm to 5 microns [16]. These particles, trademarked nanobarcodes, can be coated with analyte-specific entities such as antibodies for selective detection of complex molecules. DNA detection with these nano-scale coded particles has also been demonstrated [17, 18]. When the DNA molecules attached to the ends of the nanotubes are placed in a liquid containing DNA molecules, the DNA on the chip attaches to the target and increases its electrical conductivity. This technique is expected to reach the sensitivity of fluorescence-based detection systems and therefore may find application in the development of a portable biosensor (Fig. **2**).

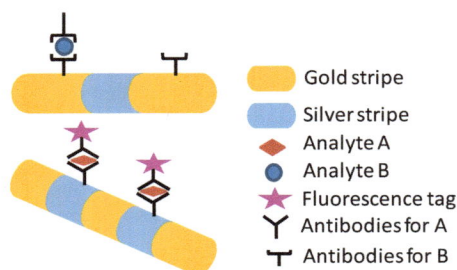

Figure 2. Semiconducting ZnO nanobelts.

3.3. Deployable Nanosensors

A different type of sensor is referred to as a deployable nanosensor. There is not a lot of research available on this type of nanosensor. These mostly refer to sensors that would be used in the military or other forms of national security. One sensor in particular is the Sniffer STAR, which is a nano-enabled chemical sensor that can be integrated into a micro unmanned aerial vehicle[19]. This sensor is a lightweight, portable chemical detection system that combines a nanomaterial for sample collection and a concentration with a Micro-Electromechanical (MEM) based "chemical lab-on-a-chip" detector. This would likely be used in homeland security and during times of war in which it could easily detect chemicals in the air without risking human lives by sending it up in the air instead.

3.4. Localized and Propagating Surface Plasmon Resonance Sensors

The last two decades has seen a tremendous advancement of optical biosensors and their applications in environmental and biotechnology fields [20]. The potential of Surface Plasmon Resonance (SPR) biosensors was realized in early 1980's by Liedberg and coworkers who were able to sense antibodies by observing the change in the critical angle when the antibodies bound selectively to an Au film [21]. Furthermore, in late 1990's, nanoparticle-based Localized Surface Plasmon Resonance (LSPR) sensors have been reported to detect biological and chemical entities (Fig. **3**) [22-24]. Hall *et al.* developed a method to amplify the wavelength shift observed from LSPR bioassays using gold nanoparticles-labeled antibodies[25]. The technique, which involved detecting surface-bound analytes using gold nanoparticles conjugated antibodies, provided a way to enhance LSPR shifts for more sensitive detection of low concentration analytes. Using the biotin and antibiotin binding pair as a model, they demonstrated up to 400% amplification of the shift upon antibody binding to analyte. In addition, the antibody-nanoparticles conjugate improved the binding constant by 2 orders of magnitude, and the limit of detection by nearly 3 orders of magnitude. This amplification strategy provided a way to improve the sensitivity of plasmon-based bioassays for the detection of single molecule and clinically relevant diagnostics.

Propagating surface plasmons are evanescent electromagnetic waves bounded by flat smooth metal-dielectric interfaces and arise from oscillations of the conduction electrons in the metal[26]. When surface plasmons are confined on either periodic or other nanosystems, localized optical modes are observed. These optical modes lead to highly localized electromagnetic fields outside the particles. Both SPR and LSPR sensors are sensitive to the local refractive index changes that occur when target analyte binds to the metal film or nanoparticles. Surface refractive index sensors have an inherent advantage over optical biosensors that require a chromophoric group or other label to transduce the binding event. Furthermore, they require very little ligand purification due to the specific ligand/receptor binding of these sensors. Also, these sensors provide real-time information on the course of binding and are applicable over a broad range of binding affinities. Additionally, LSPR sensing elements are inherently the size of a single nanoparticle, making the LSPR sensors potentially be applicable for *in situ* detection in biological systems. The sensing capability of LSPR sensors can also be tuned by changing the shape, size and material composition of the nanoparticles [27].

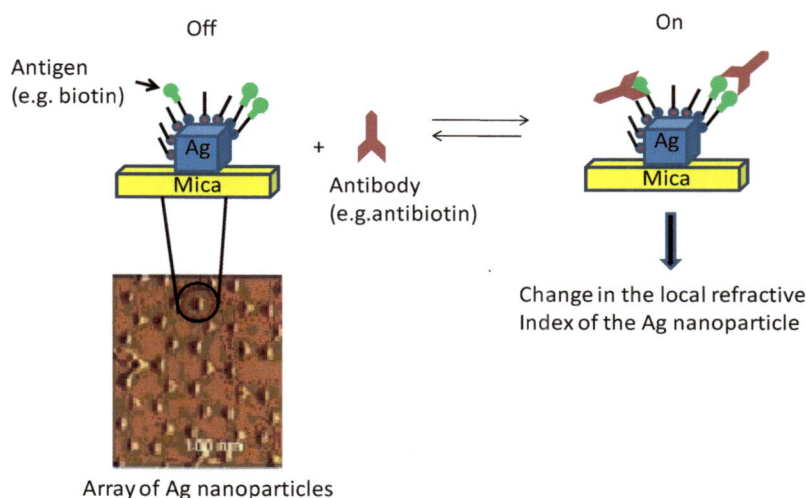

Figure 3. Biosensing mechanism of silver pyramidal nanoparticle arrays using LSPR to measure local changes in the refractive index of the Ag nanoparticles.

3.5. Sensors Based on Carbon Nanotubes (CNTs)

The electrical conductance of semiconducting Single-Walled Carbon Nanotubes (SWNTs) (band gaps ~0.5 eV) can vary by orders of magnitude under electrostatic gating or doping. This serves as the basis for nanotube field effect transistors [28-30]. SWNTs are molecular wires with every atom on the surface. Charge transfer between SWNTs and adsorbed molecules could cause drastic changes to the nanotube conductance, a chemical gating effect that can be utilized for molecular sensing, as initially demonstrated by Kong *et al.* for NO_2 and NH_3 detection [31,32]. Nanotube sensors offer significant advantages over conventional metal-oxide-based electrical sensor materials in terms of sensitivity, room temperature operation, and small sizes needed for miniaturization and construction of massive sensor arrays. Nevertheless, several outstanding issues remain. First, for sensing purposes, it is desired to reliably obtain devices consisting of only semiconductor SWNTs (one or multiple), as the electrical properties of metallic SWNTs are relatively insensitive to chemical gating [33]. This requires growing only semiconducting SWNTs through chirality control, an ability currently lacking for synthesis methods. Second, strategies should be devised to impart selectivity to nanotube sensors, a task that will ultimately determine the practical usefulness of these devices.

Recently, Qi *et al.* developed arrays of electrical devices with each comprising multiple SWNT bridging metal electrodes by Chemical Vapor Deposition (CVD) of nanotubes across prefabricated electrode arrays[34]. The ensemble of nanotubes in such a device collectively exhibits large electrical conductance changes under electrostatic gating, owing to the high percentage of semiconducting nanotubes. This leads to the fabrication of large arrays of low-noise electrical nanotube sensors with 100% yield for detecting gas molecules. Polymer functionalization is used to impart high sensitivity and selectivity to the sensors. Polyethyleneimine coating affords n-type nanotube devices capable of detecting NO_2 at less than 1 ppb (parts-per-billion) concentrations while being insensitive to NH_3. Coating Nafion (a polymeric perfluorinated sulfonic acid ionomer) on nanotubes blocks NO_2 and allows for selective sensing of NH_3. In this work, multiplex functionalization of a nanotube sensor array was carried out by microspotting. Detection of molecules in a gas mixture was demonstrated with the multiplexed nanotube sensors.

3.6. Sensors Based on Bulk Nanostructured Materials

Several properties including catalytic behaviour of nanoparticles are useful for applications in electrochemical and biosensors [35]. For instance, platinum nanoparticles supported on materials such as porous carbon or noble metals such as gold are reported for the design of gas diffusion electrodes [36]. Another property, *i.e.*, high surface area, makes nanoparticles suitable for immobilizing biomolecules, polymers and biomaterials that allow the generation of composite materials with tunable surface properties [37]. For example, modifying metal nanoparticles with predesigned receptor units and assembling them on surfaces could lead to new electrochemical sensors with tailored specificities. Simple and highly selective electroanalytical procedures can also be achieved by proper functionalization of nanoparticles [38].

3.7. Sensors Based on Porous Silicon

Nanoporous silicon consists of a complicated network of silicon threads having the thickness of 2-5 nm, displays an internal surface area-to-volume ratio of up to $500m^2/cm^3$ [39]. The extremely tiny pores give the material a strong luminescence at room temperature. This emission has the unique property that the wavelength of the emitted light depends on the porosity of the material. This property has led to the design of novel type of gas sensors that can be monitored by visually observing a change in colour.

3.8. Sensors Based on Self-Assembled Nanostructures

So far nanosensors have been developed with biomolecules (such as liposomes) using a bottom-up approach [40]. Going a further step forward of the "microarray technology", "Nanoarrays" are being developed based on the interaction between different types of receptors and ligands such as proteins or nucleic acids for sensing DNA molecules [41]. Approximately 400 nanoarray spots can be placed in the same area as a traditional microarray spot. Such DNA biosensors have been applied in environmental analysis for the quantification of genes associated to numerous environmentally prominent pathogens [42].

4. METAL OXIDE NANOWIRES AS SENSORS

The past few decades have seen the development of a multitude of simple, robust, solid-state sensors whose operation is based on the transduction of the binding of an analyte at the active surface of the sensor to a measurable signal that most often is a change in the resistance, capacitance, or temperature of the active element [43]. Molecular recognition in chemical sensing may be achieved by specific key-lock interactions, as they are known from both, natural biological systems and microstructured man-made devices. To handle the specific detection of molecules or ions reliably, a detailed understanding is required in both cases about the chemical compositions and electronic and/or optical properties of the materials involved, with particular emphasis on their interfaces. Of increasing interest are organic structures with potential future applications in chemical sensing or molecular and bioelectronic devices. Recent research and development efforts focus at proper choices of prototype materials to be microfabricated, new experimental approaches, adequate theories, reliable technologies in the preparation and modification of materials; and interfaces to produce microstructured devices down to the atomic scale [44].

Metal-oxide nanowires can function as sensitive and selective chemical or biological sensors, which could potentially be multiplexed in the small size devices. The active nanowire sensor can be configured either as resistors whose conductance is altered by charge-transfer processes occurring at their surfaces or as field-effect transistors whose properties can be controlled by applying an appropriate potential onto its gate [45]. Functionalizing the surface of the nanowires offers yet another avenue for expanding their sensing capability. The surface chemical properties and catalytic process selectivity of the functionalized nanowires can be altered due to the charge exchange between an adsorbate and the nanowire. Though research on the use of metal-oxide nanowires as sensors is still in infant stages, several encouraging results have been reported in the literature[46]. Chemical and biological sensors have a profound influence in the areas of personal safety, public security, medical diagnosis, detection of environmental toxins, semiconductor processing, agriculture, automotive and aerospace industries.

Metal oxides possess a broad range of electronic, chemical, and physical properties that are often highly sensitive to changes in their chemical environment [47]. Because of these properties, metal oxides have been widely studied, and most commercial sensors are based on appropriately structured and doped oxides. Nevertheless, much new science awaits discovery, and novel fabrication strategies remain to be explored in this class of materials by using strategies based on nanoscience and technology. Traditional sensor fabrication methods make use of pristine or doped metal oxides configured as single crystals, thin and thick films, ceramics, and powders through a variety of detection and transduction principles, based on the semiconducting, ionic conducting, photoconducting, piezoelectric, pyroelectric, and luminescence properties of metal oxides [48]. This review is limited primarily to semiconducting devices with quasi-one dimensional nanostructures such as nanowires and nanobelts and two related device configurations such as conductometric elements and field effect transistors. Numerous quasi-one-dimensional oxide nanostructures with useful properties, compositions, and morphologies have recently been fabricated using so-called bottom-up synthetic routes [49]. Some of these structures cannot be created easily or economically using top-down technologies. A few classes of these new nanostructures with potential as sensing devices are summarized in Fig. **4**.

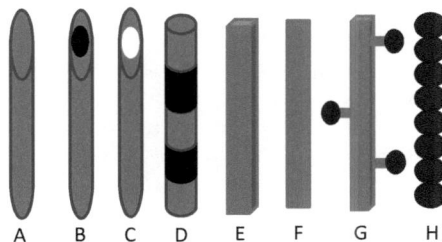

Figure 4. A schematic summary of the kinds of quasi-one-dimensional metal oxide nanostructures, **(A)** nanowires and nanorods; **(B)** core-shell structures with metallic inner core, semiconductor, or metal-oxide; **(C)** nanotubules/nanopipes and hollow nanorods; **(D)** heterostructures; **(E)** nanobelts/nanoribbons; **(F)** nanotapes; **(G)** dendrites; **(H)** nanosphere assembly.

5. GAS SENSORS BASED ON METAL OXIDES

Gas sensors have a great impact in many areas, such as environmental monitoring, domestic safety, public security, automotive applications, air conditioning in airplanes, spacecrafts and houses, sensor networks, *etc.* [50] The development of gas sensor devices with optimized stability, selectivity and sensitivity at room temperature has been gaining prominence in recent years. The gas sensing properties are related to the surface state and morphology of the material. Nanostructured materials present new opportunities for enhancing the properties and performance of gas sensors because of the much higher surface-to-volume ratio than that of coarse micro-grained materials.

Semiconductor metal oxides have attracted great attention as gas sensing materials due to their advantageous features, such as high sensitivity under ambient conditions, low cost and simplicity in fabrication. Among metal oxides, tin oxide (SnO) is the one that has received more attraction due to its high reactivity to many gaseous species. However, SnO has revealed a lack of selectivity, and thus investigation on other metal oxides has been considered necessary[51]. Among others, zinc oxide (ZnO)-based elements have concerned much attention as gas sensors because of their chemical sensitivity to different adsorbed gases, high chemical stability, amenability to doping, non-toxicity and low cost [52]. ZnO is an n-type semiconductor of wurtzite structure (lattice parameters A and C in the ratio of C/A =1.633) with a direct band gap of ~3.37eV at room temperature. The Zn atoms are tetrahedrally coordinated to four O atoms, where the Zn 3d-electrons hybridize with the O 2p-electrons [53].

A method for low-temperature synthesis of a mixture of high-density ZnO nanoflakes and nanowires is developed to produce low cost and high efficiency gas sensors with ZnO nanostructures [54, 55]. ZnO nanoflakes and nanowires are grown on glass substrates by the RF sputter deposition of Zn particles and localized oxidation at 300°C. The formed ZnO nanoflakes and nanowires are polycrystalline and they have nanometer dimensions. A gas sensor based on the mixture of ZnO nanoflakes/nanowires responded rapidly and sensitively to ethanol. The sensing properties of the ZnO nanostructure sensor was found to be ~72% for 50 ppm ethanol gas at an operating temperature of 100°C. Oxygen related gas sensing involves chemisorption of oxygen on the oxide surface, followed by charge transfer during the reaction between the reducing chemisorbed oxygen and target gas molecules [56]. This reaction changes the surface resistance of the sensor element.

5.1. NO$_x$ Sensing Properties of Metal Oxide Nanostructures

ZnO is studied intensively as a sensing material. However, there are limited reports only available on aluminium (Al)-doped ZnO as a NOx sensor. The selective NOx sensing characteristics of Al-doped ZnO porous pellets were reported [57]. The salient feature of the results is that the sensor can detect even small concentrations of NO$_x$ at lower operating temperature. It is also observed that as compared to gases such as SO$_x$, HCl, LPG, H$_2$S, H$_2$, ammonia, alcohol and acetone it selectively detects NO$_x$ due to Al-doping. In this study, the amount of Al in ZnO during synthesis was varied between 1 and 10 wt%. These sensors sensed NO$_x$ concentration as low as 20 ppm at 100°C. Of all the compositions with Al-doping, 1 wt% was found to give best results. The phase contents and lattice parameters of Al-doped ZnO porous pellets were determined by XRD and the average particle size was obtained using Scherrer formula. A probable mechanism for sensing NO$_x$ involving oxygen ion adsorption and desorption on the surface of sensor has been suggested. For gas sensing applications ZnO in pure and doped form is intensively studied [58]. Gas sensors based on ZnO make use of chemical sensitivity of the surface to different adsorbed gases, which cause change in resistance of the sensor. Appropriate doping can provide electronic defects that increase the influence of oxygen partial pressure on the conductivity. The porosity and the grain size of polycrystalline ZnO material were found to have noticeable effect on the gas sensitivity property [59].

An ultrasensitive chemiresistor based on electrospun TiO$_2$ nanofibers was fabricated by directly depositing TiO$_2$/poly(vinyl acetate) (PVAc) composite nanofiber mats onto interdigitated Pt electrode arrays with subsequent hot pressing at 120 °C and calcining at 450 °C [60]. The hot pressed TiO$_2$ fibers had much higher specific surface area (138 m^2/g) than that of unpressed fibers (31 m^2/g) and led to exceptionally high gas sensitivity. Exemplary results of the resistance response of one of the sensors upon cyclic exposure to NO$_2$ at different concentrations and operating temperatures were observed. Unusual response patterns were observed at high NO$_2$ concentrations (>12.5 ppm), consistent with n to p inversion of the surface-trap limited conduction facilitated by the high surface-to-volume ratio of this material.

A further study on TiO$_2$ nanofibers as sensing membrane to detect traces of CO and NO$_2$ in air was reported to demonstrate the processing-microstructure-properties correlation of ultrasensitive gas sensors [61]. The sensor was more sensitive at 300 °C than at 400 °C, similar to the previous report in, demonstrating exceptional sensitivity toward NO$_2$ displaying a change of 405% in the sensor's resistance on exposure to 50 ppb of NO$_2$ [60]. By extrapolating the results to lower gas concentrations they estimated the detection limit to be 5 ppb for NO$_2$ and 20 ppb for CO with operating temperature at 300 °C. Additionally, a highly sensitive and stable humidity nanosensors based on LiCl doped TiO$_2$ electrospun nanofibers was reported by Wang *et al.* [62]. The as-prepared humidity sensor exhibited excellent sensing characteristics, including ultrafast response time (≤ 3 s) and recovery time (≤7 s) for measuring relative humidity in a wide range of 11-95% in air at room temperature with the impendence changing from 107 to 104 Ω. Moreover, the nanosensor has good reproducibility, linearity, and stability.

5.2. Ozone, Methanol and Ammonia Sensing Properties of Metal Oxide Nanostructures

Ozone sensing properties of ZnO nanostructure produced by several physical techniques, such as sputtering, Pulsed Laser Deposition (PLD) and Molecular Beam Epitaxy (MBE) have been reported[63]. Aqueous Chemical Growth (ACG) is a novel and inexpensive technique for depositions of ZnO at mild temperatures. It does not require complicated setups or high-pressure containers. Furthermore, it is entirely safe and environment friendly, since water is used as solvent. In this work, ACG was applied to create ZnO nanostructures and investigated their ozone sensing characteristics. In particular, the ozone sensing properties of flowerlike ZnO nanostructures deposited on glass substrates at low temperatures was studied, in relation with deposition time and nanostructures morphology. SEM micrographs of the ZnO nanostructures formed on glass substrates at 1, 5 and 20h are presented in Fig. **5**. For 1h growth, sponge-like nanostructures were found to be formed, out of which uniform shaped nanorods were emerged.

Figure 5. SEM images of ZnO nanostructures grown by ACG at 95°C on glass for 1h (inset of **(a)**), 5h (top view **(a)**, side view **(b)**) and 20h (top view **(c)**, side view **(d)**). In the inset of **(c)** the hexagonal shape of the ZnO nanorods.

Wang *et al.* reported the fabrication of a gas sensor to detect moisture and methanol gas based on a single SnO$_2$ nanofiber made from C$_{22}$H$_{44}$O$_4$Sn/poly(ethylene oxide) (PEO) using electrospinning and metallorganic deposition techniques [64]. The sensors showed high sensitivity to both gases and response times of the complete testing system were in the range of 108-150s for moisture, and 10-38s for methanol gas, respectively. Nonwoven type fibrous SnO$_2$ gas sensors for detection of ethanol were prepared on a micro hot plate by "near-field" electrospinning PVA/SnCl$_4$ solution with a subsequent annealing at 300, 500, and 700 °C [65]. The fiber resistance response increased with a raise of operating temperature and attained a maximum at 330 °C, followed by a decrease with further increases of operating temperature. The SnO$_2$ nanofibers with an average diameter of similar to 100 nm exhibited large responses, low detection limits, fast response/recovery, and good reproducibility. The detection limit was <10 ppb and the response/recovery time towards 10 ppm ethanol was <14s. Tao *et al.* demonstrated a sensitive CO gas sensor for use at room temperature using polycrystalline SnO$_2$ and MWCNT doped SnO$_2$ nanofibers *via* electrospinning. The *n*-type MWCNT/SnO$_2$ nanofibers are able to detect CO in the concentration range of 47-547 ppm [66].

The fabrication and characterization of polycrystalline WO$_3$ nanofibers and their application for ammonia sensing were reported by Wang *et al.* [67]. Pure tungsten oxide nanofibers with controllable diameters of around 100 nm were obtained by electrospinning PVAc/tungsten isopropoxide followed by a calcinations step. The relationship between solution concentration and ceramic nanofiber morphology has been studied. It has been shown that the as-prepared tungsten oxide ceramic nanofibers have a quick response to ammonia with various concentrations at 350 °C, suggesting potential applications of the electrospun tungsten oxide nanofibers as a sensor material for gas detection. Sahner *et al.* fabricated a perovskite p-type conducting electrospun SrTi$_{0.8}$Fe$_{0.2}$O$_{3-\delta}$ nanofibrous membrane as a model sensing material for application in methanol sensing at the operating temperature of 400 °C[68]. When exposed to reducing gases, the sensor resistance increased, as expected for a p-type sensor membrane. Very good sensor characteristics such as short response and recovery times and a stable baseline can be observed. At 10 ppm methanol, sensor response R/R_0 was observed to be 5.5 (electrospun) and 2.9 (electrosprayed). In the case of the electrospun membrane, the signal for methanol with concentrations ranging from 5 to 50 ppm in dry air was found to be reproducible.

6. CONCLUSION

In this chapter, we have presented the various types of nanosensors and their production methods and applications. Since SPR and LSPR sensors are more sensitive to the local refractive index changes that occur when target analyte binds to the metal film or nanoparticles, they have potential importance in the determination of biological and chemical entities. Due to the broad range of electronic, chemical, and physical properties that are often highly sensitive to changes in their chemical environment, metal oxide nanowires have shown potential application in NO_x, methanol, ammonia and ozone sensing. We expect that this systematic review could pave the way for researchers to develop novel types of biosensors that could be utilized in various environmental and biomedical fields.

REFERENCES

[1]   Taton, T. A.; Mirkin, C. A.; Letsinger, R. L. Scanometric DNA array detection with nanoparticle probes. *J Sci.,* **2000**, *289*, 1757-1760.

[2]   Schedin, F.; Geim, A. K.; Morozov, S. V.; Hill, E. V.; Blake, P.; Katsnelson, M. I.; Novoselov, K. S. Detection of individual gas molecules adsorbed on graphene, Nat Mat., **2007**, *6(9)*, 652 - 655.

[3]   Roman, C.; Ciontu, F.; Courtois, B. *Single molecule detection and macromolecular weighting using an all-carbon-nanotube nanoelectromechanical sensor*, In: 4th IEEE Conference on Nanotechnology, **2004**, 263 - 266.

[4]   Tallury, P.; Malhotra, A.; Byrne, L. M.; Santra, S. Nanobioimaging and sensing of infectious diseases, *Adv Drug Delivery Rev.*, **2010**, *62 (4-5)*, 424 - 437.

[5]   Chrisey, L.; Lee, G.; O'Ferrall, C. Covalent attachment of synthetic DNA to self-assembled monolayer films. *J. Nucleic Acid. Res.,* **1996**, *24*, 3031-3039.

[6]   Chiesl, T. N.; Shi, W.; Barron, A. E. Poly(acrylamide-co-alkylacrylamides) for electrophoretic DNA purification in microchannels. *Anal.Chem.,* **2005**, *77*, 772-779.

[7]   Ratner, M. A.; Ratner, D.; Ratner, M. *Nanotechnology: a gentle introduction to the next big idea.* Upper Saddle River: Prentice Hall, ISBN 0131014005, **2003**.

[8]   Foster, L. E. *Medical nanotechnology: science, innovation, and opportunity.* Upper Saddle River: Pearson Education, ISBN 0131927566, **2006**.

[9]   Poncharal, P.; Wang, Z. L.; Ugarte D.; de Heer W. A. Electrostatic deflections and electromechanical resonances of carbon nanotubes, *J Sci.,* **1999**, *283*, 1513-1516.

[10]   Freitas, R. A. *Nanomedicine - basic capabilities.* Landes Bioscience, Austin, ISBN 1570596808, **1999**.

[11]   Kong, J.; Franklin N. R.; Zhou C.; Chapline M. G.; Peng S.; Cho K.; Dai H. Nanotube molecular wires as chemical sensors, *J Sci.,* **1999**, *287*, 622-625.

[12]   Modi, A.; Koratkar, N.; Lass, E.; Wei, B.; Ajayan, P. M. Miniaturized gas ionization sensors using carbon nanotubes. *J. Nat.,* **2003**, *424*, 171-174.

[13]   Watanabe, K.; Okada, T.; Choe, I.; Sato, Y. Organic vapor sensitivity in a porous silicon device. *J Sens Actuators B*, **1996**, *33*, 194-197.

[14]   Kong, J.; Franklin, N.; Zhou, C.; Chapline, M.; Peng, S.; Cho, K.; Dai, H. Nanotube molecular wires as chemical sensors. *J. Sci.,* **2000**, *287*, 622-625.

[15]   Ong, K.; Zeng, K.; Grimes, C. A. A wireless, passive carbon nanotubes-based gas sensor. *J. IEEE Sensor,* **2002**, *2*, 82-88.

[16]   Varghese, O.; Grimes C. Metal oxide nanoarchitectures for environmental sensing. *J. Nanosci. Nanotechnol.,* **2003**, *3*, 277-293.

[17]   Sedlacek, M. *Electron physics of vacuum and gaseous devices*, John Wiley & Sons: New York, **1996**.

[18]   Zhu, N.; Chang, Z.; He, R.; Rang, Y. Electrochemical DNA biosensors based on platinum nanoparticles combined carbon nanotubes. *J. Anal. Chem. Acta,* **2005**, *545*, 21.

[19]   Shimizu, Y.; Egashira, M. Basic aspects and challenges of semiconductor gas sensors. *J. Mater. Res. Soc. Bull.,* **1999**, *24*, 18-24.

[20]   Chanda, Y.; Richard, P. V. Localized and propagating surface plasmon resonance sensors: a study using carbohydrate binding protein, *Mater Res Soc Symp Proc.,* **2005,** *876E,* 731–736.

[21]   Liedberg, B.; Nylander, C.; Lundstorm, I. Surface plasmon resonance for gas detection and biosensing, *J. Sens. Actuators. B,* **1983,** *4,* 229-304.

[22]   Riboh, J. C.; Haes, A. J.; McFarland, A. D.; Yonzon, C. R.; Van Duyne, R. P. A nanoscale optical biosensor: real-time immunoassay in physiological buffer enabled by improved nanoparticle adhesion, *J. Phys. Chem. B,* **2003,** *107,* 1772-1780.

[23]   Yonzon, C. R.; Zhang, X.; Van Duyne, R. P. Localized surface plasmon resonance immunoassay and verification using surface-enhanced Raman spectroscopy, Proc SPIE, **2003,** *5224,* 78-85.

[24]   McFarland, A. D.; Van Duyne, R. P. Single silver nanoparticles as real-time optical sensors with zeptomole sensitivity, *Nano Lett.,* **2003,** *3,* 1057-1062.

[25]   Hall, W. P.; Ngatia, S. N.; Van Duyne, R. P. LSPR biosensor signal enhancement using nanoparticle–antibody conjugates, *J. Phys. Chem. C.,* **2011,** *115* (5), 1410–1414.

[26]   Reather, H. *Surface Polaritons on Smooth and Rough Surfaces and on Gratings*; Springer-Verlag: Berlin, **1988.**

[27]   Haynes, C. L.; Van Duyne, R. P. Nanosphere Lithography: A versatile nanofabrication tool for studies of size-dependent nanoparticle optics, *J. Phys. Chem. B.,* **2001,** *105,* 5599-5611.

[28]   Tans, S.; Verschueren, A.; Dekker, C. Room-temperature transistor based on a single carbon nanotube, *Nat.,* **1998,** *393,* 49-52.

[29]   Martel, R.; Schmidt, T.; Shea, H. R.; Hertel, T.; Avouris, P. Single- and multi-wall carbon nanotube field-effect transistors, *Appl. Phys. Lett.,* **1998,** *73,* 2447-2449.

[30]   Zhou, C.; Kong, J.; Dai, H. Electrical measurements of individual semiconducting single-walled carbon nanotubes of various diameters, *Appl. Phys. Lett.,* **1999,** *76,* 1597 - 1599.

[31]   Dai, H. Carbon nanotubes: synthesis, integration, and properties, *Acc. Chem. Res.,* **2002,** *35,* 1035-1044.

[32]   Kong, J.; Franklin, N.; Zhou, C.; Chapline, M.; Peng, S.; Cho, K.; Dai, H. Nanotube molecular wires as chemical sensors, *J. Sci.,* **2000,** *287,* 622-625.

[33]   Kong, J.; Dai, H. Full and modulated chemical gating of individual carbon nanotubes by organic amine compounds, *J. Phys. Chem.,* **2001,** *105,* 2890-2893.

[34]   Qi, P.; Vermesh, O.; Grecu, M.; Javey, A.; Wang, Q.; Dai, H.; Peng, S.; Cho, K. Toward large arrays of multiplex functionalized carbon nanotube sensors for highly sensitive and selective molecular detection. *Nano Lett.,* **2003,** *3,* 347-351.

[35]   Bachilo, S. M.; Strano, M. S.; Kittre, U. C.; Hauge, R. H.; Smalley, R. E.; Weisman, R. B. Structure-assigned optical spectra of single-walled carbon nanotubes. *J. Sci.,* **2002,** *298,* 2361-2366.

[36]   Htoon, H.; O'Connell, M. J.; Cox, P. J.; Doom, S. K.; Klimov, V. I. First laser-controlled antihydrogen production. *J. Phys. Rev. Lett.,* **2004,** *93,* 263401-263404.

[37]   Ma, Y. Z.; Stenger, J.; Zimmermann, J.; Bachilo, S. M.; Smalley, R. E.; Weisman, R. B.; Fleming, G. R. Ultrafast carrier dynamics in single-walled carbon nanotubes probed by femtosecond spectroscopy. *J. Chem. Phys.,* **2004,** *120,* 3368.

[38]   Saito, R.; Dresselhaus, G.; Dresselhaus, M. S. *Physical properties of carbon nanotubes*, Imperial College Press, London, **1998.**

[39]   Hagen, A.; Hertel, T. Quantitative analysis of optical spectra from individual single-wall carbon nanotubes. *Nano Lett.,* **2003,** *3,* 383-388.

[40]   Dresselhaus, M. S.; Dresselhaus, G.; Jorio, A. Unusual properties and structure of carbon nanotubes, *J. Annu. Rev. Mater. Res.,* **2004,** *34,* 247 - 278.

[41]   Wang, J. X.; Shi, J.; Li, N. Q.; Gu, Z. N. Direct electrochemistry of cytochrome C at a glassy carbon electrode modified with single-wall carbon nanotubes. *J. Anal. Chem.,* **2002,** *74,* 1993-1997.

[42]   Zhao, Y. D.; Zhang, W. D.; Chen, H.; Luo, Q. M.; Li, S. F. Y. Direct electrochemistry of horseradish peroxidase at carbon nanotube powder microelectrode. *J. Sens. Actuators. B,* **2002,** *87,* 168-172.

[43]   Mandelis, A.; Christofides, C. *Physics, chemistry and technology of solid state gas sensor devices.* New York, Wiley Interscience, **1993.**

[44]   Gopel, W. Chemical sensing, molecular electronics and nanotechnology: interface technologies down to the molecular scale. *J. Sens. Actuators. B,* **1991,** *4,* 7-21.

[45]   Moseley, P. T.; Tofield, B. C. *Solid state gas sensors.* Bristol, Philadelphia, Hilger, **1987.**

[46]   Sberveglieri, G. *Gas sensors: principles, operation and developments.* Boston, Kluwer, **1992.**

[47]   Albert, K. J.; Lewis, N. S.; Schauer, C. L.; Sotzing, G. A.; Stitzel, S. E.; *et al.* Cross-reactive chemical sensor arrays. *Chem. Rev.,* **2000**, *100*, 2595-2626.

[48]   Pearce, P. C.; Schiffman, S. S.; Nagle, H. T.; Gardner, J. W. *Handbook of machine olfaction: electronic nose technology.* Weinheim, Wiley-VCH, **2003**.

[49]   Moseley, P. T. Materials selection for semi-conducting gas sensors. *J.Sens.Actuators.B.*, **1992**, *6*, 149-156.

[50]   White, N. M.; Turner, J. D. Thick-film sensors: past, present and future. *J. Meas. Sci. Technol.,* **1997**, *8*, 1-20.

[51]   Kohl, D. One-dimensional nanostructures: synthesis, characterization, and applications. *J. Phys. D. Appl. Phys.,* **1991**, *34*, 125-149.

[52]   Henrich, V. E.; Cox, P. A. *Surface science of metal oxides,* New York, **1996**.

[53]   Xia, Y. N.; Yang, P. D.; Sun, Y. G.; Wu, Y. Y.; Mayers B. One-dimensional nanostructures: synthesis, characterization, and applications. *J. Adv. Mater.*, **2003**, *15*, 353-389.

[54]   Wang, Z. L. *Nanowires and nanobelts: materials, properties and devices*, New York, Kluwer, **2003**.

[55]   Wang, Y.; Kempa, K.; Kimball, B.; Carlson, J. B.; Benham, G.; Li, W. Z.; Kempa, T; J.; Rybczynski, A.; Herczynski, Z. F. Ren. Receiving and transmitting light-like radio waves: Antenna effect in arrays of aligned carbon nanotubes. *Appl. Phys. Lett.*, **2004**, *85*, 2607-2609.

[56]   Amao, Y.; Asai, K.; Okura, I. Optical oxygen pressure sensing based on triplet–triplet quenching of fullerene–polystyrene film using laser flash photolysis: soccerballene C_{60} versus rugbyballene C_{70}. *J. Chem. Soc.*, **1999**, *72*, 2223-2227.

[57]   Rosenblatt, S.; Yaish, Y.; Park, J.; Gore, J.; Sazonova, V.; McEuen, R. L. High performance electrolyte gated carbon nanotube transistors. *Nano Lett.*, **2002**, *2*, 869-872.

[58]   Kazaoui, S.; Minami, N.; Matsuda, N.; Kataura, H.; Achiba, Y. Electrochemical tuning of electronic states in single-wall carbon nanotubes studied by in situ absorption spectroscopy and ac resistance. *J. Appl. Rhys. Lett.*, **2001**, *78*, 3433-3435.

[59]   Kavan, L.; Rapta, R.; Dunsch, L.; Bronikowski, M. J.; Willis, R.; Smalley, R. E. Electrochemical tuning of electronic structure of single-walled carbon nanotubes: in-situ Raman and Vis-NIR Study. *J. Rhys. Chem. B,* **2001**, *105*, 10764-10771.

[60]   Kim, I.; Rothschild, A.; Lee, B.; Kim, D.; Jo, S.; Tuller, H. Ultrasensitive chemiresistors based on electrospun TiO_2 nanofibers. *Nano Lett.,* **2006**, *6*, 2009-2013.

[61]   Landau, O.; Rothschild, A.; Zussman, E. Processing-microstructure-properties correlation of ultrasensitive gas sensors produced by electrospinning. *Chem. Mater.*, **2009**, *21*, 9-11.

[62]   Wang, Y.; Ramos, I.; Santiago-Aviles, J. Detection of moisture and methanol gas using a single electrospun tin oxide nanofiber. *IEEE Sens J.,* **2007**, *7*, 1347-1348.

[63]   Bender, E.; Gagaoudakis, E.; Douloufakis, E.; Natsakou, N.; Katsarakis, V.; Cimalla, G.; Kiriakidis, E.; Fortunato, P.; Nunes, A.; Marques R. Production and characterization of zinc oxide thin films for room temperature ozone sensing. *J. Thin. Solid.Films,* **2002**, *418*, 45-50.

[64]   Li, Z.; Zhang, H.; Zheng, W.; Wang, W.; Huang, H.; Wang, C.; MacDiarmid, A.; Wei, Y. Highly sensitive and stable humidity nanosensors based on LiCl doped TiO_2 electrospun nanofibers. *J. Am. Chem. Soc.,* **2008**, *130*, 5036-5037.

[65]   Zhang, Y.; He, X.; Li, J.; Miao, Z.; Huang, F. Fabrication and ethanol-sensing properties of micro gas sensor based on electrospun SnO_2 nanofibers. *J.Sens. Actuators.B,* **2008**, *132*, 67-73. 65.

[66]   Yang, A.; Tao, X.; Wang, R. Room temperature gas sensing properties of SnO_2/multiwall-carbonnanotube composite nanofibers. *Appl. Phys. Lett.*, **2007**, *91*, 133110.

[67]   Wang, G.; Ji, Y.; Huang, X.; Yang, X.; Gouma, P.; Dudley, M. Fabrication and characterization of polycrystalline WO_3 nanofibers and their application for ammonia sensing. *J. Phys. Chem. B,* **2006**, *110*, 23777-23782.

[68]   Sahner, K.; Gouma, P.; Moos, R. Electrodeposited and sol-gel precipitated p-type $SrTi_{1-x}Fe_xO_{3-\delta}$ semiconductors for gas sensing. *Sens.,* **2007**, *7*, 1871-1886.

CHAPTER 6

Growth of CdSe Nanoparticles on Abscisic Acid Nanofibers and their Interactions with HeLa cells

Stephen H. Frayne[1], Stacey N. Barnaby[1], Areti Tsiola[2], Karl R. Fath[2,3], Evan M. Smoak[1] and Ipsita A. Banerjee[1*]

[1]*Fordham University, Department of Chemistry, 441 E. Fordham Road, Bronx, NY 10458, USA;* [2] *Biology Department, Queens College, The City University of New York, 6530 Kissena Boulevard, Flushing, NY 11367, USA and* [3]*The Graduate Center, The City University of New York, 365 Fifth Avenue, NY 10016, USA*

Abstract: Abscisic Acid (ABA) is a vital phytohormone that regulates plant elongation, fruit ripening and senescence. It also plays an important role in plant responses to environmental stress and pathogens. In this work, self-assembled ABA was utilized as template for the growth of CdSe nanoparticles. The formed assemblies were functionalized with an organic linker (ethylene diamine) to enhance the growth of the CdSe nanoparticles on the surface of the nanofibers. The nanocomposites formed were analyzed by microscopic and spectroscopic methods. It was observed that the formation of quantum dots was promoted under mild conditions, leading to the formation of uniform nanofibers coated with CdSe nanoparticles. Further, the nanocomposites were utilized for targeting HeLa cells. We believe that such nanomaterials may lead to the development of a new family of nanomaterials for bioimaging and cancer cell targeting, as well as a host of optoelectronic applications.

Keywords: Growth; CdSe nanoparitcles; self-assembly; nanofibers; interaction.

1. INTRODUCTION

In recent times, semiconductor nanoparticles (quantum dots) exhibiting Quantum Confinement Effects (QCE) have generated substantial interest due to their potential applications in various device fabrications such as sensors, detectors, light-emitting diodes, and solar cells [1-2]. Quantum Dots (QDs) have also been utilized in biomedical applications such as contrast agents and biological labeling [3], as well as for *in vivo* imaging of tumor vasculature [4], sentinel lymph nodes [5], tumor-specific receptors [6], tumor immune responses [7], and cancer cells [8]. Further, QD's have also been utilized as immunofluorescent probes for breast cancer markers [9], microbial toxins [10], cancer cell motility, and metastic potential [11]. However, concerns over the clearance of QDs from the body [12], their biocompatibility, and the toxicity associated with compounds consisting of heavy metals have limited their potential application in biology and medicine [13].

When compared to traditional organic dyes, QD's have many advantages, including tunable fluorescence, stable emission, high quantum yield, and broad excitation spectra [14-15]. In general, QD's have been synthesized on a variety of templates such as mesoporous/nanoporous supports [16], carbon nanotubes [17], peptide nanotubes [18], polystyrene microspheres [19], dendrimers [20], micelles, and polymers [21]. They have also been fabricated onto specific surfaces by colloidal lithography [22] and capillary lithography [23]. Water-in-oil (W/O) reverse microemulsion synthesis methods have also been exploited [24]. Although metal and semiconductor nanoparticles are often prepared in the presence of ligands, polymers, or other surfactants [25-31], such stabilizers may diminish the catalytic activity at the nanoparticle active sites [32-36]. Recent work has involved the use of biomolecular templates to prepare QD's, such as proteins, liposomes, and nucleic acids [37, 38], as well as bacteria, cellulose, insect wings, spider silk, wool, and wood [39]. For example, Kelley and co-workers found that RNA possesses the ability to form CdS nanocrystals during its precipitation from solution [40]. Specifically, it was found that wild

*Address correspondence to Ipsita A. Banerjee: Fordham University, Department of Chemistry, 441 E. Fordham Road, Bronx, NY 10458, USA; Tel: 718-817-4445; Fax: 718-817-4432; E-mail: banerjee@fordham.edu

Songjun Li, Yi Ge and He Li (Eds)
All rights reserved - © 2012 Bentham Science Publishers

type t-RNA exerted exceptional ability to create QD's with a uniform diameter, whereas unfolded mutant t-RNA created QD's of multiple diameters within a small range. In a separate study, peptides were used as scaffolds for the creation of QD's, where gold nanoparticle-coated viruses were connected with a second peptide that recognized CdSe nanoparticles [41]. Also, fibrous proteins and flagella with inserted histidine loops have been utilized as scaffolds for the templated assembly of QD's, as well as various metallic species [42]. In a separate study, Belcher and coworkers found that engineered viruses possess the ability to recognize semiconductor surfaces through combinatorial phage display, thus allowing the virus to organize inorganic nanocrystals [43]. Further, it was found that the M13 bacteriophage served as the basis of the self-ordering liquid crystal system for creating monodisperse ZnS crystals [44]. In another study, ZnSe quantum dots were synthesized in the cavity of apoferritin through a chemical reaction utilizing tetra-amine zinc ion and selenourea [45]. At room temperature, the synthesis yielded cubic ZnSe polycrystals, where at 500°C single-crystal ZnSe nanoparticles were formed, thus showing the ability of biological materials to control the shape of the resulting QD's.

In this work, we have investigated the use of plant based biocompatible starting materials as templates for the growth of CdSe nanoparticles. Specifically, we utilized the plant phytohormone abscisic acid (ABA), a sesquiterpenoid that is synthesized in leaves and is found in sycamore, birch, willow, and cabbage leaves, as well as in cotton balls, potatoes, avocado seeds, and lemons [46]. In plants, ABA primarily regulates physiological processes such as dormancy, acceleration of abscission, inhibition of rooting, elongation, fruit ripening, and stimulation of stomata closure [47]. It has also been reported that ABA regulates G protein signaling in *Arabidopsis* guard cells [48]. In recent times, ABA has garnered substantial interest due to its ability to respond to environmental stress, such as droughts, cold weather, and plant pathogens [49].

Furthermore, it has also been shown that ABA inhibits DNA replication in root tips and embryos of *Fraxinus excelsior* [50] and studies of the root tissue have shown that ABA allows for the inhibition of cell elongation and mitotic cell activity in the G1- phase [51, 52]. Further, it has been observed that ABA activates the signaling of G-protein pathway by producing a hyperpolarization on plasma membranes [53]. Although the exact mechanism of how ABA regulates the cell cycle is not yet unresolved, due to the complexity caused by MAP kinase (MAPK) cascades and various cell cycle components, it is possible that the mechanism may be analogous to that found in mammalian systems, and therefore may involve regulation of the retinoblastoma protein pRb as well as a MAPK pathway [54]. More recent studies by Wu and co-workers found that cyclic ADP ribose is a secondary messenger of ABA that regulates growth responses, and when cADP is inhibited, ABA signaling and cADP ribose action are both inhibited [55].

Herein, we have utilized ABA based nanostructures as templates for the growth of CdSe nanoparticles. ABA was self-assembled into nanofibers in aqueous solution at a pH value of 5. The nanostructures were then functionalized with the organic linker Ethylene Diamine (EDA) to promote the growth of CdSe nanoparticles *in situ*.

Specifically, CdSe nanoparticles are one of the most highly efficient, luminescent, semiconducting QDs [56]. Most notably, CdSe nanoparticles have been examined because of their high emission frequency and size-tuned photoluminescence (PL) [57]. In particular, nanocrystalline CdSe demonstrates quantum confinement effects that allow band-gap tuning from the near infrared to the blue-green region, making them especially strong candidates for the fabrication of a variety of nanodevices. In recent times, *in vivo* imaging using QDs has been gaining significance, particularly in the field of disease diagnostics and biosensors [58]. The use of quantum dots as bio-labeling agents, however, has proven to be a challenging task due to the known toxicity of quantum dots [59]. In order to reduce toxicity, enhance solubility, and expand their applications for bioimaging, QD's have been conjugated with a variety of materials. For example, several synthetic polymers have been utilized for bioconjugation with quantum dots, wherein specific side chains have been functionalized to enhance the attachment with QDs. In some cases, light-emitting polymers have similarly been used as probes [60]. It has also been reported that carbon nanotubes have been functionalized with streptavidin conjugated QDs for intracellular fluorescence imaging [61]. Recently, water soluble MPC (methacryloyloxyethyl phosphorylcholine) based polymers containing poly(MPC-*co-n*-butyl methacrylate (BMA)-co-ω-methacryloyloxy poly(ethylene oxide) oxycarbonyl 4-

nitrophenol (MEONP) (PMBN) have also been synthesized [62]. The preparation of artificial cell membrane-covered nanoparticles using PMBN and water-insoluble polymers such as poly(L-lactic acid) (PLA) and Polystyrene as cores as well as PMBN, PLA, and QD system as an efficient bioimaging system have also been reported [63].

In this work, CdSe nanoparticles were conjugated with the plant phytohormone ABA in order to potentially reduce the toxicity of the CdSe nanoparticles due to the benign nature of ABA, which is known to enhance cell viability [64]. Thus, the conjugation with self-assembled ABA nanostructures may potentially reduce the toxicity of CdSe nanoparticles, and such materials could be considered useful as biological templates for preparation of nanoparticles for use in fluorescence imaging. We found the ABA based nanostructures acted as efficient templates for the growth of quantum dots with an average diameter of 10-50 nm, where the quantum dots were uniformly coated on the surface of the nanofibers. Further, those quantum dot bound nanofibers were found to be potent in inhibiting the growth of HeLa cells. Thus, we have developed a new family of nano-biohybrids with potential applications such as sensors, optoelectronics and bioimaging.

2. EXPERIMENTAL

2.1. Materials

2-cis,4-trans-Abscisic acid, N-hydroxysuccinimide (NHS), N-(3-Dimethylaminopropyl)-N′-ethylcarbodiimide hydrochloride (EDAC), 1,2 – ethylene diamine, cadmium chloride, selenium (100 mesh powder), and sodium sulfite were all purchased from Sigma Aldrich. Buffer solutions of various pH values were purchased from Fisher Scientific. All compounds were used as received.

2.2 Methods

2.2.1. Self Assembly of Nanostructures

The ABA nanostructures were self-assembled in an aqueous solution at a pH value of 5 at various concentrations (0.01M-0.1M) and allowed to grow for four to five weeks. The formed nanostructures were sonicated, washed with deionized water, and centrifuged twice at 15000 rpm before further analysis.

2.2.2. Conjugation of EDA to ABA Nanostructures

Since the ABA nanostructures contain free carboxyl groups, the carboxyl group was conjugated with EDA to functionalize the nanostructures. The self-assembled ABA nanostructures (1 mL) were incubated with 1 mL of 0.25M N-hydroxysuccinimide (NHS), 1mL of 0.1M N-(3-Dimethylaminopropyl)-N′-ethylcarbodiimide hydrochloride (EDAC), followed by 2 mL of 1,2 – ethylene diamine in DMF after one hour. The solutions were allowed to incubate on a shaker for seventy-two hours. In order to remove any un-reacted materials and excess reducing agent, the samples were washed with deionized water and centrifuged twice at 15000 rpm. The excess solvent was also removed by centrifugation. The samples were dried under vacuum before further analysis. The conjugation of the linker was confirmed by FTIR spectroscopy. FTIR (KBr): $v = 3580$ cm^{-1} (O-H), 3410 cm^{-1} (– N-H), 3105 cm^{-1} (=C–H), 1647 cm^{-1}, 1552 cm^{-1} (C=O).

2.2.3. Growth of Cadmium Selenide Nanoparticle coated abscisic acid Nanostructures

The CdSe nanoparticles were grown by modification of previously established methods [65]. The self-assembled ABA nanostructures (separate vials for the ABA-amide conjugated nanofibers and ABA nanostructures) (500 μL) were incubated with a freshly prepared stock solution of 0.1M cadmium chloride (300 μL) for 24 hours. The samples were then washed with deionized water and centrifuged at 15000 rpm to remove the unbound salt. Freshly prepared sodium selenosulfate (Na$_2$SeSO$_3$) was used as the source for selenium, which was prepared by reaction between elemental selenium and sodium sulfite (0.1M) at 60°C for two hours. The as prepared Na$_2$SeSO$_3$ was then allowed to react with the ABA nanostructures, which had been incubated overnight with CdCl$_2$ solution. The reaction mixture was then heated in a dry-bath at 60°C for an additional two hours. The formation of CdSe nanoparticles was indicated by a color change to reddish-yellow. The samples were then allowed to sit undisturbed (15 hours), and were washed and centrifuged twice at 15000 rpm in order to remove un-reacted materials.

2.2.4. Cell cultures

HeLa human cervical epithelial cells (Sigma-Aldrich, St. Louis, MO) were supplemented with 10% fetal calf serum (Hyclone, Logan, UT), 100 units/mL penicillin, and 100 µg/mL streptomycin. Cells were grown as monolayers in 25 or 75 cm^2 culture flasks under a humidified atmosphere of 5% CO_2 in air at 37 °C. Cells were seeded at densities of 1-2 x 10^4 cells/mL and harvested when they reached 80-90% confluence.

Cell Proliferation

To study the effects of the ABA bound CdSe nanocomposites on cell proliferation, cells were seeded at densities of 2 x 10^4 cells/mL in twelve-well plates. Next, 25 µL of solution containing the nanocomposites were added to each well. At the designated time (24 hours, 48 hours, 4 days, 6 days) after plating, cell proliferation was calculated. The medium from each well was collected in a 15-mL centrifuge tube. Live cells were harvested after applying 500 µL 1X trypsin EDTA (Cellgro, Manassas, VA). Trypsinization was blocked with 500 µL medium and the total volume was added to the centrifuge tube. The cell suspension was centrifuged for 2 minutes and the pellet was re-suspended in 500 µL of medium. To 45 µL of the latter cell suspension, was added to 45 µl of 0.4% Trypan Blue Stain (Gibco) and the density of live and dead cells was calculated using a hemocytometer.

Imaging

Images of the plate cultures were taken before cell proliferation counts using a M165FC stereoscope equipped with a DFC400 CCD camera or a DMIL inverted microscope equipped with a DFC420C CCD camera (Leica Microsystems, Bannockburn, IL). For fluorescence imaging, cells were grown on 18 mm round glass cover slips in 12-well plates at the same concentration used in the proliferation assays. Cells were rinsed with Phosphate-Buffered Saline (PBS) and fixed with 4% paraformaldehyde in PBS pH 7.0 for 15 minutes. The fixed cells were washed with PBS and then a dip in distilled water before mounting using 90% glycerol, 10% Tris-buffered saline and 100 mg/mL of 4-diazabicyclo[2.2.2]octane (DABCO). The mounted cover slips were sealed with nail polish and the cells were imaged with a Leica TCS-SP5 laser-scanning confocal microscope. For imaging samples in the absence of cells, a drop of the sample was put on a glass microscope slide and covered with a coverslip. Images were taken using confocal microscopy.

2.3. Characterization

2.3.1. FTIR Spectroscopy

Analyses were performed using Matteson Infinity IR equipped with DIGILAB, ExcaliBur HE Series FTS 3100 software. The washed samples were dried and pressed with KBr to form flattened pellets. The measurements for the samples were carried out at 400-4000cm^{-1}.

2.3.2. Absorbance Spectroscopy

A Varian Cary 3 UV-visible Spectrophotometer was used to analyze the samples in the wavelength range of 200-800nm. Samples were washed in deionized water and centrifuged twice before analyses. Measurements were carried out in quartz cuvettes.

2.3.3. Fluorescence Spectroscopy

Analysis was carried out using a Jobin Yvon Fluoromax 3 fluorimeter. The samples were excited at appropriate wavelengths for each sample.

2.3.4. Scanning Electron Microscopy

The morphologies of the samples were analyzed by using an SEM (Hitachi S-2600N) operated at 25 kV. A few drops of the sample solution were pipetted onto MCE filter paper (5mm), which were air-dried. Portions of the stained MCE filter paper were then cut out and mounted on SEM sample stubs with double sided tape and carbon coated before analysis.

2.3.5. Transmission Electron Microscopy

The morphologies of the samples were analyzed by TEM (JEOL 1200 EX) operated at 100 KV. Samples were washed twice and air-dried on carbon-coated copper grids for analysis.

3. RESULTS AND DISCUSSION

3.1 Self-Assembly of ABA

Self-assembled nano- and microstructures have garnered substantial interest in nanotechnology as a "bottom up" approach to produce supramolecular nanostructures with a plethora of potential applications [66], such as in electronics [67], sensing [68], optics [69], transistors [70], photovoltaics [71], and drug delivery [72-73]. The formation of self-assembled structures using biological materials such as oligopeptides [74], proteins [75], DNA [76], and lipids [77] is dependent upon chemical complementarities and structural compatibility [78], and occurs due to weak non-covalent interactions such as hydrogen bonding, ionic interactions, hydrophobic interactions, van der Waals forces, and π- π stacking interactions [79-80]. Specifically, stacking interactions play a prominent role in stabilizing the helix of DNA and RNA [80, 81], packing of aromatic systems in crystals [82], aiding in the self-assembly of amyloid fibrils [79], as well as stabilizing the tertiary structure of proteins [83] and porphyrin aggregation [84]. Molecular Dynamics (MD) studies of model polypeptide multilayer nanofilms have shown that electrostatic interactions, hydrophobic interactions, and hydrogen bonding interactions influence the self-assembly of multilayer peptide films [85]. It has been shown that intra- and intermolecular hydrogen bonding interactions between NH••O=C aid in the self-assembly of Hg (II) macrocycles [86].

In this work, we utilized self-assembled ABA nanofibers as templates for the growth of CdSe nanocrystals. The ABA nanofibers were self-assembled at a pH value of 5, and the process was found to be dependent upon concentration as well as growth period. (We also examined self-assembly at other pH values, and those results will be reported in detail separately). For the purposes of this study, only nanofibers grown at pH 5 were used. Fig. **1** shows the proposed scheme for hydrogen bonding interactions between hydroxyl groups and carboxyl groups of ABA molecules.

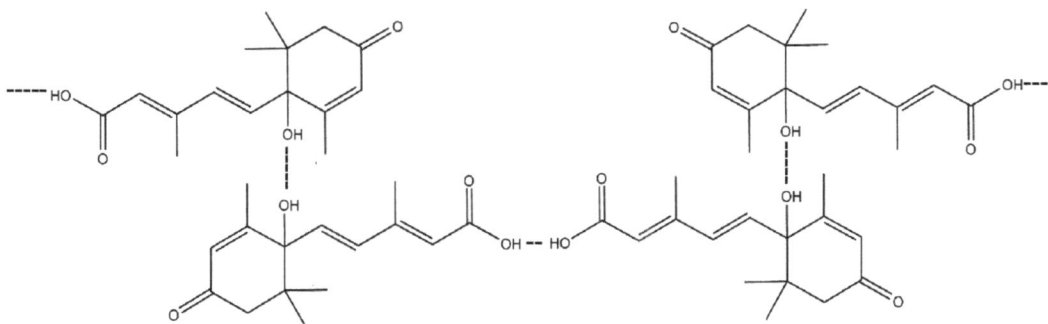

Figure 1. Scheme showing hydrogen-bonding interactions between ABA molecules.

We observed that the average time period for self-assembly of a 0.1 M solution was approximately three weeks, whereas the self-assembly took four to five weeks at lower concentrations (<0.1M). The samples were analyzed using Transmission Electron Microscopy (TEM) as shown in Fig. **2a**, where self-assembled nanofibers with a diamter of approximately 50 nm were observed after one month of growth. Because ABA possesses hydroxyl as well as carboxyl groups, it is likely that those functional groups aid in the self-assembly process under mildly acidic conditions, by hydrogen bonding interactions between adjacent moieties of ABA. Fig. **2b** shows the SEM image of the self-assembled structures after a longer period (2 months) of growth, which shows that bundles of fibers formed in the micron range. Although the exact mechanism for the fiber formation is not known, it appears that the mechanism for formation of such fibers is most likely similar to that observed in the case of amyloid fibrils, which have the capability to promote self-polymerization by seeding, where in the preformed fibers act as templates for fiber elongation [87].

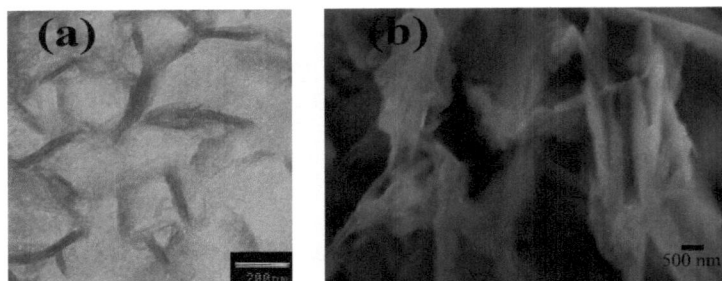

Figure 2(a). ABA nanofibers grown over a period of one month at pH 5; **(b)** SEM image of bundles of ABA nanofibers after 2 months of growth.

3.2. Formation of CdSe Nanoparticles on ABA Nanofibers

Metal ions are known to bind to biomolecules, due to the coordination with specific functional groups found in biomolecules. It has been shown that the biosorption, or binding of metal ions by the algal biomass, occurs due to the coordination of the ions to different functional groups in or on the algae [88]. The coordinating groups, often provided by proteins, lipids, and carbohydrates, include groups such as amino, sulfhydryl, carbonyl, carboxyl, imidazole, phosphate, phenolic, hydroxyl, or amide moieties. Among these, carboxyl, sulfonate, and phenolic groups have been confirmed to be prevalent in many biomaterials. The functionalities of several biopolymers, such as proteins [89], have been exploited as templates for the fabrication of semiconductor nanoparticles. DNA and RNA mediated synthesis of stable size-controlled PbS QDs have also been reported [90]. Nucleotide monophosphates, such as Guanosine 5'-Mono-Phosphate (GMP), have been used as templates for the growth of QDs such as CdS [91]. Luminescent CdSe QDs and other metal oxide nanocrystals have been successfully encapsulated in nanometric liposomes [92]. It has also been reported that $(His-Asn)_6$ domains fused to fatty acid binding proteins allows for the assembly of CdSe nanoparticles [93]. This observation is significant, since fusion proteins with this protein domain are common for affinity purification.

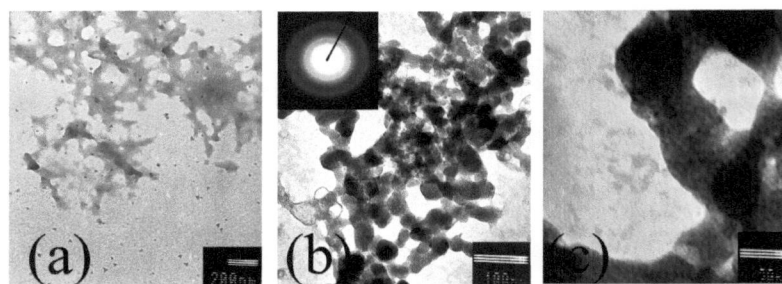

Figure 3. (a) CdSe nanoparticles on ABA nanofibers; **(b)** CdSe nanoparticles grown on ABA nanofibers conjugated with EDA; **(c)** High magnification image of CdSe nanoparticles grown on ABA nanofibers conjugated with EDA.

It has also been reported that certain metal ions such as Mn^{+2}, Gd^{+2}, and Cd^{+2} can bind to the carboxyl groups of Glu^{35} and Asp^{52} moieties of lysozyme [94]. Bacteriorhodopsin has also been reported to bind to a variety of metal ions *via* the oxygen atoms of the carboxyl groups [95]. Since ABA contains both a carboxyl as well as a carbonyl group, we suspected that it may have an affinity for metal ions, and that the self-assembled nanostructures may efficiently bind to the formed CdSe nanoparticles. The morphologies of the cadmium selenide nanoparticle coated nanostructures obtained were observed using transmission and scanning electron microscopy. As seen in Fig. **3a**, networks of ABA nanofibers sparsely coated with CdSe nanoparticles were observed. The sizes of the nanoparticles were in the range of 20-30 nm. This indicates that ABA does not have a very high affinity for CdSe nanoparticles, and consequently the CdSe nanoparticles formed are not evenly coated on the ABA nanofibers. However, upon growth in the presence of EDA-conjugated ABA nanofibers, we observed that there was a significant enhancement in the attachment of the nanofibers toward CdSe nanoparticles (Fig. **3b**), most likely due to a higher affinity of the free –NH$_2$ moiety of the EDA-ABA conjugate. The inset in Fig. **3b** shows the diffraction pattern of the CdSe nanocrystals formed, which indicates that the nanoparticles are highly crystalline, showing the (111) and (220) phases. As seen in the high magnification image (Fig. **3c**), the nanoparticles are evenly coated on the ABA-conjugated nanofibers. The sizes of the CdSe nanoparticles formed were also much smaller (~5-10 nm in diameter) compared to those formed in the presence of the non-conjugated nanofibers.

3.3. Absorbance and Fluorescence Spectroscopy

The formation of CdSe nanoparticles was also confirmed by spectroscopic methods. Fig. **4a** shows the absorbance spectrum comparing the formation of CdSe nanoparticles in the presence and absence of the ABA-EDA conjugate. In the presence of the nanoconjugate, peaks were observed at a λmax of 585 nm (2.13 eV), while when CdSe nanoparticles were synthesized in the absence of the ABA-nanoconjugate, the peak was red-shifted to 598 nm (2.07 eV). Both of the λmax peaks are significantly blue shifted from the bulk CdSe optical band gap at 713 nm (1.73 eV) [96]. These results further confirm that upon binding to the ABA-EDA nanoconjugate, the sizes of the nanoparticles obtained are more controlled and smaller (5-10 nm in diameter) compared to those in the absence of the template, which also corroborates with the TEM image obtained. The blue shift in the presence of the ABA-EDA conjugate is a result of the quantum confinement effect. The results obtained are in agreement with the sizes given in the absorption spectra of CdSe nanocrystals reported earlier [97]. The fluorescence spectra (Fig. **4b**) of the CdSe nanoparticles formed on the ABA-EDA templates showed a peak at 601 nm, while that observed in the absence of the template showed a red shift to maxima 605 nm, further confirming that the sizes of the nanoparticles obtained were larger in the absence of the template.

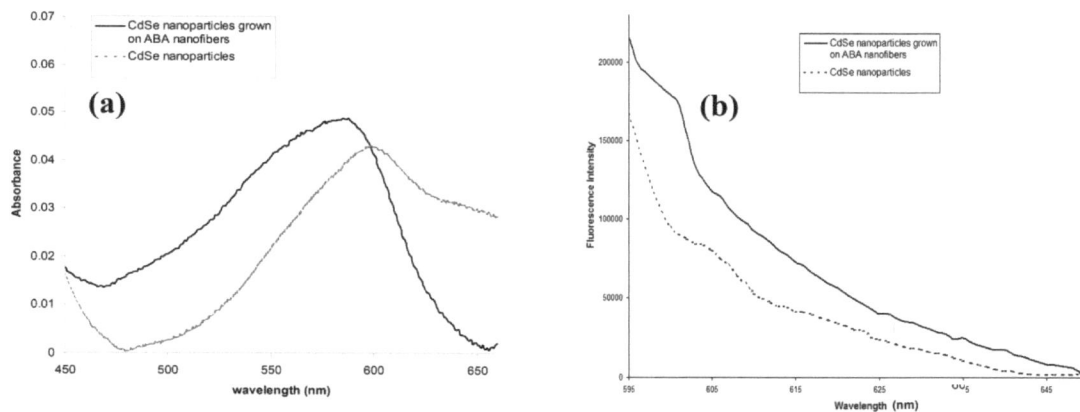

Figure 4. **(a)** Absorbance and **(b)** Fluorescence spectra of CdSe nanoparticles coated on ABA nanofibers in comparison with CdSe nanoparticles grown in the absence of the template.

3.4. Confocal Microscopy and Cell Studies

One of the goals of this study was to investigate the effect of the ABA conjugated CdSe nanoparticles on HeLa cells and their potential applications in bioimaging. Fig. **5** shows the confocal microscopy images of ABA bound CdSe nanoparticles on the ABA fibers. Fig. **5a** shows the phase contrast image, while Fig. **5b** shows the corresponding fluorescence microscopy image. The ABA bound CdSe nanoparticles were found to be fluorescent due to the incorporation of CdSe nanoparticles on the ABA fibers, as the QD's were present throughout the surface of the fibers. These results indicate that the CdSe nanoparticle bound ABA fibers could potentially be useful for purposes of bioimaging.

In a separate set of experiments, HeLa cells were allowed to grow after treatment with the ABA bound CdSe nanoparticles for different periods of time. As indicated by the MTT assay, we observed a significant reduction in HeLa cell proliferation over a period of time, when compared to untreated cells (Fig. **6**). This indicated that the ABA fibers conjugated with CdSe nanoparticles were successful in inhibiting the growth of HeLa cells.

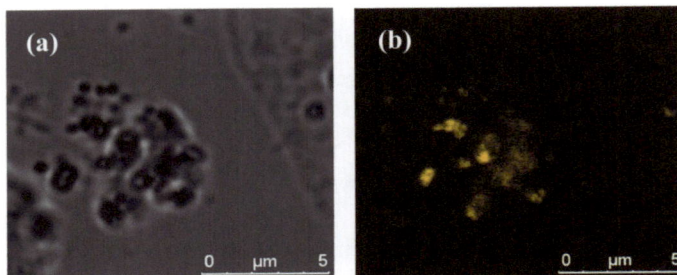

Figure 5. Confocal microscope images of ABA conjugated CdSe nanoparticles. **(a)** Phase contrast image; **(b)** Corresponding fluorescence image.

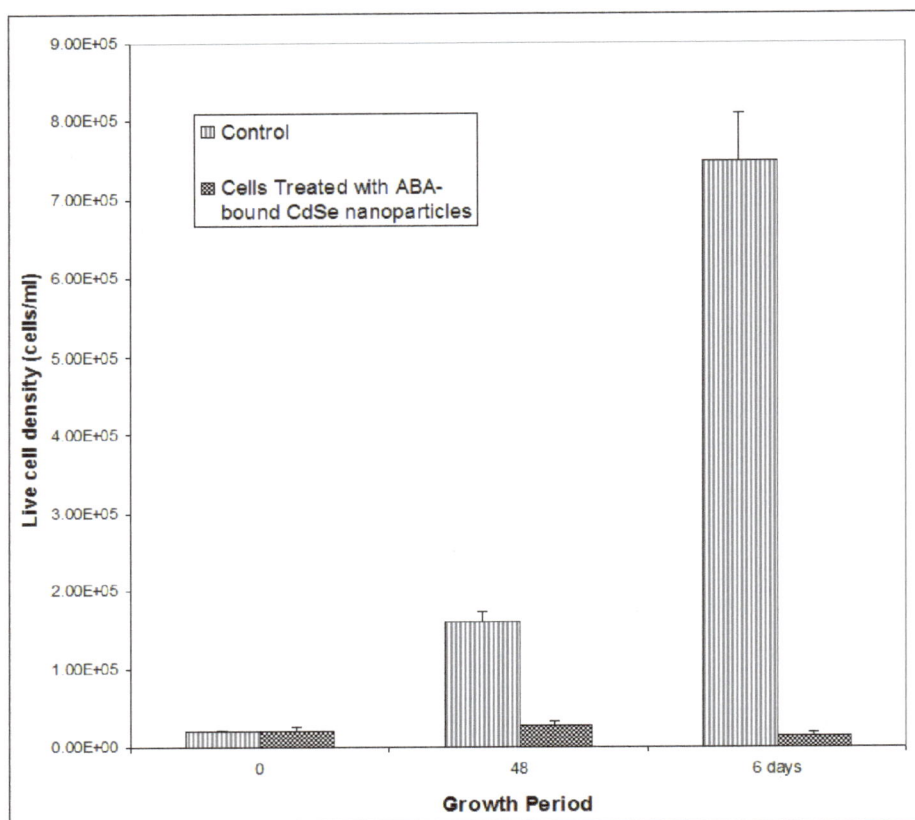

Figure 6. Cell proliferation studies of HeLa cells in the presence and absence of ABA nanofiber bound CdSe nanoparticles.

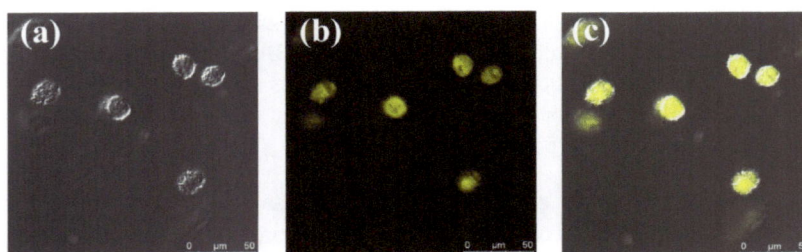

Figure 7. Confocal microscopy images. **(a)** DIC image of HeLa cells in the presence of CdSe-ABA nanocomposites; **(b)** Corresponding fluorescence image and **(c)** Overlay of fluorescence and DIC images.

Fig. **7** shows the cell attachment ability of the ABA bound CdSe nanoparticles. The constructs were plated on cover slips and allowed to grow for a period of six days. The fluorescence microscopy image of HeLa cells that had been incubated with ABA bound CdSe nanoparticles is depicted in Fig **7b**, and the corresponding DIC image (Fig. **7a**) shows that the ABA bound CdSe nanoparticles were clearly associated with the HeLa cells, and in some cases were found to be internalized. This observation was confirmed by confocal Z-stack laser scan microscopy (data not shown). The microscopy scans were made with the same settings for laser power and photomultiplier sensitivity, for direct comparison. Fig. **7c** shows an overlay of the fluorescence and the DIC images, indicating that ABA bound CdSe nanoparticle constructs were clearly associated with the HeLa cells.

CONCLUSIONS

ABA nanofibers were self-assembled at a pH value of 5 and utilized as templates for the growth of CdSe nanoparticles. It was observed that ABA templates allowed for the growth of size controlled CdSe nanoparticles when compared to those obtained in the absence of the template. The growth of CdSe nanoparticles on ABA nanofibers was examined before and after conjugation with the organic linker ethylene diamine. It was observed that the latter allowed for the formation of a more uniform coating of CdSe nanoparticles on the ABA templates. ABA was chosen as a template for the growth of QDs due to its inherent biocompatibility and its ability to self-assemble, which offers benefits, such as increased surface area, for templating the growth of QDs. Furthermore, we investigated the interactions of the CdSe nanoparticle bound ABA templates with HeLa cells for the development of a new class of chemotherapeutics and for bioimaging. Therefore, careful synthesis and targeting of nanoparticles composed of and used in conjugation with ABA as a scaffold for construction may potentially open doors to a new class of relatively less cytotoxic bioimaging probes.

ACKNOWLEDGEMENTS

SF thanks The Campion Institute Office of Prestigious Fellowships for the Matteo Ricci Summer Scholarship. SB thanks the Summer Science Internship at Fordham University for financial support. This work was conducted in part at the Core Facility for Imaging, Cell and Molecular Biology at the Queens College, Department of Biology.

REFERENCES

[1]   [a] Gill, R.; Zayats, M.; Willner, I. Angew. Semiconductor Quantum Dots for Bioanalysis. *Chem., Int. Ed.,* **2008**, *47,* 7602– 7625; [b] Sargent, E. H. Solar Cells, Photodetectors, and Optical Sources from Infrared Colloidal Quantum Dots. *Adv. Mater.,* **2008**, *20*, 3958– 3964; [c] Kamat, P. V. Quantum Dot Solar Cells. Semiconductor Nanocrystals as Light Harvesters. *J. Phys. Chem. C,* **2008**, *112*, 18737– 18753.

[2]   [a] Bawendi, M. G.; Wilson, W. L.; Rothberg, L.; Carroll, P. J.; Jedju, T. M.; Steigerwald, M. L.; Brus, L. E. Electronic structure and photoexcited-carrier dynamics in nanometer-size CdSe clusters. *Phys. Rev. Lett.,* **1990**, *65*, 1623; [b] Murray, C. B.; Norris, D. J.; Bawendi, M. G. Synthesis and characterization of nearly monodisperse CdE (E = sulfur, selenium, tellurium) semiconductor nanocrystallites. *J. Am. Chem. Soc.,* **1993**, *115*, 8706-8715; [c] Guzelian, A. A.; Banin, U.; Kadavanich, A. V.; Peng, X.; Alivisatos, A. P. Colloidal chemical synthesis and characterization of InAs nanocrystal quantum dots. *Appl. Phys. Lett.,* **1996**, *69*, 1432; [d] Alivisatos, A. P. Semiconductor Clusters, Nanocrystals, and Quantum Dots. S*cience.,* **1996**, *271*, 933-937; [e] Wu, Q.; Zheng, N.; Li, Y.; Ding, Y. Preparation of nanosized semiconductor CdS particles by emulsion liquid membrane with o-phenanthroline as mobile carrier. *J. Membr. Sci.,* 2000, *172*, 199-201; [f] Chen, S.; Liu, W. Preparation and Characterization of Surface-Coated ZnS Nanoparticles. *Langmuir.,* **1999**, *15*, 8100-8104.

[3]   [a] Prasad, P. N. *Introduction to Biophotonics*, Wiley-Interscience: Hoboken, NJ, **2003**; [b] Gao, X.; Yang, L.; Petros, J. A.; Marshall, F. F.; Simons, J. W.; Nie, S. *In Vivo* Molecular and Cellular Imaging with Quantum Dots. *Curr. Opin. Biotechnol.,* **2005**, *16*, 63–72; [c] Medintz, I. L.; Uyeda, H. T.; Goldman, E. R.; Mattoussi, H. Quantum Dot Bioconjugates for Imaging, Labelling and Sensing. *Nat. Mater.,* **2005**, *4*, 435–446; [d] Wolfgang, J. P.; Teresa, P.; Christian, P. Labeling of Cells with Quantum Dots. *Nanotechnology.,* **2005**, *16*, R9–R25.

[4]   [a] Diagaradjane, P.; Deorukhkar, A.; Gelovani, G. J.; Maru, M. D.; Krishnan, S. Gadolinium Chloride Augments Tumor- Specific Imaging of Targeted Quantum Dots *In Vivo*. ACS Nano *in press*; [b] Papagiannaros, A.; Levchenko, T.; Hartner, W.; Mongayt, D.; Torchilin, V. Quantum Dots Encapsulated in Phospholipid Micelles for Imaging and Quantification of Tumors in the Near-Infrared Region. *Nanomedicine.*, **2009**, *5*, 216–224; [c] Mulder, W. J.; Castermans, K.; van Beijnum, J. R.; Oude Egbrink, M. G.; Chin, P. T.; Fayad, Z. A.; Lowik, C. W.; Kaijzel, E. L.; Que, I.; Storm, G.; Strijkers, G. J.; Griffioen, A. W.; Nicolay, K. Molecular Imaging of Tumor Angiogenesis Using Alphavbeta3-Integrin Targeted Multimodal Quantum Dots. *Angiogenesis.*, **2009**, *12*, 17–24; [d] Chen, K.; Li, Z. B.; Wang, H.; Cai, W.; Chen, X. Dual-Modality Optical and Positron Emission Tomography Imaging of Vascular Endothelial Growth Factor Receptor on Tumor Vasculature Using Quantum Dots. *Eur. J. Nucl. Med. Mol. Imaging.*, **2008**, *35*, 2235–2244; [e] Smith, B. R.; Cheng, Z.; De, A.; Koh, A. L.; Sinclair, R.; Gambhir, S. S. Real-Time Intravital Imaging of Rgd- Quantum Dot Binding to Luminal Endothelium in Mouse Tumor Neovasculature. *Nano Lett.*, **2008**, *8*, 2599–2606; [f] Cai, W.; Shin, D. W.; Chen, K.; Gheysens, O.; Cao, Q.; Wang, S. X.; Gambhir, S. S.; Chen, X. Peptide-Labeled Near- Infrared Quantum Dots for Imaging Tumor Vasculature in Living Subjects. *Nano Lett.*, **2006**, *6*, 669–676; [g] Morgan, N. Y.; English, S.; Chen, W.; Chernomordik, V.; Russo, A.; Smith, P. D.; Gandjbakhche, A. Real Time *In Vivo* Non-invasive Optical Imaging Using near-Infrared Fluorescent Quantum Dots. *Acad. Radiol.*, **2005**, *12*, 313–323.

[5]   [a] Robe, A.; Pic, E.; Lassalle, H. P.; Bezdetnaya, L.; Guillemin, F.; Marchal, F. Quantum Dots in Axillary Lymph Node Mapping: Biodistribution Study in Healthy Mice. *BMC Cancer.*, **2008**, *8*, 111; [b] Takeda, M.; Tada, H.; Higuchi, H.; Kobayashi, Y.; Kobayashi, M.; Sakurai, Y.; Ishida, T.; Ohuchi, N. *In Vivo* Single Molecular Imaging and Sentinel Node Navigation by Nanotechnology for Molecular Targeting Drug-Delivery Systems and Tailor-Made Medicine. *Breast Cancer.*, **2008**, *15*, 145–152; [c] Kim, S.; Lim, Y. T.; Soltesz, E. G.; De Grand, A. M.; Lee, J.; Nakayama, A.; Parker, J. A.; Mihaljevic, T.; Laurence, R. G.; Dor, D. M.; Cohn, L. H.; Bawendi, M. G.; Frangioni, J. V. Near-Infrared Fluorescent Type II Quantum Dots for Sentinel Lymph Node Mapping. *Nat. Biotechnol.*, **2004**, *22*, 93–97; [d] Soltesz, E. G.; Kim, S.; Laurence, R. G.; DeGrand, A. M.; Parungo, C. P.; Dor, D. M.; Cohn, L. H.; Bawendi, M. G.; Frangioni, J. V.; Mihaljevic, T. Intraoperative Sentinel Lymph Node Mapping of the Lung Using Near-Infrared Fluorescent Quantum Dots. *Ann. Thorac. Surg.*, **2005**, *79*, 269–277, discussion 269-277; [e] Soltesz, E. G.; Kim, S.; Kim, S. W.; Laurence, R. G.; De Grand, A. M.; Parungo, C. P.; Cohn, L. H.; Bawendi, M. G.; Frangioni, J. V. Sentinel Lymph Node Mapping of the Gastrointestinal Tract by Using Invisible Light. *Ann. Surg. Oncol.*, **2006**, *13*, 386–396; [f] Parungo, C. P.; Ohnishi, S.; Kim, S. W.; Kim, S.; Laurence, R. G.; Soltesz, E. G.; Chen, F. Y.; Colson, Y. L.; Cohn, L. H.; Bawendi, M. G.; Frangioni, J. V. Intraoperative Identification of Esophageal Sentinel Lymph Nodes with Near-Infrared Fluorescence Imaging. *J. Thorac. Cardiovasc. Surg.*, **2005**, *129*, 844–850; [g] Ballou, B.; Ernst, L. A.; Andreko, S.; Harper, T.; Fitzpatrick, J. A.; Waggoner, A. S.; Bruchez, M. P. Sentinel Lymph Node Imaging Using Quantum Dots in Mouse Tumor Models. *Bioconjug. Chem.*, **2007**, *18*, 389–396.

[6]   [a] Diagaradjane, P.; Orenstein-Cardona, J. M.; Colon- Casasnovas, N. E.; Deorukhkar, A.; Shentu, S.; Kuno, N.; Schwartz, D. L.; Gelovani, J. G.; Krishnan, S. Imaging Epidermal Growth Factor Receptor Expression *In Vivo*: Pharmacokinetic and Biodistribution Characterization of a Bioconjugated Quantum Dot Nanoprobe. *Clin. Cancer Res.*, **2008**, *14*, 731–741; [b] Wu, X.; Liu, H.; Liu, J.; Haley, K. N.; Treadway, J. A.; Larson, J. P.; Ge, N.; Peale, F.; Bruchez, M. P. Immunofluorescent Labeling of Cancer Marker Her2 and Other Cellular Targets with Semiconductor Quantum Dots. *Nat. Biotechnol.*, **2003**, *21*, 41–46.

[7]   Sen, D.; Deerinck, T. J.; Ellisman, M. H.; Parker, I.; Cahalan, M. D. Quantum Dots for Tracking Dendritic Cells and Priming an Immune Response *In Vitro* and *In Vivo*. *PLoS One.*, **2008**, *3*, e3290.

[8]   Folarin Erogbogbo, Ken-Tye Yong, Indrajit Roy, GaiXia Xu, Paras N. Prasad, and Mark T. Swihart. Biocompatible Luminescent Silicon Quantum Dots for Imaging of Cancer Cells. *ACS Nano.*, **2008**, *2*, 873–878.

[9]   [a] Bentzen, L. E.; House, F.; Utley, J. T.; Crowe Jr., J. E.; Wright, W. D. Progression of Respiratory Syncytial Virus Infection Monitored by Fluorescent Quantum Dot Probes. *Nano Lett.*, **2005**, *5*, 591- 595; [b] Wu, X.; Liu, H.; Liu, J.; Haley, K. N.; Treadway, J. A.; Larson, J. P.; Ge, N.; Peale, F.; Bruchez, M. P. Immunofluorescent labeling of cancer marker Her2 and other cellular targets with semiconductor quantum dots. *Nat. Biotechol.* **2003**, *21*, 41-46.

[10]   Goldman, E. R.; Balighian, E. D.; Mattoussi, H.; Kuno, M. K.; Mauro, J. M.; Tran, P. T.; Anderson, G. P. Avidin: A Natural Bridge for Quantum Dot-Antibody Conjugates. *J. Am. Chem. Soc.* **2002**, *124*,41-46.

[11]   Parak, W. J.; Boudreau R.; Le Gros, M.; Gerion, D.; Zanchet, D.; Micheel, C. M.; Williams, S. C.; Alivisatos, A. P.; Larabell, C. Cell motility and metastatic potential studies based on quantum dot imaging of phagokinetic tracks. *Adv. Mater.* **2002**, *14(12)*, 882-885.

[12] Schipper, M. L.; Cheng, Z.; Lee, S. W.; Bentolila, L. A.; Iyer, G.; Rao, J.; Chen, X.; Wu, A. M.; Weiss, S.; Gambhir, S. S. Micropet-Based Biodistribution of Quantum Dots in Living Mice. *J. Nucl. Med.,* **2007**, *48*, 1511–1518.

[13] Michalet, X.; Pinaud, F. F.; Bentolila, L. A.; Tsay, J. M.; Doose, S.; Li, J. J.; Sundaresan, G.; Wu, A. M.; Gambhir, S. S.; Weiss, S. Quantum Dots for Live Cells, *In Vivo* Imaging, and Diagnostics. *Science.,* **2005**, *307*, 538–544.

[14] Guo, G.; Liu, W.; Liang, J.; Xu, H.; He, Z.; Yang, X. Preparation and characterization of novel CdSe quantum dots modified with poly (d, l-lactide) nanoparticles. *Mater. Lett.,* **2006**, *60*, 2565-2568.

[15] Nagasaki, Y.; Ishii, T.; Sunaga, Y.; Watanabe, Y.; Otsuka, H.; Kataoka, K. Novel Molecular Recognition *via* Fluorescent Resonance Energy Transfer Using a Biotin−PEG/Polyamine Stabilized CdS Quantum Dot. *Langmuir.,* **2004**, *20*, 6396-6400.

[16] [a] Yin, Y.; Lu, Y.; Gates, B.; Xia, Y. A Self-Assembly Approach to the Formation of Asymmetric Dimers from Monodispersed Spherical Colloids. *J. Am. Chem. Soc.* **2001**, *123*, 8718-8729; [b] Cui, Y.; Bjork, M. T.; Liddle, J. A.; Sonnichsen, C.; Boussert, B.; Alivisatos, A. P. Integration of Colloidal Nanocrystals into Lithographically Patterned Devices. *Nano Lett.,* **2004**, *4*, 1093-1098.

[17] Peng X.; Wong, S. S. Controlling Nanocrystal Density and Location on Carbon Nanotube Templates. *Chem. Mater.* **2009**, *21*, 682–694.

[18] Yan, X.; Cui, Y.; He, Q.; Wang, K.; Li, J. Organogels Based on Self-Assembly of Diphenylalanine Peptide and Their Application to Immobilize Quantum Dots. *Chem. Mater.* **2008**, *20*, 1522-1526.

[19] Chen, J.; Liao, W.S.; Chen, X.; Yang, T.; Wark, S. E.; Son, D. H.; Batteas, J. D.; Cremer, P. S. Evaporation-Induced Assembly of Quantum Dots into Nanorings. *ACS Nano.,* **2009**, *3*, 173-180.

[20] Priyam, A.; Blumling, D. E.; Knappenberger Jr., K. L. Synthesis, Characterization, and Self-Organization of Dendrimer-Encapsulated HgTe Quantum Dots. *Langmuir.,* **2010**, *26*, 10636–10644.

[21] [a] Meyer, G. J. Molecular Approaches to Solar Energy Conversion with Coordination Compounds Anchored to Semiconductor Surfaces. *Inorg. Chem.,* **2005**, *44*, 6852-6864; [b] Donega´, C. d. M.; Liljeroth, P.; Vanmaekelbergh, D. Physicochemical evaluation of the hot-injection method, a synthesis route for monodisperse nanocrystals. *Small.,* **2005**, *1*, 1152-1162.

[22] [a] Henzie, J.; Barton, J. E.; Stender, C. L.; Odom, T. W. Large- Area Nanoscale Patterning: Chemistry Meets Fabrication. *Acc. Chem. Res.,* **2006**, *39*, 249–257; [b] Aizpurua, J.; Hanarp, P.; Sutherland, D. S.; Ka´ll, M.; Bryant, G. W.; Garc´ıa de Abajo, F. J. Optical Properties of Gold Nanorings. *Phys. Rev. Lett.,* **2003**, *90*, 057401-1–057401-4; [c] Geissler, M.; McLellan, J. M.; Chen, J.; Xia, Y. Side-by-Side Patterning of Multiple Alkanethiolate Monolayers on Gold by Edge-Spreading Lithography. *Angew. Chem., Int. Ed.* **2005**, *44*, 3596–3600; [d] Yang, S.-M.; Jang, S. G.; Choi, D.-G.; Kim, S.; Yu, H. K. Nanomachining by Colloidal Lithography. *Small.,* **2006**, *2*, 458–475; [e] Li, J.-R.; Garno, J. C. Elucidating the Role of Surface Hydrolysis in Preparing Organosilane Nanostructures *via* Particle Lithography. *Nano Lett.,* **2008**, *8*, 1916–1922; [f] Hulteen, J. C.; Van Duyne, R. P. Nanosphere Lithography: A Materials General Fabrication Process for Periodic Particle Array Surfaces. *J. Vac. Sci. Technol., A.* **1995**, *13*, 1553–1558; [g] Winzer, M.; Kleiber, M.; Dix, N.; Wiesendanger, R. Fabrication of Nano-Dot- and Nano-Ring-Arrays by Nanosphere Lithography. *Appl. Phys. A: Mater. Sci. Process.,* **1996**, *63*, 617–619; [h] Garno, J. C.; Amro, N. A.; Wadu-Mesthrige, K.; Liu, G.-Y. Production of Periodic Arrays of Protein Nanostructures Using Particle Lithography. *Langmuir.,* **2002**, *18*, 8186–8192; [i] Li, J.-R.; Henry, G. C.; Garno, J. C. Fabrication of Nanopatterned Films of Bovine Serum Albumin and Staphylococcal Protein A Using Latex Particle Lithography. *Analyst.,* **2006**, *131*, 244–250; [j] Liao, W.-S.; Chen, X.; Chen, J.; Cremer, P. S. Templating Water Stains for Nanolithography. *Nano Lett.,* **2007**, *7*, 2452–2458.

[23] [a] Lu, N.; Chen, X.; Molenda, D.; Naber, A.; Fuchs, H.; Talapin, D. V.; Weller, H.; Mu¨ller, J.; Lupton, J. M.; Feldmann, J.; *et al.* Lateral Patterning of Luminescent CdSe Nanocrystals by Selective Dewetting from Self-Assembled Organic Templates. *Nano Lett.,* **2004**, *4*, 885–888; [b] Cui, Y.; Bjork, M. T.; Liddle, J. A.; Sonnichsen, C.; Boussert, B.; Alivisatos, A. P. Integration of Colloidal Nanocrystals into Lithographically Patterned Devices. *Nano Lett.,* **2004**, *4*, 1093–1098.

[24] Peng, H.; Zhang, L.; Soeller, C.; Travas-Sejdic, *J. Lumin.,* Preparation of water-soluble CdTe/CdS core/shell quantum dots with enhanced photostability. **2007**, *127*, 721-726.

[25] Mattoussi, H.; Mauro, J.M.; Goldman, E.R.; Anderson, G.P.; Sundar, V.C.; Mikulec, F.V.; Bawendi, M.G. Self-Assembly of CdSe−ZnS Quantum Dot Bioconjugates Using an Engineered Recombinant Protein. *J. Am. Chem. Soc.,* **2000**, *122*, 12142- 12150.

[26]   Wu, X.; Liu, H.; Liu, J.; Haley, K. N.; Treadway, J. A.; Larson, J. P.; Ge, N.; Peale, F.; Bruchez, M. P. Immunofluorescent labeling of cancer marker Her2 and other cellular targets with semiconductor quantum dots. *Nat. Biotechnol.,* **2003**, *21*, 41-46.

[27]   Koole, R.; van Schooneveld, M. M.; Hilhorst, J.; Donega, C. d. M.; Hart, D. C.; van Blaadersen, A.; Vanmaekelbergh, D.; Meijerink, A. On the Incorporation Mechanism of Hydrophobic Quantum Dots on Silica Spheres by a reverse Microemulsioin Method. *Chem.Mater.* **2008**, *20*, 2503-2512.

[28]   Regan, M. R.; Banerjee, I. A. Preparation of Au–Pd bimetallic nanoparticles in porous germania nanospheres: A study of their morphology and catalytic activity. *Scr. Mater.,* **2006**, *54*, 909–914.

[29]   Reetz M. T.; Lohmer G. Propylene carbonate stabilized nanostructured palladium clusters as catalysts in Heck reactions. *Chem Commun.,* **1996**, *16*, 1921-1922.

[30]   Mayer, A. B. R.; Mark, J. E.; Hausner, S. H. Palladium nanocatalysts protected by polyacids. *J. Appl. Polym. Sci.,* **1998**, *70*, 1209-1219.

[31]   Zhao, Z. M.; Crooks, R. M. Homogeneous hydrogenation catalysis with monodisperse, dendrimer-encapsulated Pd and Pt nanoparticles. *Angew. Chem. Int. Ed.,* **1999**, *38*, 364-366.

[32]   Monnier, A.; Schüth, F.; Huo, Q.; Kumar, D.; Margolese, D.; Maxwell, R. S.; Stucky, G. D.; Krishnamurty, M.; Petroff, P.; Firouzi, A.; Janicke, M.; Chmelka, B. F. Cooperative Formation of Inorganic-Organic Interfaces in the Synthesis of Silicate Mesostructures. *Science,* **1993**, *261*, 1299-1303.

[33]   Cai W.; Zhong, H.; Zhang, L. Optical measurements of oxidation behavior of silver nanometer particle within pores of silica host. *J. Appl. Phys.,* **1998**, *83*, 1705-1711.

[34]   Mulukurtla, R. S.; Asakura, K.; Kogure, T.; Namba, S.; Iwasawa, Y. Synthesis and characterization of rhodium oxide nanoparticles in mesoporous MCM-41. *Phys. Chem. Chem. Phys.,* **1999**, *1*, 2027-2032.

[35]   Aronson, B. J.; Blanford, C. E.; Stein A. Solution-Phase Grafting of Titanium Dioxide onto the Pore Surface of Mesoporous Silicates: Synthesis and Structural Characterization. *Chem Mater.,* **1997**, *9*, 2842-2851.

[36]   Raevskaya, A. E.; Stroyuk, A. L.; Kuchmiy, S. Y. Preparation of colloidal CdSe and CdS/CdSe nanoparticles from sodium selenosulfate in aqueous polymers solutions. *J. Colloid Interf. Sci.,* **2006**, *302*, 133-141.

[37]   [a] Nan, M.; Dooley, C. J.; Kelley, S. O. RNA-Templated Semiconductor Nanocrystals. *J. Am. Chem. Soc.,* **2006**, *128*, 12598-12599; [b] Seeman, N. C.; Belcher, A. M. Emulating biology: building nanostructures from the bottom up. *Proc. Natl. Acad. Sci. U.S.A.,* **2002**, *99*, 6451-6455; [c] Lee, S.-W.; Mao, C.; Flynn, C. E.; Belcher, A. M. Ordering of quantum dots using genetically engineered viruses. *Science.,* **2002**, *296*, 892-895; [d] Hinds, S.; Taft, B. J.; Levina, L.; Sukhovatkin, V.; Dooley, C. J.; Roy, M. D.; MacNeil, D. D.; Sargent, E. H.; Kelley, S. O. Nucleotide-Directed Growth of Semiconductor Nanocrystals. *J. Am. Chem. Soc.,* **2006**, *128*, 64-65; [e] Gugliotto, L. A.; Feldheim, D. L.; Eaton, B. E. RNA-Mediated Metal-Metal Bond Formation in the Synthesis of Hexagonal Palladium Nanoparticles. *Science.,* 2004, *304*, 850-852; [f] Petty, J. T.; Zheng, J.; Hud, N. V., Dickson, R. M. DNA-templated Ag nanocluster formation. *J. Am. Chem. Soc.,* **2004**, *126*, 5207-5212; [g] Coffer, J. L.; Bigham, S. R.; Pinozzotto, R. F.; Yang, H. *Nanotechnology,* **1992**, *3*, 69-76.

[38]   Bergkvist, M.; Mark, S. S.; Yang, X.; Angert, E. R.; Batt, C. A. Bionanofabrication of Ordered Nanoparticle Arrays: Effect of Particle Properties and Adsorption Conditions. *J. Phys. Chem. B,* **2004**, *108*, 8241-8248.

[39]   [a] Sotiropoulou, S.; Sierra-Sastre, Y.; Mark, S. S.; Batt, C. A. Biotemplated Nanostructured Materials. *Chem. Mater.,* **2008**, *20*, 821–834; [b] Caruso, R. A. Micrometer-to-nanometer replication of hierarchical structures by using a surface sol-gel process. *Angew. Chem., Int. Ed.,* **2004**, *43*, 2746–2748; [c] Kumara, M. T.; Tripp, B. C.; Muralidharan, S. Exciton Energy Transfer in Self-Assembled Quantum Dots on Bioengineered Bacterial Flagella Nanotubes. *J. Phys. Chem. C,* **2007**, *11*, 5276–5280.

[40]   [a] Feldheim, D. L.; Eaton, B. E. Selection of Biomolecules Capable of Mediating the Formation of Nanocrystals. *ACS Nano.,* **2007**, *1*, 154-159; [b] Ma, N.; Dooley, C. J.; Kelley, S. O. RNA-Templated Semiconductor Nanocrystals. *J. Am. Chem. Soc.,* **2006**, *128*, 12598–12599.

[41]   Q. Huang, Y.; Chiang, C. Y.; Lee, S. K.; Gao, Y.; Hu, E. L.; De Yoreo, J.; Belcher, A. M. Programmable Assembly of Nanoarchitectures Using Genetically Engineered Viruses. *Nano. Lett.,* **2005**, *5*, 1429–1434.

[42]   [a] Pejova, B.; Tanusˇevski, A.; Grozdanova, I. Semiconducting thin films of zinc selenide quantum dots. *J. Solid State Chem.,* **2004**, *177*, 4785–4799; [b] Empedocles, S. A.; Norris, D. J.; Bawendi, M. G. Photoluminescence Spectroscopy of Single CdSe Nanocrystallite Quantum Dots. *Phys. Rev. Lett.,* **1996**, *77*, 3873-3876.

[43]   Whaley, S. R.; English, D. S.; Hu, E. L.; Barbara, P. F.; Belcher, A. M. Selection of peptides with semiconductor binding specificity for directed nanocrystal assembly. *Nature,* **2000**, *405*, 665.

[44]   Lee, S.W.; Mao, C.; Flynn, C. E.; Belcher, A. M. Ordering of Quantum Dots Using Genetically Engineered Viruses. *Science 3.,* **2002**, *296*, 892 – 895.

[45]   [a] Feldheim, D. L.; Eaton, B. E. Selection of Biomolecules Capable of Mediating the Formation of Nanocrystals. *ACS Nano.,* **2007,** *1,* 154-159; [b] Ma, N.; Dooley, C. J.; Kelley, S. O. RNA-Templated Semiconductor Nanocrystals. *J. Am. Chem. Soc.,* **2006,** *128,* 12598–12599.

[46]   [a] Roberts, D. L.; Heckman, R. A.; Hege, B. P.; Bellin, S. A. Synthesis of (RS)-Abscisic Acid. *J. Org. Chem.,* **1968,** *33,* 3566–3569; [b] Addicott, F. T.; Lyon, J. L.; Ohkuma, K.; Thiessen, W. E.; Carns, H. R.; Smith, O. E.; Cornforth, J. W.; Milborrow, B. V.; Ryback, G.; Wareing, P. F. Abscisic Acid: A New Name for Abscisin II (Dormin). *Science.,* **1968,** *29,* 1493; [c] Addicott, F. T.; Cams, H. R.; Lyon, J. L.; Smith, O. E.; MoeMeans, J. L. Regulateurs Naturels de la Croissanoe VBgBtale. Centre National de la Recherche Scientifique, Paris, **1964,** 687-703; [d] Ohkuma, K.; Addicott, F. T.; Smith, O. E.; Thies- sen, W. E. Structure of abscisin II. *Tetrahedron Lett.,* **1965,** *29,* 2529-2535; [e] Addicott, F. T.; Ohkuma, K.; Smith, O. E.; Thiessen, W. E. Natural Pest Control Agents. *Advances in Chemistry Series.; No 53* Washington D.C. **1966,** Chp. 9, 97-105;[f] Cornforth, J. W.; Milborrow, B. V.; Ryback, G.; Wareing, P. F. Chemistry and physiology of dormins in sycamore. Identity of sycamore dormin with abscisin II. *Nature,* **1965,** *406,* 1269-1270; [g] Cornforth, J. W.; Milborrow, B. V.; Rybaok, G. Identification and Estimation of (+-)-abscisin II (dormin) in plant extracts by spectropolarimetry. *Nature.,* **1966,** *210,* 627-628; [h] Dörffling, K. (+)-Abszisin II (Dormin) im Ätherextrakt von Pisumsprossen. *Naturwissenschaften.,* **1967,** *64,* 23-24; [i] El-Antably, H. M. M.; Wareing, P. F.; Hillman, J. Some physiological responses to d,l abscisin (dormin). *Planta,* **1967,** *78,* 74-90; [j] Hoad, G. V. (+)-Abscisin II, ((+)-dormin) in phloem exudate of willow. *Life Sci.,* **1967,** *6,* 1113-1118; [k] Milborrow, B. V. The effects of synthetic dl-dormin (Abscisin II) on the growth of the oat mesocotyl. *Planta,* **1966,** *70,* 155-171;

[47]   [a] Kim, B. T.; Min, Y. K.; Asami, T.; Park, N. K.; Kwon, O. Y.; Cho, K. Y.; Yoshida, S. 2-Fluoroabscisic Acid Analogues: Their Synthesis and Biological Activities. *J. Agric. Food Chem.,* **1999,** *47,* 313−317; [b] Ohkuma, K.; Lyon, J. L.; Addicot, F. D.; Smith, O. E. Abscisin II, an abscission-accelerating substance from young cotton fruit. *Science,* **1963,** *142,* 1592-1593.

[48]   Zhao, Z.; Stanley, B. A.; Zhang, W.; Assmann, S. M. ABA-Regulated G Protein Signaling in Arabidopsis Guard Cells: A Proteomic Perspective. *J. Proteome Res.,* **2010,** *9,* 1637–1647.

[49]   [a] Prasad, T. K.; Anderson, M. D.; Stewart, C. R. Acclimation, hydrogen peroxide, and abscisic acid protect mitochondria against irreversible chilling injury in maize seedlings. *Plant Physiol.,* **1994,** *105,* 619-627; [b] Davies, W. J.; Tardieu, F.; Trejo, C. L. *In Plant Adaptation to Environmental Stress;* Fowden, L.; Mansfield, T.; Stoddart, J., Eds, Chapman and Hall Ltd: London, **1993;** 209-222; [c] Anderson, M. D.; Prasad, T. K.; Martin, B. A.; Stewart, C. R. Differential gene expression in chilling-acclimated maize seedlings and evidence for the involvement of abscisic acid in chilling tolerance. *Plant Physiol.,* **1994,** *105,* 331-339.

[50]   [a] Himmelbach, A.; Iten, M.; Grill, E. Signalling of abscisic acid to regulate plant growth. *Phil. Trans. R. Soc. Lond. B,* **1998,** *353,* 1439-1444; [b] Overbeek, J. V.; Loe¥er, J. E.; Mason, M. I. R. Dormin (abscisin II), inhibitor of plant DNA synthesis. *Science,* **1967,** *156,* 1497-1499; [c] Villiers, T. A. An autoradiographic study of the elect of the plant hormone abscisic acid on nucleic acid and protein metabolism. *Planta,* **1968,** *82,* 342-354.

[51]   [a] Nagl, W. Selective inhibition of cell cycle stages in the Allium root meristem by colchicine and growth regulators. *Am. J. Bot.,* **1972,** *59,* 346-351; [b] de la Torre, C.; Diez, J. L.; Lopez-Saez, J. F.; Gimenez-Martin, G. Elect of abscisic acid on the cytological components of the root growth. *Cytologia,* **1972,** *37,* 197-205.

[52]   [a] Yin, H.; Zhang, X.; Liu, J.; Wang, Y.; He, J.; Yang, Y.; Hong, X.; Yang, Q.; Gong, Z. Epigenetic Regulation, Somatic Homologous Recombination, and Abscisic Acid Signaling Are Influenced by DNA Polymerase ε Mutation in *Arabidopsis. Plant Cell,* **2009,** *21,* 386-402; [b] Swiatek, A.; Lenjou, M.; Van Bockstaele, D.; Inze, D.; Van Onckelen, H. Differential effect of jasmonic acid and abscisic acid on cell cycle progression in tobacco BY-2 cells. *Plant Physiol.,* **2002,** *128,* 201-221.

[53]   Wang, X.Q.; Ullah, H.; Jones, A. M.; Assmann, S. M. G Protein Regulation of Ion Channels and Abscisic Acid Signaling in Arabidopsis Guard Cells. *Science,* **2001,** *292,* 2070-2072.

[54]   [a] Doonan, J.; Fobert, P. Conserved and novel regulators of the plant cell cycle. *Curr. Opin. Cell Biol.,* **1997,** *8,* 824-830; [b] Mizoguchi, T.; Irie, K.; Hirayama, T.; Hayashida, N.; Yamaguchi-Shinozaki, K.; Matsumoto, K.; Shinozaki, K. A gene encoding a mitogen-activated protein kinase kinase kinase is induced simultaneously with genes for a mitogen-activated protein kinase and an S6 ribosomal protein kinase by touch, cold, and water stress in Arabidopsis thaliana. *PNAS,* **1996,** *93,* 765-769; [c] Mizoguchi, T.; Ichimura, K.; Shinozaki, K. Environmental stress response in plants: the role of mitogen- activated protein kinases. *Trends Biotechnol.,* **1997,** *15,* 15-19; [d] Machida, Y.; Nishihama, R.; Kitakura, S. Progress in studies of plant homologs of Mitogen-Activated Protein (MAP) kinase and potential upstream components in kinase cascades. *Crit. Rev. Pl. Sci.,* **1997,**

16, 481-496; [e] Ligterink, W.; Kroj, T.; Zurnieden, U.; Hirt, H.; Scheel, D. Receptor-mediated activation of a MAP kinase in pathogen defense of plants. *Science,* **1997**, *276*, 2054-2057.

[55] [a] Armstrong, F.; Leung, J.; Grabov, A.; Brearley, J.; Giraudat, J.; Blatt, M. R. Sensitivity to abscisic acid of guard-cell K^+ channels is suppressed by abi1-1, a mutant Arabidopsis gene encoding a putative protein phosphatase. *PNAS,* **1995**, *92*, 9520-9524; [b] Wu, Y.; Kuzma, J.; Marechal, E.; Grae¡, R.; Lee, H. C.; Foster, R.; Chua, N.-H. Abscisic acid signaling through cyclic ADP-ribose in plants. *Science,* **1997**, *278*, 2126-2130.

[56] Mauro, J. M.; Mattoussi, H.; Medintz, I. L.; Goldman, E. R.; Tran, P. T.; Anderson, G. P. In *Defense Applications of Nanomaterials*; Miziolek, A. W.; Karna, S. P.; Mauro, J. M.; Vaia, R. A.; Eds. American Chemical Society: Washington, DC, **2005**, pp. 16-30.

[57] [a] Chong, S. V.; Suresh, N.; Xia, J.; Al-Salim, N.; Idriss, H. TiO$_2$ Nanobelts/CdSSe Quantum Dots Nanocomposite. *J. Phys. Chem. C,* **2007**, *111*, 10389– 10393; [b] Lee, J. C.; Sung, Y. M.; Kim, T. G.; Choi, H. J. TiO$_2$-CdSe nanowire arrays showing visible-range light absorption. *Appl. Phys. Lett.,* **2007**, *91*, 113104/1-113104/3; [c] Mora-Sero, I.; Bisquert, J.; Dittrich, T.; Belaidi, A.; Susha, A. S.; Rogach, A. L. Photosensitization of TiO$_2$ Layers with CdSe Quantum Dots: Correlation between Light Absorption and Photoinjection. *J. Phys. Chem. C,* **2007**, *111*, 14889– 14892; [d] Danek, M.; Jensen, K. F.; Murray, C. B.; Bawendi, M. G. Synthesis of Luminescent Thin-Film CdSe/ZnSe Quantum Dot Composites Using CdSe Quantum Dots Passivated with an Overlayer of ZnSe. *Chem. Mater.,* **1996**, *8*, 173-180; [e] Rodriguez-Viejo, J.; Jensen, K. F.; Mattoussi, H.; Michel, J.; Dabbousi, B. O.; Bawendi, M. G. Cathodoluminescence and photoluminescence of highly luminescent CdSe/ZnS quantum dot composites. *Appl. Phys. Lett.,* **1997**, *70*, 2132-2134; [f] Peng, X.; Schlamp, M. C.; Kadavanich, A. V.; Alivisatos, A. P. Epitaxial Growth of Highly Luminescent CdSe/CdS Core/Shell Nanocrystals with Photostability and Electronic Accessibility. *J. Am. Chem. Soc.,* **1997**, *119*, 7019-7029; [g] Hines, M. A.; Guyot-Sionnest P. Synthesis and Characterization of Strongly Luminescing ZnS-Capped CdSe Nanocrystals. *J. Phys. Chem. B,* **1996**, *100*, 468-471.

[58] [a] Bertolini, G.; Paleari, L.; Catassi, A.; Roz, L.; Cesario, A.; Sozzi, G.; Russo, P. *In vivo* Cancer Imaging with Semiconductor Quantum Dots. *Curr. Pharm. Anal.* **2008**, *4*, 197–205; [b] Cai, W. B.; Hsu, A. R.; Li, Z. B.; Chen, X. Y. Are quantum dots ready for *in vivo* imaging in human subjects? *Nanoscale Res. Lett.,* 2007, *2*, 265–281.

[59] [a] Hardman, R. A toxicologic review of quantum dots: toxicity depends on physicochemical and environmental factors. *Environ Health Perspect.* **2006**, *114*, 165-172; [b] Larson, D. R.; Zipfel, W. R.; Williams, R. M.; Clark, S. W.; Bruchez, M. P.; Wise, F. W.; Webb, W. W. Water-Soluble Quantum Dots for Multiphoton Fluorescence Imaging *in Vivo. Science,* **2003**, *300*, 1434–1436; [c] Kim, S.; Lim, Y. T.; Soltesz, E. G.; De Grand, A. M.; Lee, J.; Nakayama, A.; Parker, J. A.; Mihaljevic, T.; Laurence, R. G.; Dor, D. M.; Cohn, L. H.; Bawendi, M. G.; Frangioni, J. V. Near-infrared fluorescent type II quantum dots for sentinel lymph node mapping. *Nat. Biotechnol.,* **2004**, *22*, 93–97; [d] Derfus, A. M.; Chan, W. C. W.; Bhatia, S. N. Probing the Cytotoxicity of Semiconductor Quantum Dots. *Nano Lett.,* **2004**, *4*, 11–18; [e] Hoshino, A.; Fujioka, K.; Oku, T.; Suga, M.; Sasaki, Y. F.; Ohta, T.; Yasuhara, M.; Suzuki, K.; Yamamoto, K. Physiochemical Properties and Cellular Toxicity of Nanocrystal Quantum Dots Depend on Their Surface Modification. *Nano Lett.,* **2004**, *4*, 2163–2169; [f] Liang, J. G.; He, Z. K.; Zhang, S. S.; Huang, S.; Ai, X. P.; Yang, H. X.; Han, H. Y. Study on DNA damage induced by CdSe quantum dots using nucleic acid and molecular "light switches" as probe. *Talanta,* **2007**, *71*, 1675–1678; [g] Anas, A.; Akita, H.; Harashima, H.; Itoh, T.; Ishikawa, M.; Biju, V. Photosensitized Breakage and Damage of DNA by CdSe-ZnS Quantum Dots. *J. Phys. Chem. B,* **2008**, *112*, 10005–10011.

[60] [a] Oikawa, H.; Oshikiri, T.; Kasai, H.; Okada, S.; Tripathy, S. K.; Nakanishi, H. Various types of polydiacetylene microcrystals fabricated by reprecipitation technique and some applications. *Polym. Advan. Technol.,* **2000**, *11*, 783–790; [b] Landfester, K.; Montenegro, R.; Scherf, U.; Guntner, R.; Asawapirom, U.; Patil, S.; Neher, D.; Kietzke, T. Semiconducting Polymer Nanospheres in Aqueous Dispersion Prepared by a Miniemulsion Process. *Adv. Mater.,* **2002**, *14*, 651–655.

[61] Bottini, M.; Cerignoli, F.; Dawson, M. I.; Magrini, A.; Rosato, N.; Mustelin, T. Full-Length Single-Walled Carbon Nanotubes Decorated with Streptavidin-Conjugated Quantum Dots as Multivalent Intracellular Fluorescent Nanoprobes. *Biomacromolecules,* **2006**, *7*, 2259-2263.

[62] [a] Ishihara, K.; Iwasaki, Y.; Nakabayashi, N. Polymeric lipid nanosphere consisting of water-soluble poly(2-methacryloyloxyethyl phosphorylcholine-co-n-butyl methacrylate). *Polym. J.,* **1999**, *31*, 1231–1236; [b] Konno, T.; Watanabe, J.; Ishihara, K. Conjugation of Enzymes on Polymer Nanoparticles Covered with Phosphorylcholine Groups. *Biomacromolecules* **2004**, *5*, 342–347; [c] Takei, K.; Konno, T.; Watanabe, J.; Ishihara, K. Regulation of Enzyme-Substrate Complexation by a Substrate Conjugated with a Phospholipid

Polymer. *Biomacromolecules* **2004**, *5*, 858–862; [d] Konno, T.; Watanabe, J.; Ishihara, K. Enhanced solubility of paclitaxel using water-soluble and biocompatible 2-methacryloyloxyethyl phosphorylcholine polymers *J. Biomed. Mater. Res.,* **2003**, *65A*, 209–214.

[63] [a] Goto, Y.; Matsuno, R.; Konno, T.; Takai, M.; Ishihara, K. Polymer Nanoparticles Covered with Phosphorylcholine Groups and Immobilized with Antibody for High-Affinity Separation of Proteins. *Biomacromolecules* **2008**, *9*, 828–833; [b] Goto, Y.; Matsuno, R.; Konno, T.; Takai, M.; Ishihara, K. Artificial cell membrane-covered nanoparticles embedding quantum dots as stable highly sensitive fluorescence bioimaging probes. *Biomacromolecules*, **2008**, *9*, 3252-3257.

[64] Zhang, C.; Fevereiro, P. S. The effect of heat shock on paclitaxel production in Taxus yunnanensis cell suspension cultures: Role of abscisic acid pretreatment. *Biotechnol. Bioeng.,* **2007**, *96,* 506–514.

[65] [a] Raevskaya, A. E.; Stroyuk, A. L.; Kuchmiv, S. Ya.; Azhnjuk, Yu. M.; Dzhagan, V. M.; Yukhymchuk, V. O.; Valakh, M. Ya. Growth and spectroscopic characterization of CdSe nanoparticles synthesized from CdCl$_2$ and Na$_2$SeSO$_3$ in aqueous gelatine solutions. *Eng. Aspects,* **2006**, *290*, 304-309; [b] Zhang, S.; Yu, J.; Li, X.; Tian, W. Photoluminescence properties of mercaptocarboxylic acid-stabilized CdSe nanoparticles covered with polyelectrolyte. *Nanotechnology,* **2004**, *15*, 1108-1112.

[66] Zhang, Y.; Kuang, Y.; Gao, Y.; Xu, B. Versatile Small-Molecule Motifs for Self-Assembly in Water and the Formation of Biofunctional Supramolecular Hydrogels. *Langmuir*, **2011**, *27(2)*, 529-537.

[67] [a] Chen, X.; Rogach, A. L.; Talapin, D. V.; Fuchs, H.; Chi, L. Hierarchical Luminescence Patterning Based on Multiscaled Self-Assembly. *J. Am. Chem. Soc.,* **2006**, *128*, 9592-9593; [b] Colfen, H.; Mann, S. Higher-order organization by mesoscale self-assembly and transformation of hybrid nanostructures *Angew. Chem. Int. Ed.,* **2003**, *42*, 2350-2365; [c] Whang, D.; Jin, S.; Lieber, C. M. *Nano Lett.* Nanolithography Using Hierarchically Assembled Nanowire Masks. **2003**, *3*, 951-954; [d] Zhang, X.; Shen, Z.; Feng, C.; Yang, D.; Li, Y.; Hu, J.; Lu, G.; Huang, X. PMHDO-g-PEG Double-Bond-Based Amphiphilic Graft Copolymer: Synthesis and Diverse Self-Assembled Nanostructures. *Macromolecules.,* **2009**, *42*, 4249–4256; [e] Whitesides, G. M.; Grzybowski, B. Self-Assembly at All Scales. *Science,* **2002**, *295*, 2418–2421; [f] Ruokolainen, J.; Makinen, R.; Torkkeli, M.; Makela, T.; Serimaa, R.; Brinke, G. T.; Ikkala, O. Switching Supramolecular Polymeric Materials with Multiple Length Scales. *Science,* **1998**, *280*, 557–560.

[68] [a] Sahoo, S.; Husale, S.; Colwill, B.; Lu, T-M.; Nayak, S.; Ajayan, P. M. Electric Field Directed Self-Assembly of Cuprous Oxide Nanostructures for Photon Sensing. *ACS Nano,* **2009**, *3*, 3935–3944; [b] Lu, G.; Chen, Y.; Zhang, Y.; Bao, M.; Bian, Y.; Li, X.; Jiang, J. Morphology Controlled Self-Assembled Nanostructures of Sandwich Mixed (Phthalocyaninato)(Porphyrinato) Europium Triple-Deckers. Effect of Hydrogen Bonding on Tuning the Intermolecular Interaction. *J. Am. Chem. Soc.,* **2008**, *130*, 11623–11630; [c] Liu, R.; Holman, M. W.; Zang, L.; Adams, D. M. Single-Molecule Spectroscopy of Intramolecular Electron Transfer in Donor-Bridge-Acceptor Systems *J. Phys. Chem. A,* **2003**, *107*, 6522–6526; [d] Holman, M. W.; Liu, R.; Zang, L.; Yan, P.; DiBenedetto, S. A.; Bowers, R. D.; Adams, D. M. Studying and Switching Electron Transfer: From the Ensemble to the Single Molecule. *J. Am. Chem. Soc.,* **2004**, *126*, 16126– 16133; [e] Sauer, M. Single-molecule-sensitive fluorescent sensors based on photoinduced intramolecular charge transfer. *Angew. Chem., Int. Ed.,* **2003**, *42*, 1790–1793.

[69] Grimsdale, A. C.; Müllen, K. The chemistry of organic nanomaterials. *Angew. Chem., Int. Ed.,* **2005**, *44*, 5592–5629.

[70] Xu, Q.; Bao, J.; Rioux, R. M.; Perez-Castillejos, R.; Capasso, F.; Whitesides, G. M. Fabrication of Large-Area Patterned Nanostructures for Optical Applications by Nanoskiving. *Nano Lett.,* **2007**, *7*, 2800–2805.

[71] [a] Xu, B. Q.; Xiao, X.; Yang, X.; Zang, L.; Tao, N. J. Large Gate Modulation in the Current of a Room Temperature Single Molecule Transistor. *J. Am. Chem. Soc.,* **2005**, *127*, 2386–2387; [b] Li, X.; Xu, B. Q.; Xiao, X.; Yang, X.; Zang, L.; Tao, N. J. Controlling charge transport in single molecules using electrochemical gate. *Faraday Discuss.,* **2006**, *131*, 111–120; [c] Chen, Y.; Su, W.; Bai, M.; Jiang, J.; Li, X.; Liu, Y.; Wang, L.; Wang, S. High Performance Organic Field-Effect Transistors Based on Amphiphilic Tris(phthalocyaninato) Rare Earth Triple-Decker Complexes. *J. Am. Chem. Soc.,* **2005**, *127*, 15700–15701; [d] Li, R.; Ma, P.; Dong, S.; Zhang, X.; Chen, Y.; Li, X.; Jiang, J. Synthesis, Characterization, and OFET Properties of Amphiphilic Heteroleptic Tris(phthalocyaninato) Europium(III) Complexes with Hydrophilic Poly(oxyethylene) Substituents. *Inorg. Chem.,* **2007**, *46*, 11397–11404.

[72] [a] Schmidt-Mende, L.; Fechtenkotter, A.; Mullen, K.; Moons, E.; Friend, R. H.; MacKenzie, J. D. Self-Organized Discotic Liquid Crystals for High-Efficiency Organic Photovoltaics. *Science,* **2001**, *293*, 1119–1122; [b] Gregg, B. A. Excitonic Solar Cells. *J. Phys. Chem. B,* **2003**, *107*, 4688–4698; [c] Gregg, B. A. Evolution of

Photophysical and Photovoltaic Properties of Perylene Bis(phenethylimide) Films upon Solvent Vapor Annealing. *J. Phys. Chem.,* **1996**, *100*, 852–859; [d] Tamizhmani, G.; Dodelet, J. P.; Cote, R.; Gravel, D. Photoelectrochemical characterization of thin films of perylenetetracarboxylic acid derivatives. *Chem. Mater.,* **1991**, *3*, 1046–1053; [e] Li, Y.; Xiao, S.; Li, H.; Li, Y.; Liu, H.; Lu, F.; Zhuang, J.; Zhu, D. Self-Assembly and Characterization of A Novel Hydrogen-Bonded Nanostructure. *J. Phys. Chem. B,* **2004**, *108*, 6256–6260; [f] Peeters, E.; Van Hal, P. A.; Meskers, S. C. J.; Janssen, R. A. J.; Meijer, E. W. Photoinduced Electron Transfer in a Mesogenic Donor-Acceptor-Donor System. *Chem. Eur. J.,* **2002**, *8*, 4470–4474.

[73]　[a] Pan, D.; Turner, J. L.; Wooley, K. L. Folic Acid-conjugated Nanostructured Materials Designed for Cancer Cell Targeting. *Chem. Commun.,* **2003**, 2400–2401.; [b] Kataoka, K. J. Design of nanoscopic vehicles for drug targeting based on micellization of amphiphilic block copolymers. *Macromol. Sci., Pure Appl. Chem.,* **1994**, *A31*, 1759–1769; [c] Rosler, A.; Vandermeulen, G. W. M.; Klok, H. A. Advanced drug delivery devices *via* self-assembly of amphiphilic block copolymers. *Adv. Drug Delivery Rev.,* **2001**, *53*, 95–108; [d] Ahmed, F.; Discher, D. E. Self-porating polymersomes of PEG-PLA and PEG-PCL: hydrolysis-triggered controlled release vesicles. *J. Control. Release.,* **2004**, *96*, 37–53.

[74]　[a] Zhao, X.; Zhang, S. Molecular Designed Self- Assembling Peptides. *Chem. Soc. Rev.,* **2006**, *35*, 1105- 1110; [b] Ku, S.H.; Park, C.B. Highly Accelerated Self- Assembly and Fibrillation of Prion Peptides on Solid Surfaces. *Langmuir,* **2008**, *24*, 13822-13827; [c] Aulisa, L.; Dong, H.; Hartgerink, J.D. Self- Assembly of Multidomain Peptides: Sequence Variation Allows Control over Cross-linking and Viscoelasticity. *Biomacromolecules,* **2009**, *10*, 2694-2698; [d] Yang, H.; Pritzker, M.; Fung, S.Y.; Sheng, Y.; Wang, W.; Chen, P. Anion Effect on the Nanostructure of a Metal Ion Binding Self- Assembling Peptide. *Langmuir,* **2006**, *22*, 8553-8562; [e] Rajagopal, K.; Lamm, M.S.; Haines- Butterick, L.A.; Pochan, D.J.; Schneider, J.P. Tuning the pH Responsiveness of β-Hairpin Peptide Folding, Self- Assembly, and Hydrogel Material Formation. *Biomacromolecules,* **2009**, *10*, 2619-2625; [f] Zhao, Y.; Yokoi, H.; Tanaka, M.; Kinoshita, T.; Tan, T. Self- Assembled pH-Responsive Hydrogels Composed of the RATEA16 Peptide. *Biomacromolecules,* **2008**, *9*, 1511-1518.

[75]　[a] Sapsford, K.E.; Pons, T.; Medintz, I.L.; Higashiya, S.; Brunel, F.M.; Dawson, P.E.; Mattoussi, H. Kinetics of Meal- Affinity Driven Self- Assembly between Proteins or Peptides and CdSe-ZnS Quantum Dots. *J. Phys. Chem. C,* **2007**, *111*, 11528-11538; [b] Liu, J.; Mao, Y.; Lan, E.; Banatao, D.R.; Forse, G.J.; Lu, J.; Blom, H.O.; Yeates, T.O.; Dunn, B.; Chang, J.P. Generation of Oxide Nanopatterns by Combining Self- Assembly of S-Layer Proteins and Area- Selective Atomic Layer Deposition. *J. Am. Chem. Soc.,* **2008**, *130*, 16908-16913; [c] Kainz, B.; Steiner, K.; Moller, M.; Pum, D.; Schaffer, C.; Sleyter, U.B.; Toca-Herrera, J. L. Absorption, Steady-State Fluorescence, Fluorescence Lifetime, and 2D Self-Assembly Properties of Engineered Fluorescent S-Layer Fusion Proteins of *Geobacillus stearothermophilus* NRS 2004/3a. *Biomacromolecules,* **2010**, *11*, 207-214.

[76]　[a] Nguyen, H.D.; Reddy, V.S.; Brooks III, C.L. Deciphering the Kinetic Mechanism of Spontaneous Self-Assembly of Icosahedral Capsaids. *Nano Lett.,* **2007**, *7*, 338-344; [b] Sun, X.; Ko, S.H.; Zhang, C.; Ribbe, A.E.; Mao, C. Surface- Mediated DNA Self- Assembly. *J. Am. Chem. Soc.,* **2009**, *131*, 13248-13249; [c] Zhang, C.; He, Y.; Chen, Y.; Ribbe, A.E.; Mao, C. Aligning One- Dimensional DNA Duplexes into Two- Dimensional Crystals. *J. Am. Chem. Soc.,* **2007**, *129*, 14134-14135; [d] Johnson, R.R.; Johnson, A.T.C.; Klein, M.L. Probing the Structure of DNA- Carbon Nanotube Hybrids with Molecular Dynamics. *Nano Lett.,* **2008**, *8*, 69-75.

[77]　[a] Li, H.; Park, S.H.; Reif, J.H.; LaBean, T.H.; Yan, H. DNA- Templated Self- Assembly of Proteins and Nanoparticle Linear Arrays. *J. Am. Chem. Soc.,* **2004**, *126*, 418-419; [b] Qiao, R.; Ke, P.C. Lipid-Carbon Nanotube Self- Assembly in Aqueous Solution. *J. Am. Chem. Soc.,* **2006**, *128*, 13656-13657.

[78]　Menzenski, M.Z.; Banerjee, I.A. Self- assembly of supramolecular nanostructures from phenylalanine derived bolaamphiphiles. *New. J. Chem.,* **2007**, *31*, 1674- 1680.

[79]　[a] Bayburt, T. H.; Grinkova, Y. V.; Sligar, S. G. Self- Assembly of Discoidal Phospholipid Bilayer Nanoparticles with Membrane Scaffold Proteins. *Nano Lett.,* **2002**, *2*, 853-856; [b] Gazit, E. A Possible Role for π- Stackking in the Self- Assembly of Amyloid Fibrils. *FASEB J.,* **2002**, *16*, 77- 83; [c] Smoak, E.M.; Carlo, A.D.; Fowles, C.C.; Banerjee, I.A. Self- Assembly of Gibberellic Amide Assemblies and their Applications in the Growth and Fabrication of Ordered Gold Nanoparticles. *Nanotechnol.,* **2010**, *21*, 1- 10; [d] Hunter, C. A.; Lawson, K. R.; Perkins, J.; Urch, C. J. Aromatic Interactions. *J. Chem. Soc.,* **2001**, *2*, 651- 669.

[80]　[a] Kotera, M.; Lehn, J.M.; Vigneron, J.-P. Design and synthesis of complementary components for the formation of self-assembled supramolecular rigid rods. *Tetrahedron,* **1995**, *51*, 1953-1972; [b] Sessler, J. L.; Wang, B.; Harriman, A. Photoinduced energy transfer in associated, but noncovalently-linked photosynthetic model systems. *J. Am. Chem. Soc.,* **1995**, *117*, 704-714; [c] Springs, S. L.; Andrievsky, A.; Kral, V.; Sessler, J. L. Energy transfer in a supramolecular complex assembled *via* sapphyrin dimer-mediated dicarboxylate anion

chelation. *J. Porphyrins Phthalocyanines,* **1998**, *2,* 315-325; [d] Arimura, T.; Ide, S.; Sugihara, H.; Murata, S.; Sessler, J. L. A non-covalent assembly for electron transfer based on a calixarene-porphyrin conjugate: tweezers for a quinine. *New J. Chem.,* **1999**, *23,* 977-979; [e] Berg, A.; Shuali, Z.; Asano-Someda, M.; Levanon, H.; Fuhs, M.; Mo¨bius, K.; Wang, R.; Brown, C.; Sessler, J. L. A First High-Field EPR Study of Photoinduced Electron Transfer in a Base-Paired Porphyrin−Dinitrobenzene Supramolecular Complex. *J. Am. Chem. Soc.,* **1999**, *121,* 7433-7434; [f] Hunter, C. A.; Sanders, J. K. M. The nature of pi.-pi. interactions. *J. Am. Chem. Soc.* **1990**, *112,* 5525-5534; [g] Shimomura, M.; Olaf, K.; Ijiro, K. Tailoring of stacked π-electron arrays from electron—and/or energy donor—acceptor molecules based on two-dimensional supramolecular assemblies. *Synth. Methods,* **1996**, *81,* 251-257; [h] Zangmeister, R. A. P.; Smolenyak, P. E.; Drager, A. S.; O'Brien, D. F.; Armstrong, N. R. Transfer of Rodlike Aggregate Phthalocyanines to Hydrophobized Gold and Silicon Surfaces: Effect of Phenyl-Terminated Surface Modifiers on Thin Film Transfer Efficiency and Molecular Orientation. *Langmuir,* **2001**, *17,* 7071-7078.

[81]   [a] Guckian, K. M.; Schweitzer, B. A.; Ren, R. X. F.; Sheils, C. J.; Tahmassebi, D. C.; Kool, E. T. Factors Contributing to Aromatic Stacking in Water: Evaluation in the Context of DNA. *J. Am. Chem. Soc.,* 2000, *122,* 2213-2222; [b] Saenger, W. in *Principles of Nucleic Acid Structure,* Springer-Verlag: New York, **1984.**

[82]   Desiraju, G. R.; Gavezzotti, A. From molecular to crystal structure; polynuclear aromatic hydrocarbons. *J. Chem. Soc., Chem. Commun.,* **1989**, 621-623.

[83]   Burley, S. K.; Petsko, G. A. Weakly Polar Interactions In Proteins. *Adv. Protein Chem.,* **1988,** *39,* 125-192.

[84]   [a] Alexander, A. E. Monolayers of porphyrins and related compounds. *J. Chem. Soc.,* **1937,** 1813-1816; [b] Abraham, R.J.; Eivazi, F.; Pearson, H.; Smith, K. M. Mechanisms of aggregation in metalloporphyrins: demonstration of a mechanistic dichotomy. *Chem. Commun.,* **1976**, 698-699.

[85]   [a] Zhao, W.; Zheng, B.; Haynie, D. T. A Molecular Dynamics Study of the Physical Basis of Stability of Polypeptide Multilayer Nanofilms. *Langmuir,* **2006**, *22,* 6668–6675; [b] Zhang, L.; Zhao, W.; Rudra, J. S.; Haynie, D. T. Context Dependence of the Assembly, Structure, and Stability of Polypeptide Multilayer Nanofilms. *ACS Nano,* **2007**, *1,* 476-486.

[86]   Burchell, T. J.; Eisler, D. J.; Puddephatt, R. J. Self-Assembly Using Dynamic Coordination Chemistry and Hydrogen Bonding: Mercury(II) Macrocycles, Polymers and Sheets. *Inorg. Chem.,* **2004**, *43,* 5550-5557.

[87]   Wang, X.; Hammer, N. D.; Chapman, M. R. The Molecular Basis of Functional Bacterial Amyloid Polymerization and Nucleation. *J. Biol. Chem.,* **2008**, *283,* 21530-21539.

[88]   [a] Fourest, E.; Volesky, B. Contribution of Sulfonate Groups and Alginate to Heavy Metal Biosorption by the Dry Biomass of *Sargassum fluitans. Environ. Sci. Technol.,* **1996**, *30,* 277-282; [b] Ke, H. Y.; AndersonW.L.; Moncrief, R. M.; Rayson, G. D.; Jackson, P. J. Luminescence Studies of Metal Ion-Binding Sites on Datura innoxia Biomaterial. *Environ. Sci. Technol.,* **1994**, *28,* 586-591; [c] Schiewer, S.; Volesky, B. Modeling of the Proton-Metal Ion Exchange in Biosorption. *Environ. Sci. Technol.* **1995**, *29,* 3049-3058.

[89]   [a] Kumar, A.; Jakhmola, A. RNA-Templated Fluorescent Zn/PbS (PbS + Zn^{2+}) Supernanostructures. *J. Phys. Chem. C,* **2009**, *113, 9553–9559;* [b] Srivastava, S.; Verma, A.; Frankamp, B. L.; Rotello, V. M. Controlled assembly of protein-nanoparticle composites through protein surface recognition. *Adv. Mater.,* **2005**, *17,* 617–621.

[90]   [a] Kumar, A.; Jakhmola, A. RNA-Mediated Fluorescent Q-PbS Nanoparticles. *Langmuir,* **2007**, *23,* 2915–2918; [b] Feldheim, D. L.; Eaton, B. E. Selection of Biomolecules Capable of Mediating the Formation of Nanocrystals. *ACS Nano,* **2007**, *1,* 154-159.

[91]   [a] Dooley, C. J.; Rouge, J.; Ma, N.; Invernale, M.; Kelly, S. O. Nucleotide-stabilized cadmium sulfide nanoparticles. *J. Mater. Chem.,* **2007**, *17,* 1687–1691; [b] Green, M.; Symeth-Boyle, D.; Harries, J.; Taylor, R. Nucleotide passivated cadmium sulfide quantum dots. *Chem. Commun.,* **2005**, 4830–4832; [c] Kumar, A,; Kumar, V. Synthesis and Optical Properties of Guanosine 5'-Monophosphate-Mediated CdS Nanostructures: An Analysis of their Structure, Morphology, and Electronic Properties. *Inorg. Chem.,* **2009**, *48,* 11032–11037.

[92]   [a] Barbara-Guillem, E. US Patent 2002001716 A, 2001; [b] Chen, Y.; Rosenzweig, Z. Luminescent CdSe Quantum Dot Doped Stabilized Micelles. *Nano Lett.,* **2002**, *2,* 1299-1302.

[93]   Aryal, B. P.; Benson, D. E. Polyhistidine Fusion Proteins Can Nucleate the Growth of CdSe Nanoparticles, *Bioconj. Chem.,* **2007**, *18(2),* 585-589.

[94]   Kurachi, K.; Sieker, L. C.; Jensen, L. H. Metal ion binding in triclinic lysozyme. *J. Biol. Chem.,* **1975**, *250,* 7663-7667.

[95]   Ariki, M.; Lanyi, J. K. Characterization of Metal Ion-binding sites in Bacteriorhodopsin. *J. Biol. Chem.,* **1986**, *261,* 8167-8174.

[96] [a] Kutschabsky, L; Adam, G. Molecular and crystal structure of the phytohormone gibberellin A3. *J. Chem. Soc. Perkin. Trans.,* **1983**, 1653–1655; [b] Smith, K. L.; Winslow, A. E.; Petersen, D. E. Association reactions for poly(alkylene oxides) and polymeric poly(carboxylic acids). *Ind. Eng. Chem.,* **1959**, *51*, 1361-1364.

[97] [a] Sapra, S.; Sarma, D. D. Evolution of the electronic structure with size in II-VI semiconductor nanocrystals. *Phys. Rev. B: Condens. Matter.,* **2004**, *69*, 125304; [b] Sapra, S.; Rogach, A. L.; Feldmann, J. Phosphine-free synthesis of monodisperse CdSe nanocrystals in olive oil *J. Mater. Chem.,* **2006**, *16*, 3.

Smart Nanomaterials for Sensor Application, 2012, 111-125

Fabrication and Optimization of a Hydrogel Drug Delivery System for a Potential Wound Healing Application

Thomas J. Smith, James E. Kennedy and Clement L. Higginbotham[*]

Materials Research Institute, Athlone Institute of Technology, Dublin Rd, Athlone, Co. Westmeath, Ireland

Abstract: In this work, a two phase hydrogel was prepared by physically imbedding a xerogel in the core of a hydrogel that was subsequently freeze thawed. The outer hydrogel was prepared by freeze thawing poly (vinyl alcohol) (PVA) and poly (acrylic acid) (PAA) while the xerogels were prepared by UV polymerization of 1-vinyl-2-pyrrolidinone (NVP), acrylic acid (AA) and various percentages of an Active Pharmaceutical Ingredient (API). Attenuated total reflectance Fourier transform infrared spectroscopy (ATR-FTIR) confirmed that hydrogen bonding had occurred between the constituents of the two phase hydrogels, while swelling experiments in distilled water indicated that the swelling of the gels is temperature dependent. The rheological studies confirmed that, by incorporating PAA into the two phase hydrogel system the strength increased significantly. However, at temperatures under the same nominal force the hydrogels were observed to lose their physical structure. Thermal analysis suggested that the incorporation of API into the xerogel reduced the T_g by approximately 13 °C, thus suggesting that the API acts as a plasticizer within the xerogel matrix. In all cases, drug dissolution showed that the API was released at a slower rate from hydrogels that contained poly (acrylic acid).

Keywords: Hydrogels; poly (vinyl alcohol); poly (acrylic acid); drug dissolution; wound healing.

1. INTRODUCTION

Hydrogels are polymeric networks, which absorb and retain large amounts of water. This network contains hydrophilic groups or domains which become hydrated in an aqueous environment thereby creating the hydrogel structure [1]. Hydrogels resemble natural living tissue more than any other class of synthetic biomaterial due to their high water content and soft consistency [2]. The two phase hydrogels characterised in this chapter are physically crosslinked systems. These physical gels are networks comprised of an amorphous hydrophilic polymer phase held together by highly ordered aggregates of polymer chain segments arising from secondary molecular forces in conjunction with other types of molecular interaction. Unlike chemical gels, these types of hydrogel systems will eventually dissolve in water or solvents and can be melted by applying heat.

A novel approach to preparing physical crosslinks is the freeze thawing technique [3-6]. Smith *et al.* [3]. described a novel method of preparing a strong PVA/PAA hydrogel without utilisation of chemical crosslinking or reinforcing agents *via* a densification of the macro molecular structure. It was noted in Smith's contribution that the strength, stability and swelling ratio of the gels were a function of the solution concentration and freeze thaw profile. Therefore, this method of preparing hydrogels avoids the toxicity issues inherent in hydrogels prepared using chemical crosslinking agents.

In recent years, much research has been carried out on drug delivery devices capable of releasing an active agent at a constant rate over a long period of time [7]. However, not all therapeutic agents need to be released at a constant rate. Instead, it is essential to release certain drugs in a patterned manner, responding to the body's need for the drug. Controlled release devices can prolong the release of rapidly metabolised drugs, allowing effective dosages. They can also protect sensitive drugs, such as proteins, until they are delivered to the desired point of action.

*Address correspondence to Clement L. Higginbotham: Materials Research Institute, Athlone Institute of Technology, Dublin Rd, Athlone, Co. Westmeath, Ireland; Tel: 00353 90 6468050; Fax: 00353 90 6424493; E-mail: chigginbotham@ait.ie

Songjun Li, Yi Ge and He Li (Eds)
All rights reserved - © 2012 Bentham Science Publishers

In order to achieve an effective hydrogel delivery system, the swelling behaviour triggers the drug diffusion as water is absorbed to obtain zero order release [8, 9]. Physical entrapment is one of the simplest methods used for incorporating active agents into hydrogels that are intended for controlled drug delivery applications. With physical entrapment, the active agent is contained within the hydrogel structure to inhibit diffusion of the drug into the surrounding environment, i.e. there must be sufficient crosslinking or entanglements to ensure the solute remains in the hydrogel [10]. The Active Pharmaceutical Ingredient (API) chosen for this chapter was Diclofenac Sodium (DS). Diclofenac sodium is a potent nonsteroidal drug which has anti-inflammatory, analgesic and antipyretic properties. It is used for the treatment of wound healing and degenerative joint diseases such as rheumatoid arthritis and osteoarthritis [11, 12]. In this work Poly (1-vinyl-2-pyrrolidinone) (PVP) / Poly Acrylic Acid (PAA) was used to encapsulate the API within the hydrogel and to enhance the physical strength of the hydrogel. The toxicity of PVP has been extensively studied in a variety of species, including humans and other primates and has been proved to be of a very low order [6]. The objective of the present chapter was to develop a two phase hydrogel for a potential wound healing application. The inner gel (PVP and PAA) containing API was prepared by UV polymerisation and the outer gel (PVA/PAA) was prepared by freeze thawing. The primary objective for using a two phase hydrogel is to improve the mechanical properties of the hydrogel and to control the release of the API from the two phase hydrogel.

2. EXPERIMENTAL

2.1. Synthesis of Xerogels and Incorporation of Active Agent

The xerogels investigated in this work were prepared by free-radical polymerisation using ultra violet light. The monomers used were 1-vinyl-2-pyrrolidinone (NVP, Lancaster synthesis) and acrylic acid (AA, Merck-Schuchardt, Germany). The polymers tested had monomeric feed ratios of 100 wt.% NVP, 90 wt.% NVP/10 wt.% AA and 80 wt.% NVP/20 wt.% AA, where 100wt% equates to 1g. To initiate the reactions, 1-hydroxycyclohexylphenylketone (Irgacure® 184, Ciba speciality chemicals) was used as a UV-light sensitive initiator at 3 wt% of the total monomer weight. This was added to a 20 mL NVP/AA mixture and stirred continuously for 1 h. The API under investigation in this chapter was diclofenac sodium. Drug loadings (1wt%, 5wt%, 10wt%, and 15wt%) of the co-monomer content were added and stirred continuously for 30 min. Finally the solution was pipetted into a silicone mould that contained 20 disk impressions (1mL of the monomeric mixture was placed into each impression which had a diameter of 15mm). The mould was positioned horizontally to the gravity direction under two UVA 340 UV lamps (Q-panel products) and the solution was cured for 1 h in an enclosed environment. Samples were then dried in a vacuum oven at 40 °C, 500 mm Hg for 24 h prior to use.

2.2. Preparation of the Outer Hydrogel

The outer hydrogels were prepared by heating a known quantity of PVA (weight average molecular weight 146,000-186,000) mixture in 40 mL of H_2O to 80 °C for 1 h, while slowly stirring until the polymer is no longer apparent, after which time PAA (weight average molecular weight 3,000,000) was slowly added for 30 min. Both polymers used in this chapter were supplied by Aldrich. From previous studies the ratio for viable hydrogels was found to be 85% PVA and 15% PAA, where 85:15 equates to 1g PVA + 0.2g PAA. All outer hydrogels in this current chapter were synthesised in this ratio. To remove air bubbles the solution was placed in an ultra sonic bath for 5-10 min.

2.3. Preparation of the Two Phase Hydrogel

The two phase hydrogels system was prepared by placing 20 mL of the PVA/PAA solution into a polystyrene mould. The polystyrene mould was placed in approximately 300 mL of liquid nitrogen for a period of 10 min. The xerogel was then placed on top of the frozen hydrogel and the remaining 20 mL of the PVA/PAA solution was poured into the mould. The polystyrene mould was returned to the liquid nitrogen bath, again for a period of 10 min. Solidified solutions were then placed in a fridge to thaw at 3 °C for 24 h.

2.4. Differential Scanning Calorimetry

A DSC (2920 TA Instruments) containing a refrigerated cooling system was used to evaluate the gels. Approximately 8-12 mg samples were weighed out using a Sartorius scale capable of reading up to 5 decimal places. Aluminium pans were crimped before testing, with an empty crimped aluminium pan being used as the reference cell. Calorimetry scans were carried out from 20 to 220 °C at a scanning rate of 2 °C/min. Samples were taken from the xerogels and the hydrogels that had being dried in an oven at 37 °C for 24 h. The instrument was calibrated using indium as standard.

2.5. Parallel Plate Rheometry

Rheological measurements were performed using an Advanced Rheometer AR1000 (TA instruments) fitted with a Peltier temperature control. The samples were tested using a 40 mm diameter steel plate. The swollen two phase gels containing 98% water (diameter 40 mm and thickness 20 mm) were placed on the Peltier plate, and tests were carried out over a temperatures range of 30 to 80 °C at two-degree intervals. The tests were performed in an oscillation mode with a strain sweep of 1 Hz, 5 Hz, and 10 Hz. A normal force of 0.3 N was applied to the surface of the samples in order to avoid the slipping of the gel from the Peltier plate. In all cases, the force resulted in a slight compression of the samples.

2.6. Swelling Studies

The swelling characteristics of both xerogels and two phase hydrogels were investigated at 37 °C. Samples were placed in Petri dishes and were then filled with distilled water and placed in a fan oven at 37 °C. Petri dish lids were placed on the Petri dishes to prevent evaporation. Periodically, the gels were removed after predetermined time intervals. The samples were then blotted free of surface water with filter paper and the wet weight of the gel sample (in grams) was measured using a Sartorius scales capable of reading up to 5 decimal places. The xerogels and the two phase hydrogel samples were re-submerged in fresh distilled water (40-150mL) and returned to the oven at 37 °C. Swelling percentages were calculated using the formula, as outlined in Equation 1.

$$\text{Swelling} = \frac{W_t - W_0}{W_0} \times 100 \qquad\qquad \textbf{Eq. (1)}$$

where W_t is the mass of the gel at a predetermined time and W_0 is the initial weight of the gel. This process was continued for up to 16 h.

2.7. Attenuated Total Reflectance Fourier Transform Infrared Spectroscopy (ATR-FTIR)

The ATR-FTIR spectroscopy was carried out using the Attenuated Total Reflectance (ATR) mode on a Nicolet Avator 360 (FTIR) with a 32 scan per sample cycle and a resolution of 8 cm^{-1}. The samples were scanned from 400 to 4000 cm^{-1}.

2.8. Drug Dissolution Analysis

Drug dissolution studies were conducted on selected samples using a Sotax® AT7 Smart on-line dissolution system (Carl Stuart Ltd). The tests were carried out in triplicate using the Paddle method (test speed 100rpm) at 37 °C. The wavelength and absorption of a 100% drug concentration was determined using a Perkin Elmer Lambda 40 UV/Vis spectrometer. These values were entered into software calculations prior to commencement of testing. The UV absorbance of each sample was recorded at specific pre-programmed times which generated the release profile of the drug eluted sample. The tests were repeated in triplicate.

3. RESULTS AND DISCUSSION

Due to the similarities in results, the DSC, rheometry, ATR-FTIR and swelling results were discussed on the gels that contained 10% diclofenac sodium.

3.1. Thermal Analysis of the Hydrogel and the Two Phase Hydrogel System

Due to the high water content (94-98 wt %) in the hydrogel samples, the T_g values of the polymers was masked by the presence of water. Therefore, all samples were dried at 37 °C for 48 h before testing.

Moisture testing confirmed the water content in the hydrogels was 5% (± 0.5%) after drying. As presented in Fig. 1 (**A** and **B**) are the thermal transitions for the neat PVA and PAA respectively. The transitions obtained in Fig. 1 (**A**) confirmed the T_g and T_m of PVA at 80.56 °C and 203.18 °C, while the transition observed in Fig. 1 (**B**) confirmed the T_g of PAA at 78.89 °C. As PAA degrades before it reaches its melting temperature no value was obtained [14]. As illustrated in Fig. 2, three transitions were evident for the freeze thawed hydrogel containing 1g PVA + 0.2g PAA. Their transitions correspond to the T_g (80.62 °C) for PVA, a secondary relaxation (124.06 °C) which is due to the relaxation of the crystalline domains and the melting point (204.49 °C) for PVA [13].

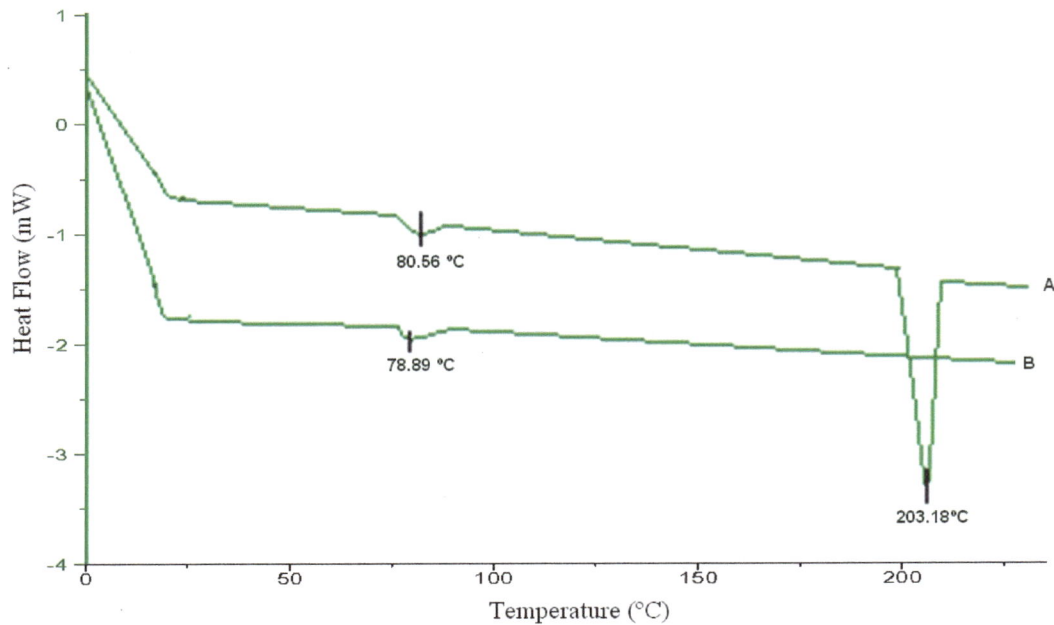

Figure 1. Thermal transition for (**A**) neat PVA and (**B**) neat PAA.

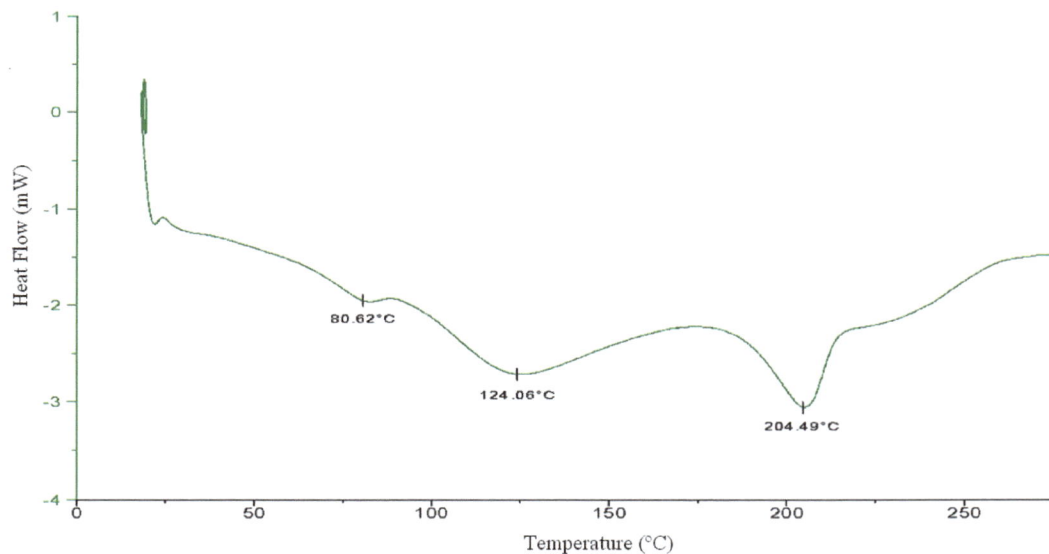

Figure 2. Thermal transition for the hydrogel in the ratio of 85:15% PVA/PAA.

DSC analysis was also performed in order to determine the effect the addition of diclofenac sodium (see Fig. **3**) had on the properties of PVP. Yaung and Kwei [15] performed DSC thermograms on PVP complexes and found a glass transition (T_g) in the order of 146 °C. As illustrated in Fig. **3(A)** the T_g of 100% PVP was found to be 146 °C correlating well with the results obtained by Yaung[15]. It can be seen from the DSC thermograms obtained, the T_g of each of the samples were above 130 °C which potentially has no impact on the behaviour of the gel at body temperature, this does however, give an indication of the stiffness of the dry polymer. As presented in Fig. **3 (D, E and F)** the DSC analysis confirmed the incorporation of API into the xerogels decreased the T_g of PVP by approximately 13 °C when compared to Fig. **3 (A, B and C)** which contained no API, thus indicating the API acts as a plastciser. It can also be seen that there appears to be no secondary transitions in any of these polymers in the region tested *i.e.* 20 – 200 °C. In an earlier study carried out by Wu & McGinity [16] a plasticising effect was observed when API's were incorporated into the polymeric system resulting in the T_g being lowered. Nair *et al.* [17] incorporated an API into a solvent casted PVP homopolymer and found a linear decrease in T_g as the drug concentrations were increased.

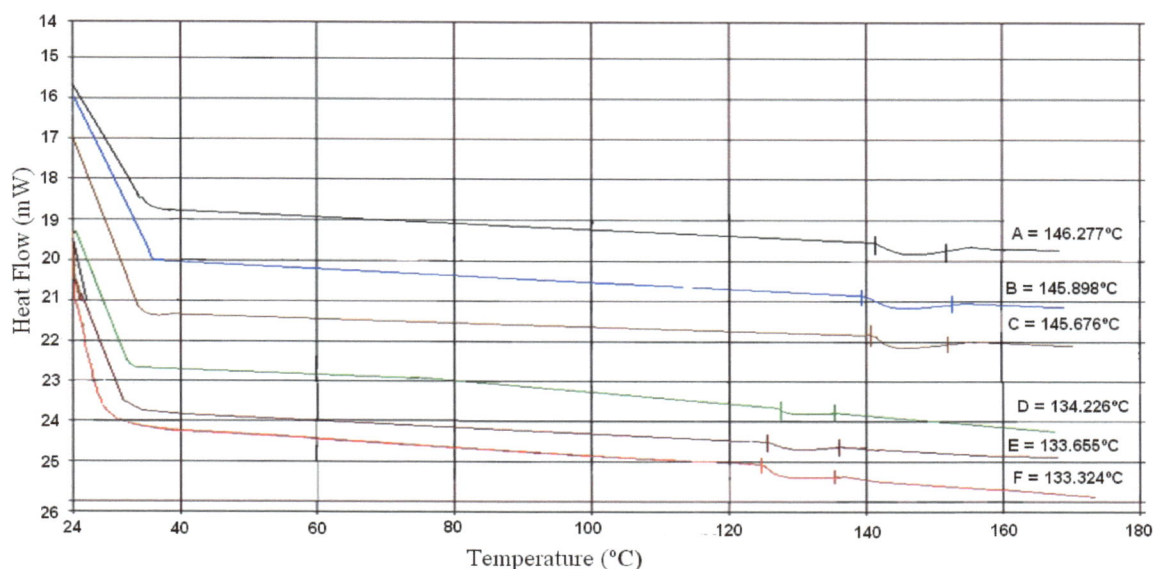

Figure 3. Thermal transition for the xerogel containing **(A)** 100% PVP with no drug, **(B)** 90/10 PVP/PAA with no drug, **(C)** 80/20 PVP/PAA with no drug, **(D)** 100% PVP with 10% diclofenac sodium, **(E)** 90/10 PVP/PAA with 10% diclofenac sodium and **(F)** 80/20 PVP/PAA with 10% diclofenac sodium.

3.2. Rheological Analysis of the Two Phase Hydrogel System

The temperature performance of the hydrogels is critical in determining the viscoelastic behaviour with respect to the storage modulus (G') and loss modulus (G"). G' is in phase with the solid and is called the storage modulus because it defines the energy stored in the specimen due to the applied strain. G" which is $\pi/2$ out of phase with the strain defines the dissipation of energy and is called the loss modulus [18]. This chapter was concerned only with the energy (G') stored in the sample, and observing how this energy affects the strength of the hydrogels. As presented in Fig. **4**, the gel containing ratios of 85:15% PVA/PAA confirmed that G' is much greater than G", when analysed between 30 and 75 °C. The gel showed the G' value, which corresponds to the strength of the hydrogel ranging between 150 and 200 Pa at a frequency of 1, 5, and 10 Hz.

It is evident that as the temperature is increased, the gels weaken which corresponds with the findings of Smith *et al.* [3].

As illustrated in Fig. **6**, the two phase gel containing ratios of 85:15% PVA/PAA and the inner gel containing 100% PVP with 10% diclofenac sodium, shows an increase in storage modulus of 300 Pa when compared to the two phase gel that contained 100% PVP with no drug (Fig. **5**). The strength of the gel gradually decreased as the temperature increased from 30 to 80 °C. At this point, the gel had completely broken down into an aqueous solution, which is due to heat energy breaking the crosslinks present in the hydrogel. The solubility of PVA/PAA in water increases greatly as its degree of hydrolysis increases.[19] Properties such as water solubility and high elasticity make these hydrogels useful as skin adhesives for the development of a wound healing device. With the incorporation of PAA within the xerogel, the strength of the two phase system increased (0-200 Pa) as presented in Fig. **7**. By varying the PAA concentration (Fig. **8**) the strength was further increased to over 700 Pa. In comparison to the pure PVA/PAA hydrogel (Fig. **4**), the two phase hydrogel system (Fig. **8**) increased in strength by approximately 1400 Pa. Hydrogen bonding between the carboxylic acid group of the API and the carboxylic acid group of PAA in the copolymer is believed to be the primary cause of the increase in strength.

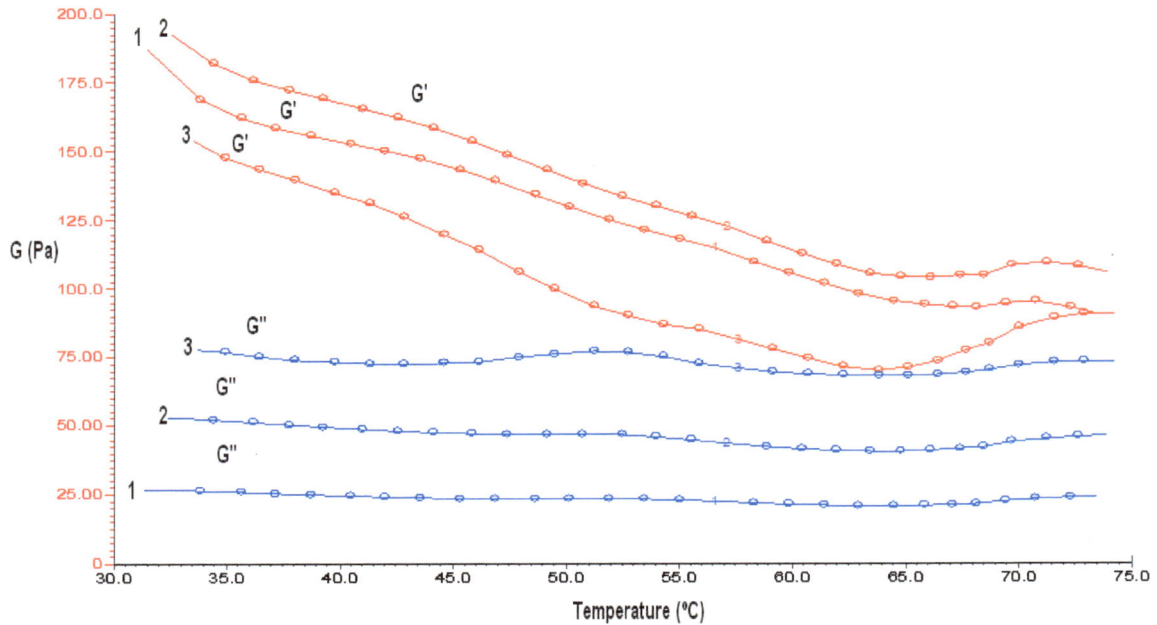

Figure 4. Rheological results for the hydrogel in the ratio of 85:15% PVA/PAA, where the storage modulus (G') 1, 2, and 3 represents 1, 5 and 10 Hz respectively.

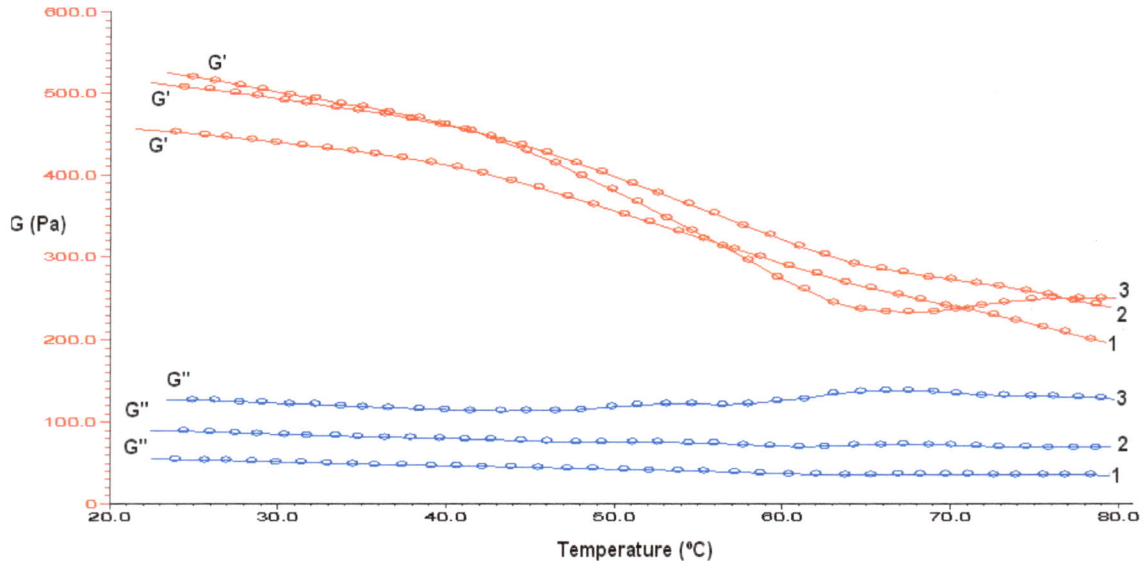

Figure 5. Rheological results for the two phase hydrogel. The outer gel contains the ratio 85:15% PVA/PAA and the inner gel contains 100% PVP, where the storage modulus (G') 1, 2, and 3 represents 1, 5 and 10 Hz respectively.

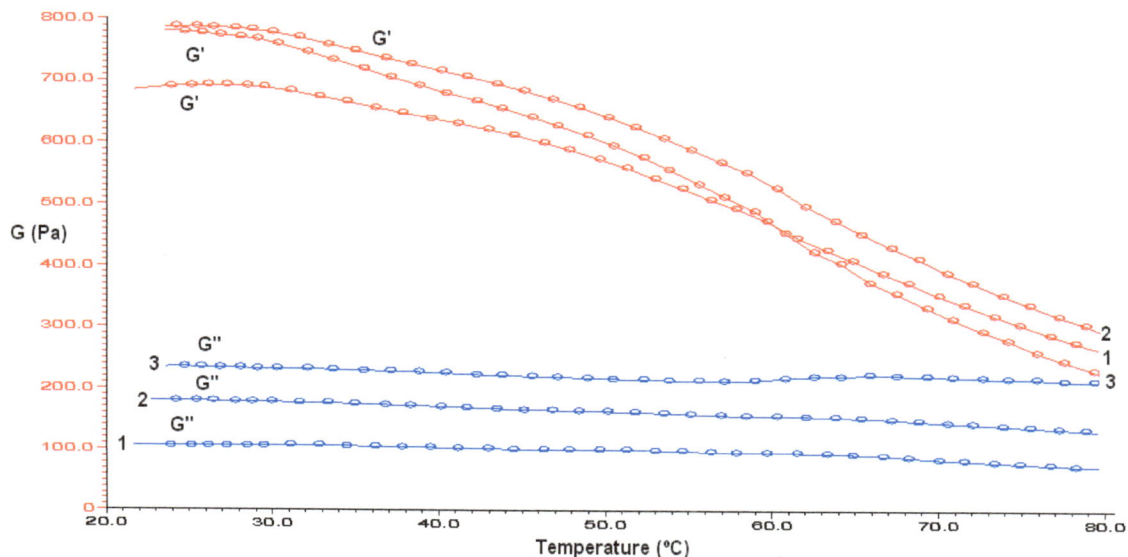

Figure 6. Rheological results for the two phase hydrogel. The outer gel contains the ratio 85:15% PVA/PAA and the inner gel contains 100% PVP with 10% diclofenac sodium, where the storage modulus (G') 1, 2, and 3 represents 1, 5 and 10 Hz respectively.

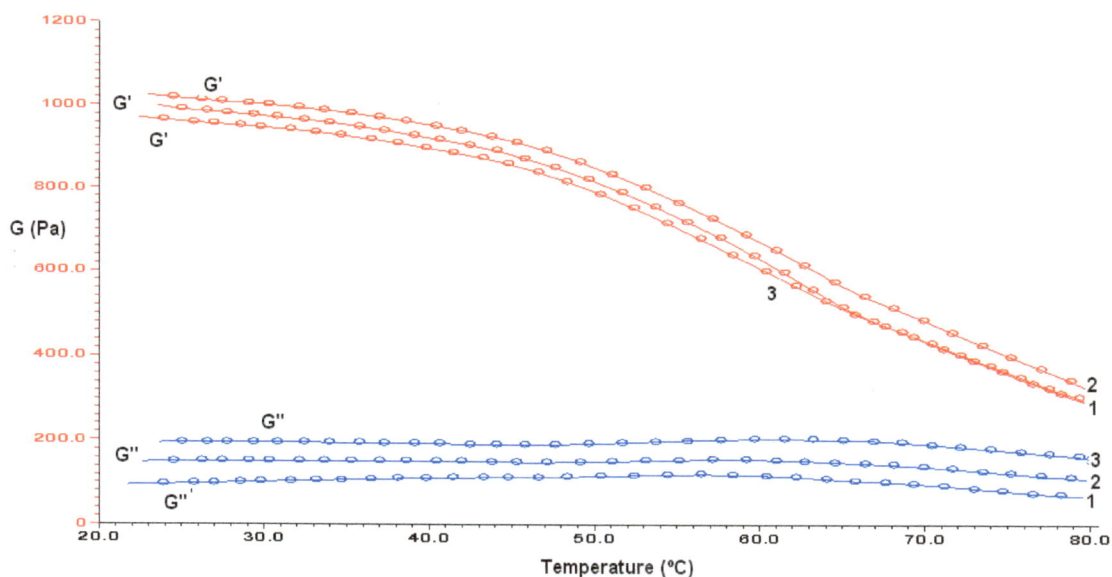

Figure 7. Rheological results for the two phase hydrogel. The outer gel contains the ratio 85:15% PVA/PAA and the inner gel contains (90/10) PVP/PAA with 10% diclofenac sodium, where the storage modulus (G') 1, 2, and 3 represents 1, 5 and 10 Hz respectively.

3.3. Attenuated Total Reflectance Fourier Transform Infrared Spectroscopy (ATR-FTIR) for the Two Phase Hydrogels

IR spectroscopy has been used to characterise both the polymerisation systems used to produce hydrogels, as well as the resulting hydrogel polymers [5]. In this chapter ATR-FTIR was used to confirm the hydrogen bonding in the two phase hydrogel. With reference to Fig. **9 (A)**, for the pure PVA/PAA hydrogel, the peak at 3,311 cm^{-1} can be attributed to the O-H bond of PVA, the strong peak at 1,645 cm^{-1} shows the C=O stretching bonds of PAA [20]. As presented in Fig. **9 (B)**, the ATR-FTIR result for the two phase hydrogel containing 100% PVP with no drug confirmed a peak at approximately 1,670 cm^{-1} which was associated

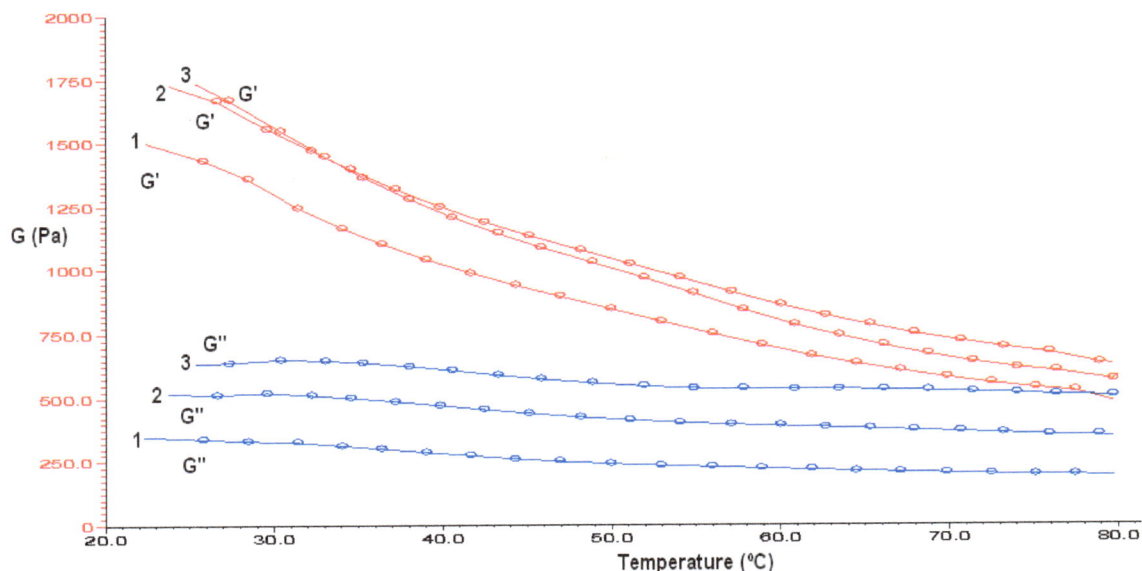

Figure 8. Rheological results for the two phase hydrogel. The outer gel contains the ratio 85:15% PVA/PAA and the inner gel contains (80/20) PVP/PAA with 10% diclofenac sodium, where the storage modulus (G') 1, 2, and 3 represents 1, 5 and 10 Hz respectively.

with the carbonyl group of PVP, while the peak at 3,093 cm^{-1} represents the O-H bond of PVA. As illustrated in Fig. **9 (C)** the two phase gel containing 100% PVP with 10% diclofenac sodium, confirms a peak at 3,387 cm^{-1} which represent the N-H stretching of the secondary amine for diclofenac sodium Shivakumar *et al.* [21]. The peak at 3,088 cm^{-1} can be attributed to the O-H bond of PVA; this peak appears at 3,475 cm^{-1} for pure PVA [22]. Hassan *et al.* [5] found that PVA/PAA gels possessed an absorption peak at 1,198 – 1,225 cm^{-1}, which confirms crystallinity within the PVA/PAA structure, this is due to the C-C stretching mode increasing with an increase in the degree of crystallinity. The peak observed at 1,222 cm^{-1} in the two phase system tested indicates that crystallinity occurred. The peak observed at 1,721 cm^{-1} represents the C=O for diclofenac sodium, again showing excellent correlation with Shivakumar *et al.* [21]. As presented in Fig. **9 (D),** the two phase gel containing 90/10 PVP/PAA with 10% diclofenac sodium confirms a peak at 3,079 cm^{-1} which is characteristic of the stretching frequency of the O-H bond of PVA, while the peak observed at 2,304 cm^{-1} indicates the O-H overtone vibrations of PAA. The peak observed at 3,385 cm^{-1} again confirms the N-H stretching of the secondary amine for diclofenac sodium. Formation of hydrogen bonding is exhibited on the IR spectrum as a negative shift of the stretching vibration of the functional group involved in the hydrogen bond, which is typically a carbonyl group. Yaung and Kwei [23] describe how the frequency of the PVP carbonyl group shifts from 1,670 to 1,640 cm^{-1} when it forms hydrogen bonds with the carboxyl group of the PAA. In this work, the PVP carbonyl group was found to be at 1,650 cm^{-1} with 90/10 PVP/PAA and shifted to 1,622 cm^{-1} with 80/20 PVP/AA, thus proving hydrogen bonding occurred in the complexes. There is also evidence that the PAA molecules formed cyclic rings through hydrogen bonding of the carboxylic acid side groups. An increase in intermolecular hydrogen bonding between the carboxylic acid groups of the PAA segments was observed as the percentage of PAA increased. The peaks observed in Fig. **9 (D & E)** (1,700 and 1,696 cm^{-1}) respectively represents the C=O for diclofenac sodium. It is evident that with the incorporation of PAA into the two phase hydrogel these peaks move to lower wave numbers, suggesting bonding interactions with the polymers. Therefore, only the unbound or weakly bound molecules are released from the hydrogel. A shift to lower wave numbers for the N-H stretching of the secondary amine is again evident in Fig. **9 (E)**, for the two phase gel containing 80/20 PVP/PAA. The peak observed at 3,384 cm^{-1} represents the N-H stretching of the secondary amine [21] while the peak at 3,031 cm^{-1} is characteristic of the O-H bonds of PVA. Finally the peak at 2,299 cm^{-1} is attributed to the O-H overtone of PAA.

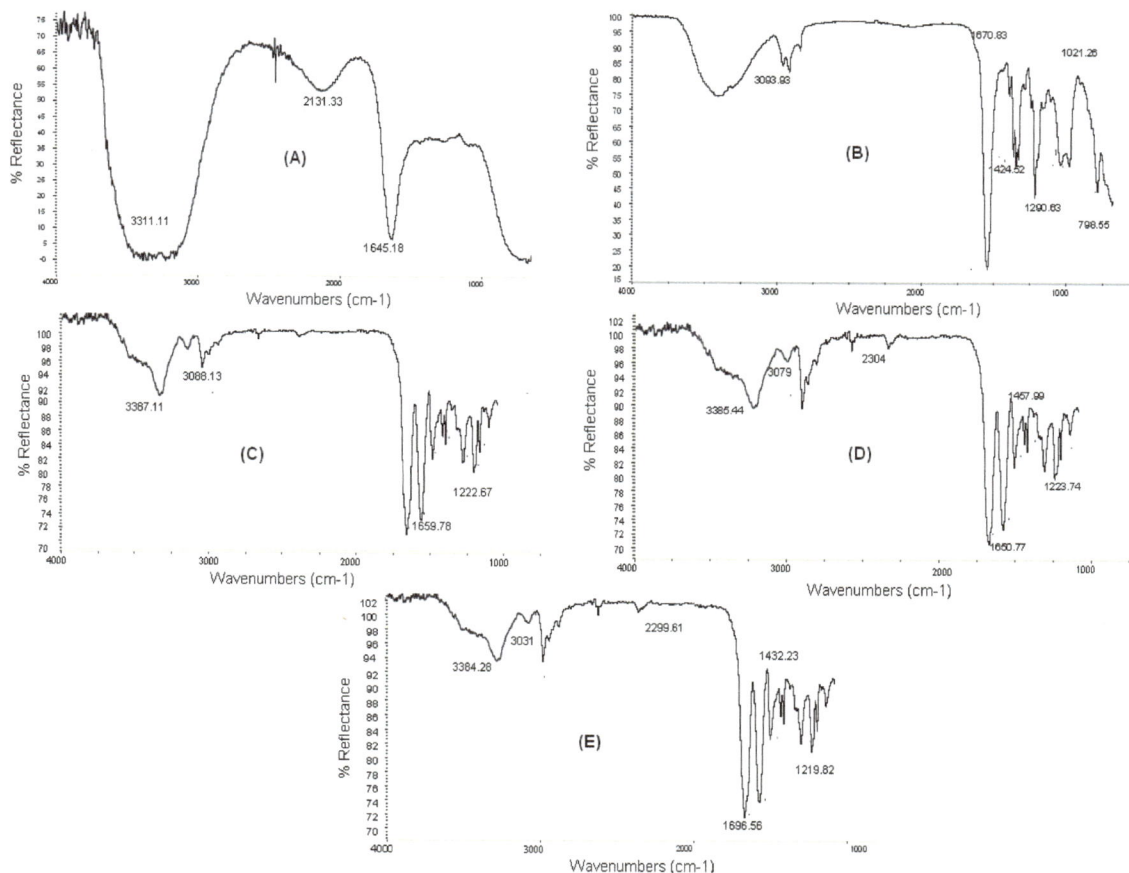

Figure 9. The ATR-FTIR results for the hydrogel and the two phase hydrogels containing **(A)** 85:15% PVA/PAA,**(B)** 85:15% PVA/PAA with 100% PVP **(C)** 85:15% PVA/PAA with 100% PVP and 10% diclofenac sodium, **(D)** 85:15% PVA/PAA with 90/10 % PVP/PAA and 10% diclofenac sodium and **(E)** 85:15% PVA/PAA with 80/20 % PVP/PAA and 10% diclofenac sodium.

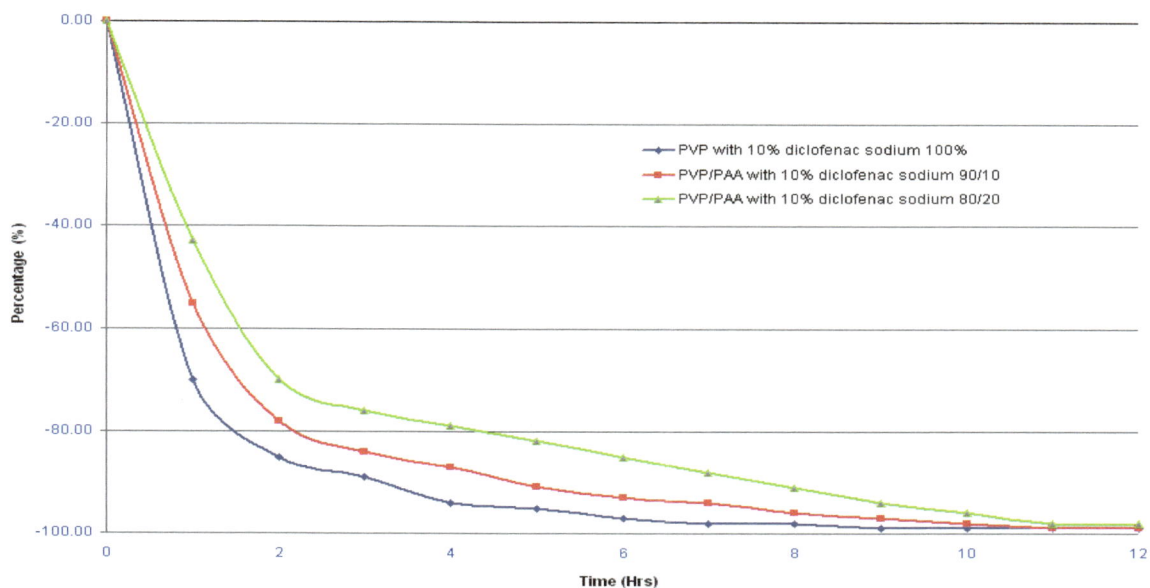

Figure 10. Dissolution results for the xerogels that contain various concentrations of PVP/PAA and 10% diclofenac sodium.

3.4. Swelling Behavior of the Xerogels and the Two Phase Hydrogels at 37 °C

Due to the hydrogels been prepared in distilled water (pH 6.9) and the pH of blood ranging from 6.9 to 7.2, it was decided to carry out the experiments using distilled water. As illustrated in Fig. **10**, dissolution experiments were carried out on the xerogels that contained various concentrations of PVP and PVP/PAA. On inspection of the samples in distilled water it was found that the sample prepared from PVP and PVP/PAA dissolved rapidly and returned to an aqueous solution within 6 h. The samples that contained PAA appeared to dissolve slower than would be expected. In an earlier study carried out by Devine *et al.* [24] rapid dissolution of the PVP and PVP/PAA gels within 2 - 6 h was reported. In that study, it was found that as the PAA content was increased the hydrogels did not dissolve as quickly. This is due to the increase in intermolecular bonding caused by the addition of PAA. These findings were supported by the rheological studies, where, by increasing the doses of PAA in the two phase hydrogels, the hydrogels increased in strength. With reference to Fig. **11**, it is evident that the PAA slows the degree of swelling when compared to the xerogels which contained no concentration of PAA. With respect to the gel that contained 100% PVP, a degree of swelling of 64% was achieved over the16 h period, however, the gel prepared using 80/20 PVP/PAA confirmed a degree of swelling of 58% over the 16 h period. It is thought that an increase in hydrogen bonding between the carboxylic acid groups of the PAA and the diclofenac sodium was the primary cause for the slow swelling profile.

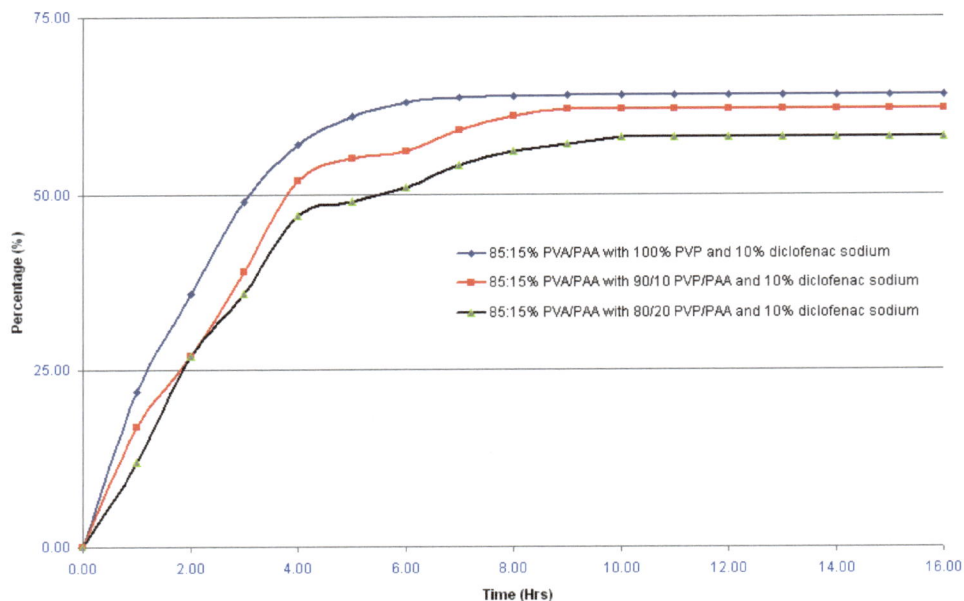

Figure 11. Swelling results for the two phase hydrogels. The outer gel contains the ratio 85:15% PVA/PAA and the inner gel contains various concentrations PVP/PAA with 10% diclofenac sodium.

3.5. Drug Dissolution Analysis of the Xerogels and Two Phase Hydrogel System

The objective of an ideal drug release system is to generate a zero order release profile where the amount of drug released is independent of the amount of drug in the system [25]. Within the context of this chapter, zero order release was obtained when diclofenac sodium (1, 5, 10 and 15wt %) was used in the two phase hydrogel system. Diclofenac sodium in powder form was initially tested to establish the dissolution properties in distilled water. The API was found to be water-soluble after 10 min (see Fig. **12**). The xerogels loaded with 1, 5, 10 and 15wt% diclofenac sodium were first analysed and found to release 100% API after approximately 5 h as can be seen in Fig. **13-15**. From further analysis it was evident that the release of the diclofenac sodium had become retarded by the incorporation of PAA. As illustrated in Fig. 13, the xerogel containing 100% PVP confirmed a 100% release of API after approximately 90 min. In comparison (Fig. **15**), the xerogel containing 80/20 PVP/PAA confirmed a slower release rate of up to 5 h. The primary cause of the retardation is believed to be due to a level of bonding between the diclofenac sodium and the PAA in the copolymer.

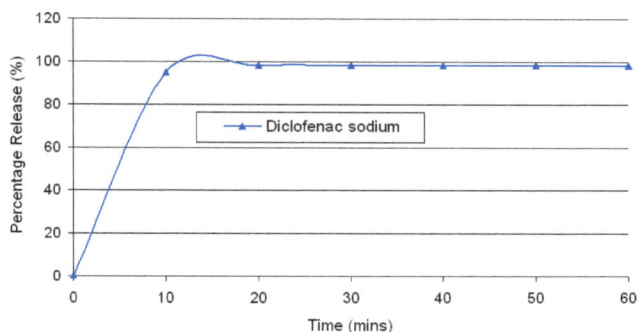

Figure 12. Dissolution behaviour of diclofenac sodium powder in distilled at 37 °C.

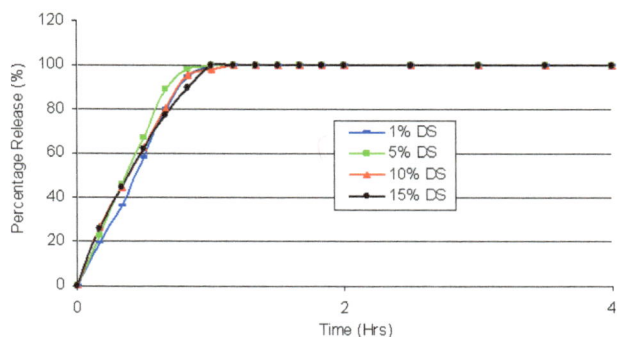

Figure 13. Drug dissolution results for the xerogel containing 100% PVP with various percentages diclofenac sodium.

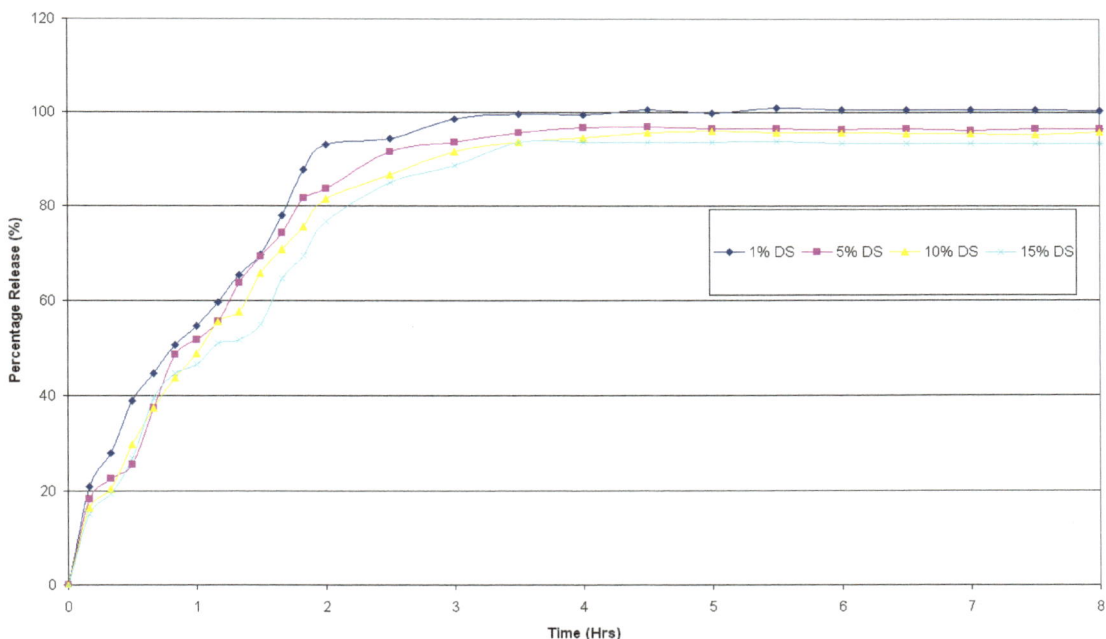

Figure 14. Drug dissolution results for the xerogel containing 90/10 PVP/PAA with various percentages of diclofenac sodium.

As presented in Fig. **16–18**, the drug dissolution results for the two phase hydrogel systems, confirmed a more controlled release of API over a 24 h period. The two phase hydrogel system which contained 100% PVP (Fig. **16**) showed 100% release of API after approximately 8 h. This was a significant reduction in release time when compared to the xerogel with the same compositions of API (see Fig. **13**). As presented in Fig.**17** and **18** the addition of PAA (90/10 PVP/PAA and 80/20 PVP/PAA) illustrated a release up to 100% after 10 and 12 h respectively. Again, this release delay can most probably be attributed to the aforementioned hydrogen bonding between the carboxylic acid group of the diclofenac sodium and the carboxylic acid group of poly acrylic acid. The increase in hydrogen bonding was confirmed in the ATR-FTIR results which exhibited a negative shift in the IR spectra.

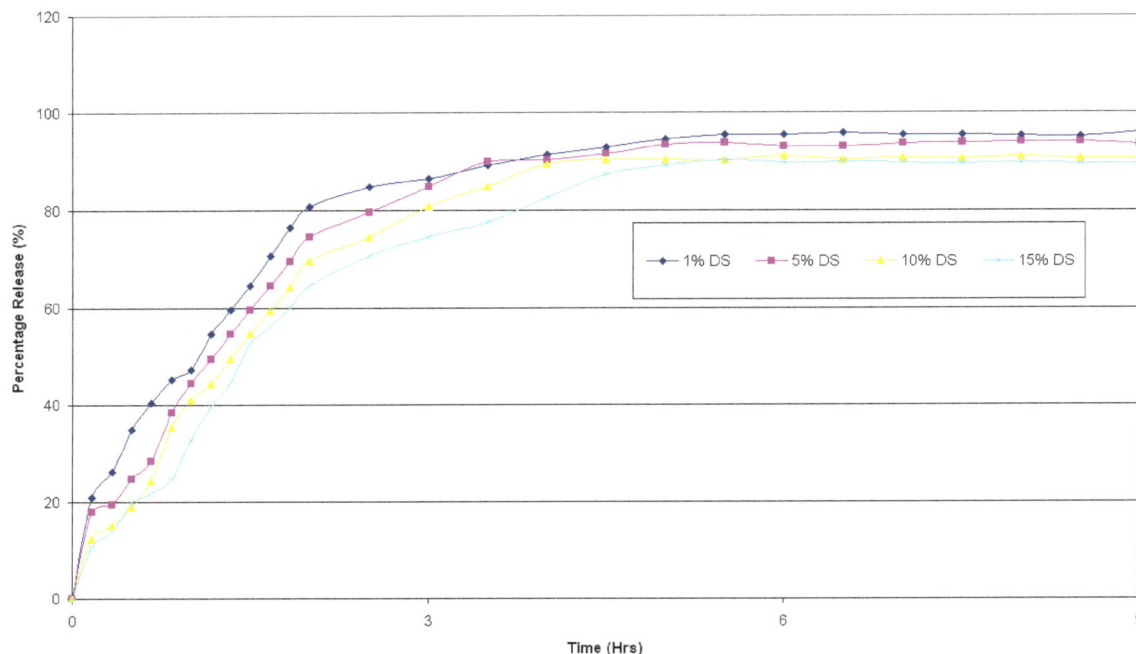

Figure 15. Drug dissolution results for the xerogel containing 80/20 PVP/PAA with various percentages of diclofenac sodium.

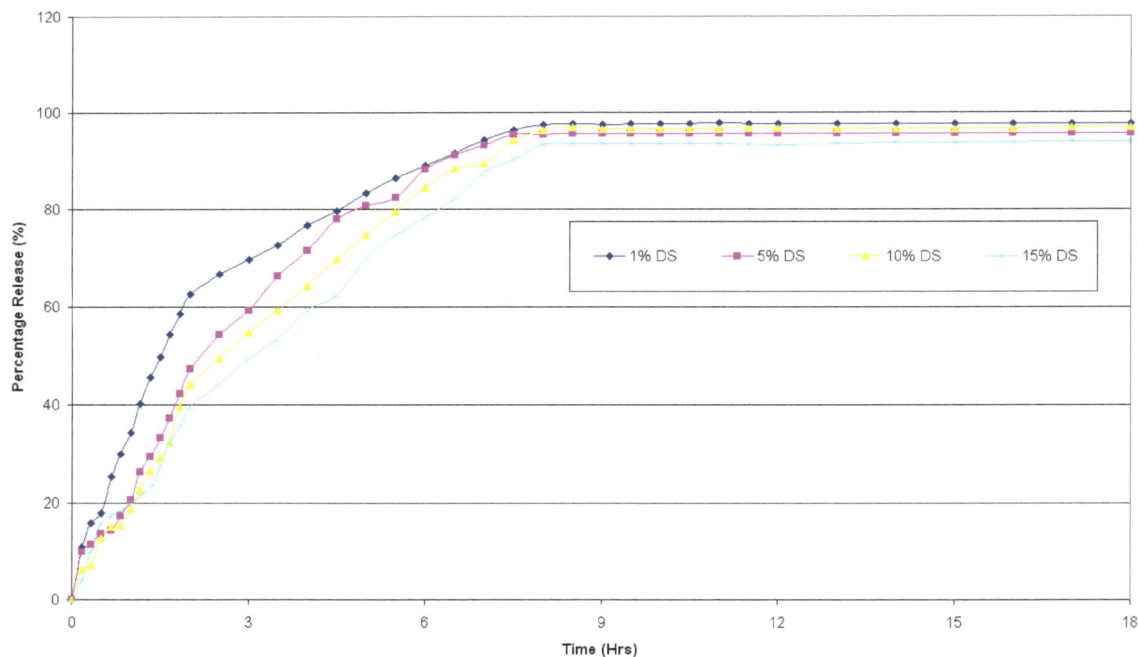

Figure 16. Drug dissolution results for the two phase hydrogel. The outer gel contains the ratio 85:15% PVA/PAA and the inner gel contains 100% PVP with various percentages of diclofenac sodium.

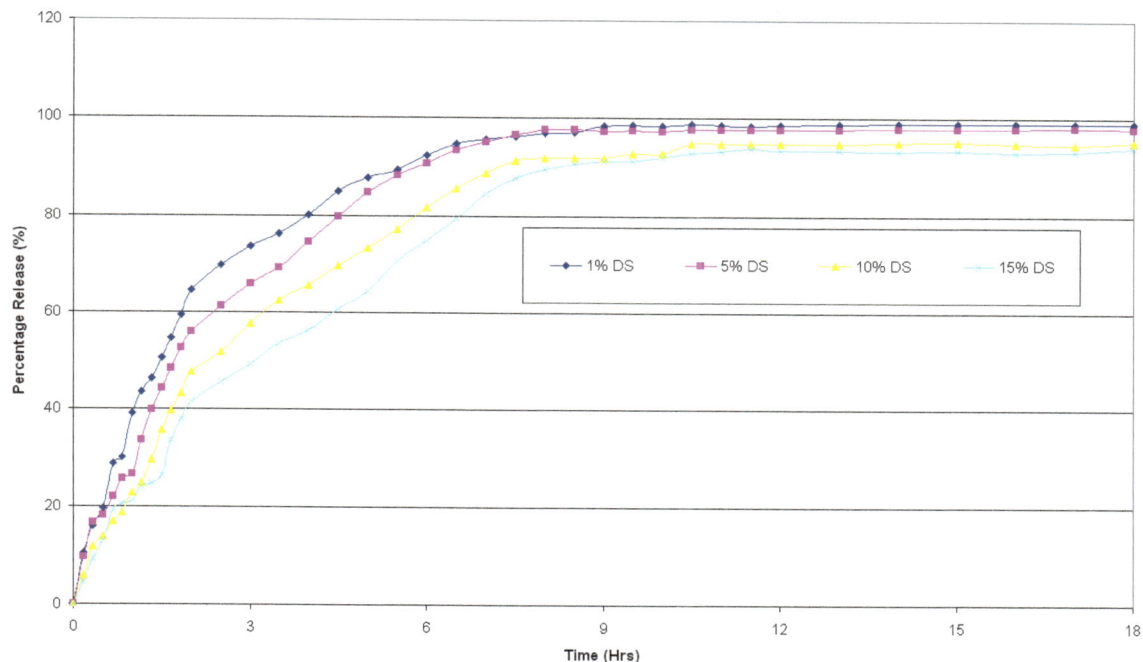

Figure 17. Drug dissolution results for the two phase hydrogel. The outer gel contains the ratio 85:15% PVA/PAA and the inner gel contains 90/10 PVP/PAA with various percentages of diclofenac sodium.

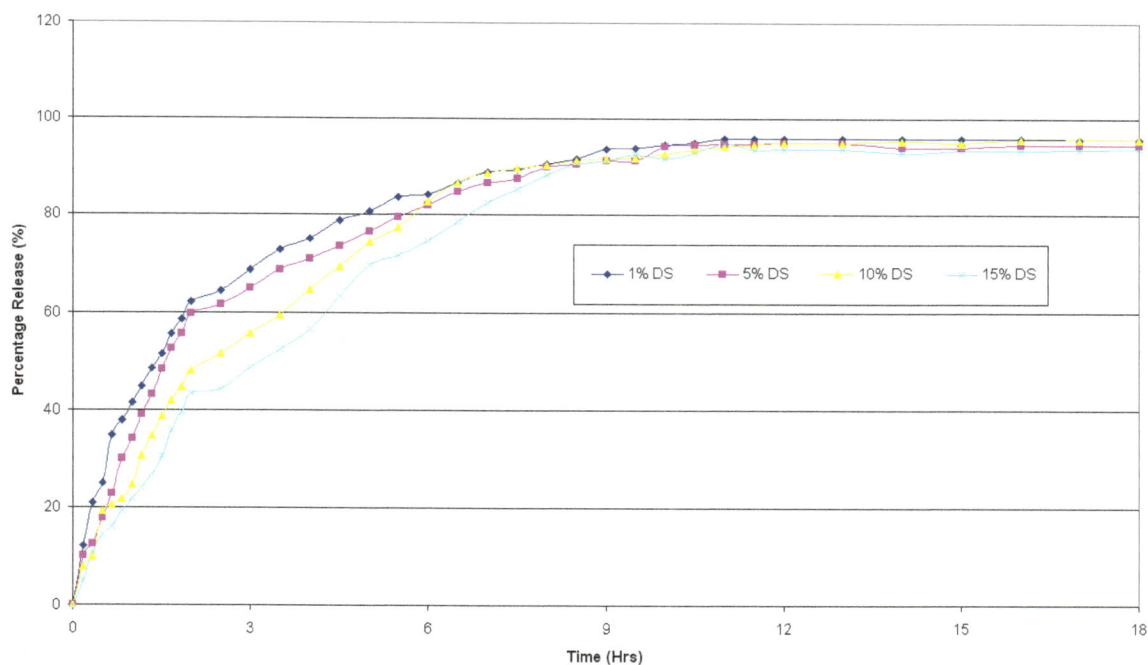

Figure 18. Drug dissolution results for the two phase hydrogel. The outer gel contains the ratio 85:15% PVA/PAA and the inner gel contains 80/20 PVP/PAA with various percentages of diclofenac sodium.

4. CONCLUSIONS

In this chapter a novel two phase hydrogel has been synthesised and evaluated for use as a potential wound healing device. The ATR-FTIR spectroscopy results confirmed hydrogen bonding occurred between the carboxylic acid group of the diclofenac sodium and the carboxylic acid group of poly acrylic acid which, resulted in a negative shift of the carbonyl group. The rheological studies confirmed that, by incorporating PAA into the two phase hydrogel system the strength increased by approximately 1400 Pa. However, at temperatures above 35 °C, under the same nominal force, the hydrogels were observed to lose their physical structure.

The swelling profiles confirmed the xerogels samples dissolved rapidly and returned to an aqueous solution within 6 h, while the two phase hydrogel systems swelled by approximately 60%. Drug dissolution analysis showed that the incorporation of PAA had obvious effects on the rate of drug release from the xerogels and the two phase hydrogel systems. The release rates showed 100% release of API after 5 h in the xerogels while the two phase hydrogel system confirmed a 100% release after 12 h. The work has indicated that the addition of PAA to the two phase hydrogels is an effective technique for improving mechanical properties and controlling the release of the API from the two phase hydrogel system. Additionally due to the biocompatible nature of both polymeric components, these hydrogels have potential in biomedical applications particularly in the area of wound healing.

ACKNOWLEDGEMENTS

This study was supported in parts by grants from both Enterprise Ireland and the Athlone Institute of Technology research and development fund.

REFERENCES

[1]   Hennink, W.E.; Van Nostrum, C.F. Novel crosslinking methods to design hydrogels. *Adv. Drug. Deliv. Rev.*, **2002**, *54*, 13–36.
[2]   Ratner, B.D.; Hoffman, A.S. Hydrogels for medical and related applications. In: ACS Symposium Series, American Chemical Society, Washington, DC, **1976**, *31*.
[3]   Smith, T.; Kennedy, J.; Higginbotham, C. The rheological and thermal characteristics of freeze-thawed hydrogels containing hydrogen peroxide for potential wound healing applications. *J. Mech. Behav*, **2009**, 264-271.
[4]   Stauffer, S.R.; Peppas, N.A. Poly (vinyl alcohol) hydrogels prepared by freezing-thawing cyclic processing. *Polymer*, **1992**, *33*, 3932–3936.
[5]   Hassan, C.H.; Stewart, J.E.; Peppas, N.A. Diffusional characteristics of freeze/thawed poly (vinyl alcohol) hydrogels: applications to protein controlled release from multilaminate devices. *Eur J Pharm Biopharm*, **2000**, *49*, 161–165.
[6]   Hassan, C.M.; Peppas, N.A. Structure and applications of poly (vinyl alcohol) hydrogels produced by conventional crosslinking or by freezing/thawing methods. *Adv. Polym. Sci.*, **2000**, *153*, 37–65.
[7]   Baker, R.W. *Controlled Release of Biologically Active Agents*, John Wiley and Sons, New York; **1987**, pp. 75-78, 179-183.
[8]   Hoffman, A. Conventional and environmentally-sensitive hydrogels for medical and industrial uses: a review paper, *ACS. Polym. Preprints,* **1991**, *31*, 289-297.
[9]   Am Ende, M.T.; Hariharan, D.; Peppas N.A. Factors influencing drug and protein transport and release from ionic hydrogels, *React. Polym.,* **1995**, *25*, 127-137.
[10]   Ward, J.W.; Peppas, N.A. Preparation of controlled release systems by free-radical UV polymerizations in the presence of a drug, *J. Control. Release*, **2001**, *71*, 183-192.
[11]   Kendall, M.; Thornhill D.; Willis J. Factors affecting the pharmacokinetics of diclofenac sodium. *Rheumatol Rehabil Suppl.,* **1979**, *2*, 38-43.
[12]   Willis, J.; Kendall, M.; Flinn, R.; Thornhill, D.; Welling P. The pharmacokinetics of Diclofenac sodium following intravenous and oral administration. *Eur. J. Clin. Pharmacy.,* **1979**, *16*, 405-413.
[13]   Park, J.S.; Park, J.W.; Ruckenstein, E. Thermal and dynamic mechanical analysis of PVA/MC blend hydrogels. *Polymer,* **2001**, *42*, 4271-4280.
[14]   Kaczmarek, H.; Kaminska, A.; Swiatek, M.; Rabek, J.F. Photo-oxidative degradation of some water-soluble polymers in the presence of accelerating agents. *Angew. Macromol. Chem.*, **1998**, *261/262*, 109–121.
[15]   Yaung, J.F.; Kwei, T.K. pH sensitive hydrogels based on polyvinylpyrrolidinone-polyacrylic acid (PVP-PAA) semi interpenetrating networks (semi-IPN): swelling and controlled release. *J Appl Poly Sci,*. **1997**, *69*, 921-930.

[16]   Wu, C., Mc Ginity, J.W. Non-traditional plasticization of polymeric films. *Int. J.Pharm*, **1999**, *177*, 15-27.

[17]   Nair, R.; Nyamweya, N.; Gönen, S.; Martínez-Miranda, L.J.; Hoag, S.W. Influence of various drugs on the glass transition temperature of poly (vinylpyrrolidone): a thermodynamic and spectroscopic investigation. *Int. J.Pharm*, **2001**, *225*, 83-96.

[18]   Ward, I.M.; Hadley, D.W. An introduction to the mechanical properties of solid polymers hichester, John Wiley & Son, **1992**.

[19]   Peppas, N.A. Hydrogels of poly (vinyl alcohol) and its copolymers in hydrogels in Medicine and Pharmacy. **1987**, *2*, Polymers. Boca Raton, FL, CRS Press.

[20]   Peppas, N.A.; Bures, P.; Leobandung, W.; Ichikawa, H. Hydrogels in pharmaceutical formulations. *Eur. J. Pharmaceut. Biopharmaceut.*, **2000**, *50*, 27-46.

[21]   Shivakumar, H.N.; Desai, B.G.; Deshmukh, G. Design and optimization of diclofenac sodium controlled release solid dispersions by response surface methodology. *Indian. J. Pharm. Sci.*, **2008**, *70*, 22-30.

[22]   Yeom, C.K.; Huang, R.Y.M. *Angew, Makromol Chem.*, **1991**, *184*, 27-40.

[23]   Yaung, J.F.; Kwei, T.K. pH-sensitive hydrogels based on polyvinylpyrrolidone–polyacrylic acid (PVP–PAA) semi-interpenetrating networks (semi-IPN) swelling and controlled release. *J Appl Polym. Sci.*, **1998**, *69*, 921-30.

[24]   Devine, D.M.; Devery, S.M.; Lyons, J.G.; Geever, L.M.; Kennedy, J.E.; Higginbotham, C.L. The synthesis of a physically crosslinked NVP based hydrogel. *International Journal of Pharmaceutics*, **2003**, *44*, 7851-7860.

[25]   Sershen, S.; West J. Implantable polymeric systems for modulated drug delivery. *Adv. Drug. Deliv. Rev.*, **2002**, *54*, 1225-1235.

Smart Nanomaterials for Sensor Application, 2012, 126-148

Advanced Carbon Nanotubes and Carbon Nanotube Fibers for Biosensing Applications

Zhigang Zhu[1*], Andrew J. Flewitt[1], William I. Milne[1] and Francis Moussy[2]

[1]*Electrical Engineering Division, Department of Engineering, University of Cambridge, 9 JJ Thomson Avenue, Cambridge, CB3 0FA, UK and* [2]*Brunel Institute for Bioengineering, Brunel University, Uxbridge, Middlesex, UB8 3PH, UK*

Abstract: Since the discovery of Carbon Nanotubes (CNTs) by Iijima in 1991[1, 2], there has been an explosion of research into the physical and chemical properties of this novel material. CNT based biosensors can play an important role in amperometric, immunosensor and nucleic-acid sensing devices, *e.g.* for detection of life threatening biological agents in time of war or in terrorist attacks, saving life and money for the NHS. CNTs offer unique advantages in several areas, like high surface-volume ratio, high electrical conductivity, chemical stability and strong mechanical strength, and CNT based sensors generally have higher sensitivities and lower detection limit than conventional ones. In this review, recent advances in biosensors utilising carbon nanotubes and carbon nanotube fibres will be discussed. The synthesis methods, nanostructure approaches and current developments in biosensors using CNTs will be introduced in the first part. In the second part, the synthesis methods and up-to-date progress in CNT fibre biosensors will be reviewed. Finally, we briefly outline some exciting applications for CNT and CNT fibres which are being targeted. By harnessing the continual advancements in micro and nano- technology, the functionality and capability of CNT-based biosensors will be enhanced, thus expanding and enriching the possible applications that can be delivered by these devices.

Keywords: Advanced carbon nanotubes; fibres; biosensors.

1. INTRODUCTION OF CARBON NANOTUBES

Biosensors are a type of bio-molecular probe that measure the concentration of biological molecules by transducing a biochemical interaction into a quantifiable electrical signal. Biosensors hold the promise of early diagnosis of diseases and genetic disorders through detection of associated molecules such as DNA, enzymes, proteins and peptide aptamers. There is intensive interest in the use of nanomaterials for such applications, and CNTs are one of the most promising materials due to the following properties: i) large length-to-diameter aspect ratios produce high surface-to-volume ratios, and are thus responsible for the efficient capture and promotion of electron transfer reactions from analytes [3]; ii) carbon nanotubes can be functionalized by most biomolecules both at the tube ends and on the walls, which allows the design of novel biosensors or enhances the solubility of the tubes in a polymer matrix through adjusting the hydrophobicity or hydrophilicity of the surface; iii) the conductivity of the tubes is critical in controlling electron transfer kinetic for use in biosensors as an electrode. For example, Multi-Wall Carbon Nanotubes (MWNTs) are metallic conductors and can be used directly as electrodes in electrochemical biosensors; however, the semi-conductor Single-Wall Carbon Nanotubes (SWNTs) are ideal For Nano-Scale Field-Effect transistors (FETs) to detect single molecules [4, 5]. But the separation and purification of CNTs are real problems and haven't been completely solved.

CNTs are well-ordered, hollow graphitic nanomaterials made of cylinders of sp^2- hybridized carbon atoms, and they have two main structural types. SWNTs consist of a single cylindrical tube of graphite sheet, while MWNTs comprise an array of such nanotubes that concentrically nest like rings of a tree trunk to share a common longitudinal axis [6-8]. As shown in Fig. **1**, SWNTs can be either metallic or semiconducting,

---

*Address correspondence to Zhigang Zhu: Electrical Engineering Division, Department of Engineering, University of Cambridge, 9 JJ Thomson Avenue, Cambridge, CB3 0FA, UK; Tel: +44 01223 748304; Fax, +44 1223 748348; Email: zz259@cam.ac.uk

Songjun Li, Yi Ge and He Li (Eds)
All rights reserved - © 2012 Bentham Science Publishers

depending on the direction in which the sheets roll to form the tube cylinders. The direction and the nanotube diameter are obtainable from a pair of integers (n, m) that denote the nanotube type. Depending on the appearance of a belt of carbon bonds around the nanotube diameter, the nanotube is either of the achiral armchair $(n = m)$, achiral zigzag $(n = 0$ or $m = 0)$, or chiral (any other n and m) type. All achiral armchair SWNTs are metals; while chiral and achiral zigzag tubes are semi-conducting unless the vector indices give a whole number when the calculation (n-m)/3 is performed [9]. Thus, with small diameter SWNTs, approximately two-thirds are semi-conducting and one-third are metallic. The electrical properties between MWNTs and SWNTs are similar, since the coupling between the cylinders is weak in MWNTs. In the meantime, electrical transport in metallic SWNTs and MWNTs occurs ballistically (*i.e.*, without scattering) along the nearly one-dimensional electronic structure, enabling them to carry high currents with essentially no heating [3].

Apart from the conductivity, another important structural aspect of CNTs is their local anisotropy. The sidewalls, comprising the vast majority of the tubes, are a relatively inert layer of sp^2 hybridized carbon atoms and are highly hydrophobic. The open ends of CNTs have carbon atoms bonded to oxygen to give far more reactive species, which are quite hydrophilic and critical for the good electrochemical properties of nanotubes. This hydrophobicity presents a major challenge when it comes to dispersing and manipulating carbon nanotubes to give controlled modification of electrode surfaces [10]. There is a tendency to rapidly coagulate both in aqueous solution and polar solvents, although Sano *et al.* have reported that acid shortening can extend the storage time for homogenous CNT solution [11]. Meanwhile, dispersing CNTs is usually performed in non-polar organic solvents such as in Dimethylformamide (DMF) [12, 13] or with the aid of surfactants or polymers [14]. The difficulty in dispersing nanotubes in aqueous solutions has sometimes been used as an advantage in preparing nanotube modified electrodes, where nanotubes dispersed in an organic solvent are dropped onto an electrode surface following solvent evaporation [10, 11].

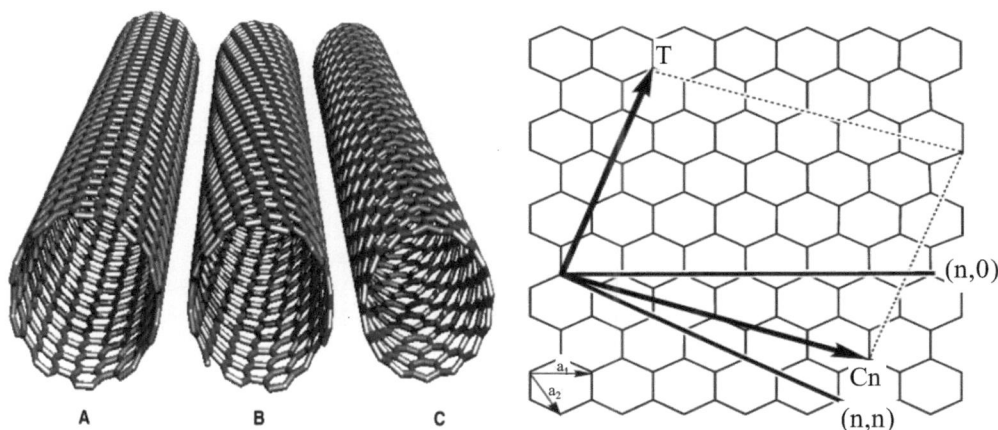

Figure 1. Idealized representation of defect-free (n,m) SWNTs with open ends. **A)** A metallic conducting (10,10) tube ("armchair"), **B)** a chiral, semiconducting (12,7) tube, and **C)** a conducting (15,0) tube ("zigzag"). **D)** Schematic representation of a 2D graphite layer with the lattice vectors a_1 and a_2, and the roll-up vector $C_h = na_1 + ma_2$. Achiral tubes exhibit roll-up vectors derived from (n,0) (zigzag) or (n,n) (armchair). The translation vector T is parallel to the tube axis and defines the 1D unit cell. The rectangle represents an unrolled unit cell, defined by T and Ch. In this example, (n,m) =(4,2), Reprinted with permission from [9].

1.1 CNTs Synthesis Methods

Three main methods are used to synthesize SWNTs and MWNTs: arc-discharge, laser-ablation and Chemical Vapour Deposition (CVD), and each method will be considered here.

1.1.1. Arc-Discharge

Arc-discharge is the easiest and most common way to produce CNTs [15]. Indeed, CNTs were firstly discovered by Japanese scientist, Iijima, who was planning to utilize an arc-discharge method to synthesize fullerenes [1]. As shown in the Fig. **2a**, the chamber consists of one carbon stick at the cathode and the other at the anode, where inert gases are helium or argon. The gap between the anode and cathode can be reduced to less than 1 mm by movement of the anode, and creates plasma by 100 A current passing through the electrode. This process typically causes the temperature of the plasma to be as high as 4000 K, and thus carbon on the anode is vaporized and deposited on the cathode or the chamber wall.

MWNT synthesis by the arc discharge technique is straightforward if two graphite electrodes are introduced, however, a great amount of side products, like fullerenes, amorphous carbon and graphite sheets, are simultaneously formed, which cause difficulty and increase cost to meet the need for purification. SWNTs can be produced by incorporating a metal catalyst on the cathode or anode [2, 16]. The metal concentration, type and pressure of inert gas, current *etc.* will greatly affect the quality of tubes. The arc-discharge is the simplest and most common method to synthesize CNTs, especially for MWNTs. However, the purification of tubes is costly and time-consuming and may damage the CNTs.

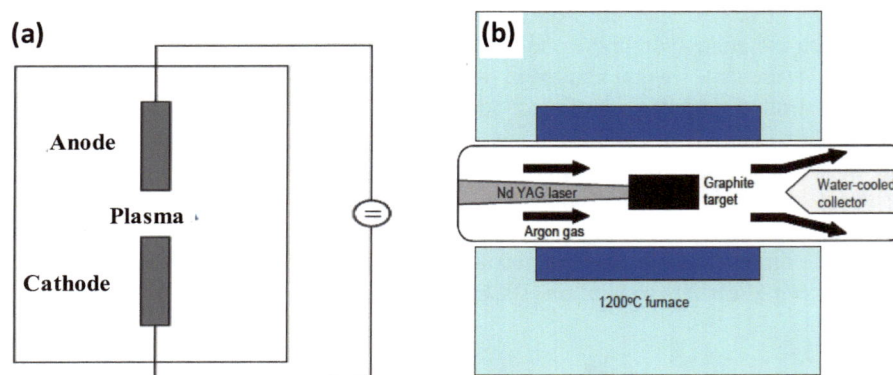

Figure 2. (a) Schematic diagram for Arc-discharge setup, two graphite electrodes are used to produce a dc electric arc-discharge in inert gas atmosphere. **(b)** Schematic diagram for laser ablation apparatus. Reprinted with permission from [15].

1.1.2. Laser-Ablation

This was first introduced by Guo *et al.* [17] when direct laser vaporization of transition-metal/graphite composite rods produced Single-Walled Carbon Nanotubes (SWNTs) in the condensing vapour in a heated flow tube. An oven laser-vaporization apparatus is illustrated in Fig. **2b**. A pulsed or continuous laser beam was introduced into a 1200 °C furnace to vaporize a target, which is made of graphite and metal catalysts (cobalt or nickel), in the presence of helium or argon gas. The growing mechanism was suggested by Scott and co-workers in 2001 [18]. A very hot vapour plume is formed that expands and cools rapidly. As the vaporized species cool, small carbon molecules and atoms quickly condense to form larger clusters, possibly including fullerenes. The catalysts also begin to condense, but more slowly at first, and attach to carbon clusters and prevent them closing into cage structures. Catalysts may also open cage structures when they attach to them. From these initial clusters, tubular molecules grow into single-wall carbon nanotubes until the catalyst particles become too large, or until conditions have cooled sufficiently that carbon no longer can diffuse through or on the catalyst particles. It is also possible that the particles coated with a carbon layer can no longer absorb more carbon and nanotube growth ceases.

The merit of the laser-ablation method can be summarized as following: i) relatively high purity SWNTs can be synthesized, ii) a lower temperature furnace can be used with a CO_2 infrared laser system, iii) the quality of CNTs is tunable by adjusting the nature of gas and its pressure. On the other hand, it only suitable for the growth of SWNTs, and is limited to laboratory scale system. Compared with the arc-discharge method, both techniques produce reasonable yields (*ca.* 70%) of CNTs with impurities in the form of amorphous carbon and catalyst particles because of the high temperature of the heated source. The CNTs obtained from both these techniques are tangled and thus post growth purification is essential.

1.1.3. Chemical Vapour Deposition (CVD) method

Chemical Vapour Deposition (CVD) synthesis is a better technique for high yield and low impurity production of CNT arrays at moderate temperature, which can be achieved by putting a carbon source in the gas phase and transferring energy to the gaseous carbon molecule by selecting an energy source, such as a plasma or resistively heated coil. A typical CVD system is illustrated in Fig. **3**.

The CVD method was firstly used by Yacaman *et al.* [19] and Ivanov *et al.* [20] to produce MWNTs, and involves the decomposition of gaseous carbon sources, including methane, carbon monoxide and acetylene, which are transformed into reactive atomic carbon. Then, the carbon diffuses towards the substrate, which is heated and coated with a catalyst. The most common catalysts used for CVD are iron, nickel and cobalt. Carbon has a low solubility in these metals at high temperature and thus CNTs arrays with excellent alignment can be grown perpendicular to the substrate. There are two main types of CVD: thermal CVD and plasma enhanced CVD (PECVD).

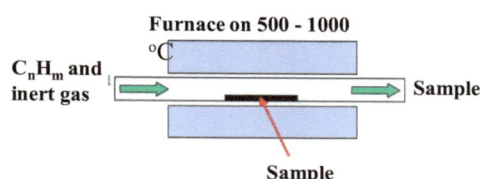

Figure 3. Schematic diagrams for CVD apparatus, Hydrocarbon gas is decomposed in a quartz tube in a furnace at 550–750 °C over a transition metal catalyst (a CVD reactor). Reprinted with permission from [15].

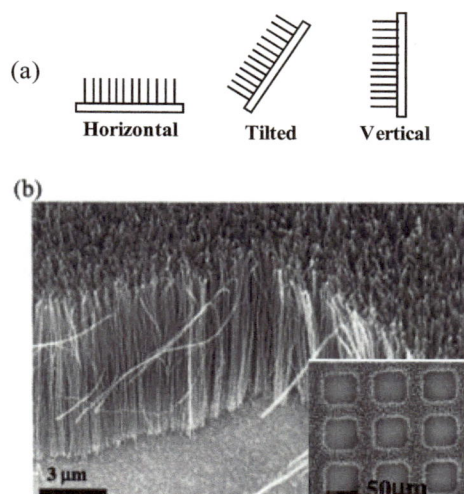

Figure 4. (a) Schematics showing the perpendicular alignment of nanotube growth regardless of the substrate position. **(b)** An SEM image that shows the growth of well aligned nanotubes. The inset shows patterned nanotube growth. Reprinted with permission from [25].

In thermal CVD, there is essentially a two-step process consisting of a catalyst preparation step followed by the actual synthesis of the nanotube. The catalyst is generally prepared by sputtering a transition metal onto a substrate and then using either chemical etching or thermal annealing to induce catalyst particle nucleation. Thermal annealing results in cluster formation on the substrate, from which the nanotubes will grow. Ammonia may be used as the etchant. The temperatures for the synthesis of nanotubes by CVD are generally within the 650–1100 °C range. A higher CVD temperature has positive effects on cyrstallinity of the structure produced and can also result in higher growth rates. However, if the CVD temperature is too high, pyrolytic decomposition becomes significant, which is detrimental to CNT formation as amorphous carbon is deposited, and this can result in the undesirable and excessive deposition of amorphous carbon either on the catalyst particles or on the growing carbon structures. In some applications, well aligned carbon nanotubes on substrates are desired in this regard CVD is uniquely superior to the other methods described above. It is general believed that the vertical alignment is due to a crowding effect, where nanotubes support each other by van der Waals interaction and the nanotubes are so closed packed that the only possible growth direction is upward [21-23].

In a PECVD growth of CNTs, the decomposition of hydrocarbon gas is largely assisted or solely achieved by the plasma energy, thus growth can occur at a considerably lower temperature than in thermal CVD. The self-bias electrical field associated with the plasma can induce CNT alignment along the direction perpendicular to the substrates and plasma depositions are very stable which leads to highly controllable and reproducible growth conditions. The carbon nanotubes growth by PECVD have also been studied using some new technologies, like hot filament assisted PECVD [24, 25], microwave PECVD [26], dc glow discharge PECVD [27], inductively coupled plasma PECVD [28] and rf PECVD [29]. Uniform films of well-aligned carbon nanotubes have been grown using microwave plasma-enhanced chemical vapor deposition, as shown in Fig. **4**. It is shown that nanotubes can be grown on contoured surfaces and aligned in a direction always perpendicular to the local substrate surface. The alignment is primarily induced by the electrical self-bias field imposed on the substrate surface from the plasma environment. It is found that switching the plasma source off effectively turns the alignment mechanism off, leading to a smooth transition between the plasma-grown straight nanotubes and the thermally grown "curly" nanotubes [25].

There are great concerns about the impurities in SWNTs synthesized by the above methods. These impurities are typically removed by acid treatments. However, these acid treatments can introduce other type of impurities, which degrade the nanotube length and perfection, and also increase nanotube costs. The mixture of semi-conducting and metallic tubes in the grown SWNTs is another concern when trying to make electronic devices. A brief summary of the three mentioned methods are listed in the Table **1** [30].

Table 1: Summary and comparison of CNTs synthesis methods, Reprint with the permission from [30].

Method	Arc-Discharge	Laser-Ablation	CVD
Pioneer	Iijima (1991)	Guo *et al.* (1995)	Yacaman *et al.* (1993)
Method	CNT growth on graphite electrodes during direct current arc-discharge evaporation of carbon in presence of an inert gas	Vaporization of a mixture of carbon (graphite) and transition metals located on a target to form CNTs	Fixed bed methods: acetylene decomposition over graphite-supported iron particles at 700°C
Yields	<75%	<75%	>75%
SWNTs or MWNTs	Both	Only SWNTs	Both
Advantages	Simple, cheap	Relatively high purity CNTs, room temperature synthesis option with continuous laser	Simple, low cost, low temperature, high purity and high yields, aligned growth is possible, fluidized bed technique for large-scale
Disadvantages	Purification of crude product is required, method cannot be scaled up, must being processed at high temperature	Cannot produce MWNTs, only adapted to lab-scale, crude product purification required	CNTs usually are MWNTs, parameters must closely be watched to obtain SWNTs

1.2 Nanostructure Approaches Using CNTs

1.2.1. CNT Paste Electrodes

Carbon Nanotube Paste Electrodes have been achieved by mixing CNT powder with deionised water, bromoform or mineral oil. Rubianes *et al.* have reported the advantages of Carbon Nanotubes Paste Electrodes (CNTPE) on the electrochemical behaviour of dopamine, ascorbic acid, uric acid, hydrogen peroxide, guanine, adenine and nucleic acids [31-33]. The resulting CNTPE retains the properties of the classical Carbon Paste Electrode (CPE), such as the viability to incorporate different substances, the low background currents, the renewability and composite nature. Therefore, this new composite electrode combines the advantages of the efficient electrocatalytic activity of CNTs with the excellent properties of composite materials.

1.2.2. CNT Solution-Based Electrodes

One method to produce CNT-modified electrodes is to cast a solution of CNTs onto a conventional Glassy Carbon Electrode (GCE) or metal electrode. The introduction of CNTs can overcome most disadvantages of the above conventional electrodes, such as poor sensitivity and stability, low reproducibility, large response time and a high overpotential for electron transfer reactions [34]. The dispersion of CNTs in an aqueous solution can be obtained by use of a specific surfactant, although strong ultrasonication is required during all the processes. Acetone [35], DMF [12, 13] and Hexadecyl Hydrogen Phosphate (DHP) [36] have also been used as solvents to fabricate electrodes. Nafion, which has been extensively used in amperometric biosensors due to its unique ion-exchange, discriminative and biocompatibility properties, has commonly been used to strengthen the mechanical connection of nanotubes and electrodes [37-39].

1.2.3. Nanotube Electrode Arrays

With the ability of promoting redox reactions of hydrogen peroxide and Nicotinamide Adenine Dinucleotide (NADH), the fabrication of vertically aligned CNTs is an effective way to produce a molecular wire and allow electrical communication between the underlying electrode and a redox enzyme, and thus could be used as amperometric biosensors associated with oxidase and dehydrogenase enzymes [40]. For example, direct electron transfer between the Flavin Adenine Dinucleotide (FAD) of an enzyme and a CNT electrode eliminates the requirement of mediators and is thus attractive for the development of reagentless biosensors. Another benefit for the nanotube electrode arrays is their high signal-to-noise ratio and low detection limits (in ppb), owing to the size reduction of each individual electrode and the increased total number of the electrodes. The schematic diagrams for the fabrication of CNT Nanoelectrode Arrays (NEAs) are depicted in Fig. **5**.

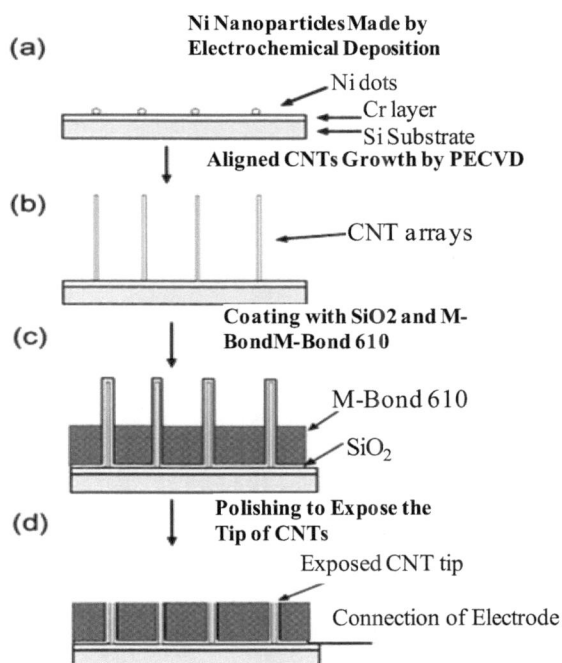

Figure 5. Fabrication scheme of the NEAs. **(a)** Ni Nanoparticles electrodeposition; **(b)** aligned carbon nanotube growth; **(c)** coating of SiO_2 and M-Bond; and **(d)** polishing to expose CNTs. Reprinted with permission from [42].

Lin *et al.* [41, 42] at Pacific Northwest National Laboratory have developed a PECVD method that allows the fabrication of low-site density-aligned carbon nanotubes with an interspacing of more than several micrometers. From this low-site density CNTs, NEAs consisted of millions of nanoelectrodes per cm^2 where each electrode is less than 100 nm in diameter. They developed these NEAs into mediator-free and membrane-free glucose biosensors. Glucose oxidase was covalently immobilized on CNT NEAs *via* carbodiimide chemistry by forming amide linkages between their amine residues and carboxylic acid groups on the CNT tips.

Li *et al.* [43] have demonstrated a new electrochemical platform based on CNT nanoelectrode arrays for ultrasensitive chemical and DNA detection. The use of aligned MWNTs provides a new bottom-up scheme for fabricating reliable nanoelectrode arrays. Fig. **6a** and **6b** indicate Scanning Electron Microscope (SEM) images of a 3 × 3 array of individually addressed electrodes on a Si (100) wafer, and each electrode can be varied in area from 2 × 2 to 200 × 200 μm^2. Each electrode consists of a vertically aligned MWNT array grown by PECVD on a 10 to 20 nm thick Ni catalyst film. Fig. **6c** and **6d** show MWNT arrays grown on 2 μm and 200 nm diameter Ni spots defined by UV and e-beam lithography, respectively. The spacing and spot size can be precisely controlled. The diameter of the MWNTs is uniform over the whole chip and can be controlled between 30 and 100 nm by PECVD conditions. Single nanotubes can be grown at each catalyst spot if their size is reduced below 100 nm. Combining such a nanoelectrode platform with Ru(bpy)$_3^{2+}$ mediated guanine oxidation, a detection limit lower than a few attomoles of oligonucleotide targets has been achieved.

Highly ordered Porous Anodic Alumina (PAA) films were first reported by Masuda *et al.* [44], and since then, PAA has been widely used as a template to prepare various nanowire and nanotube arrays, including those of CNTs, metals, semi-conductors and polymers [45-47]. PAA has several advantages. Its parallel nature allows for the simultaneous preparation of nanostructured arrays over very large areas at low cost; the pore diameter and length, as well as the interpore distance, can be easily tuned by varying the applied potential and choosing an appropriate acidic electrolyte, *etc.* One possible method for controlled vertical synthesis of CNT arrays is the use of a nanoporous template (PAA), which acts as a spatial constraint and allows growth to proceed through a narrow vertical channel [48, 49]. Maschmann *et al.* [49] grew vertical single-walled and double-walled carbon nanotubes through a PAA template by inserting a catalyst embedded layer, as shown in Fig. **7**. The Al and Fe insert layers were simultaneously anodized to create a vertically oriented pore structure through the film stack. CNTs were synthesized from the catalyst layer by plasma-enhanced chemical vapour deposition.

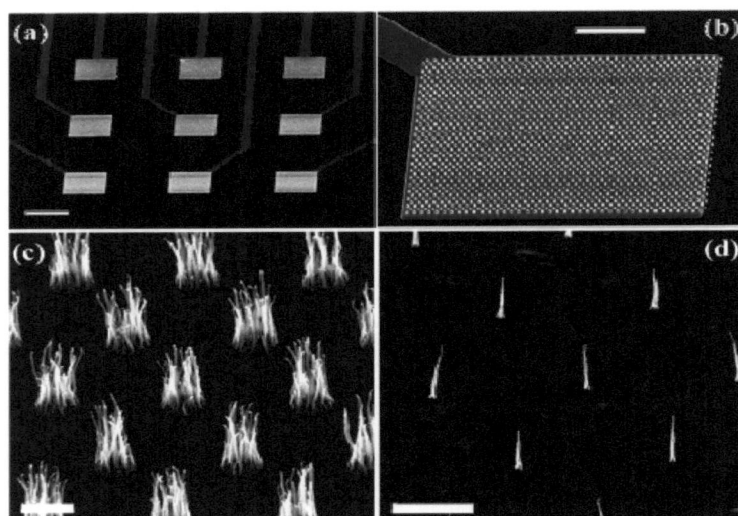

Figure 6. SEM images of **(a)** 3 × 3 electrode arrays, **(b)** arrays of MWNT bundles on one of the electrode pads, **(c)** and **(d)** arrays of MWNTs at UV-lithography and e-beam patterned Ni spots, respectively. Reprinted with permission from [43].

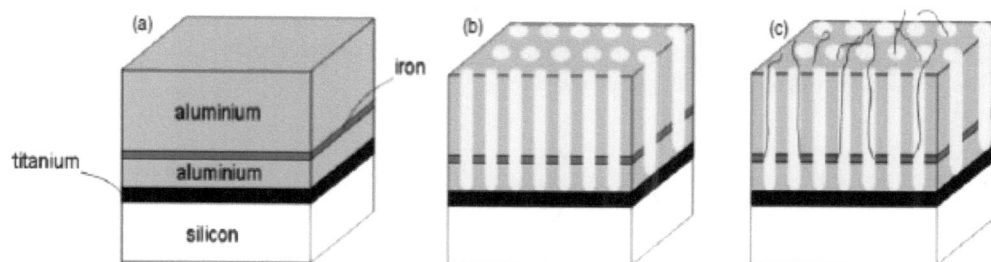

Figure 7. Schematic diagrams of catalyst preparation and CNT synthesis procedure. **(a)** Initial deposited film structure. **(b)** Anodized film structure. **(c)** CNTs synthesized from pore channels. Reprinted with permission from [49].

1.2.4. Immobilization of Enzyme

The selectivity and sensitivity of CNT-modified electrodes can be improved through the immobilization of enzymes. Various methods are used for immobilizing enzymes in biosensors, including (1) adsorption; (2) electropolymerization; (3) chemical cross-linking; (4) encapsulation, (5) covalent coupling; and (6) electrochemical co-deposition [50-52]. As for the first method, CNTs were dispensed into DMF with ultrasonic agitation, and then the suspension was cast or evaporated on the Glass Carbon Electrode (GCE) to make CNT-modified electrodes. The desired enzyme solution was dropped on the top of electrodes and allowed to evaporate. A tyrosinase-SWNTs sensor covered with Nafion film was successfully constructed for the detection of the phenolic compounds. The sensor displayed good response to phenol and other phenolic derivative compounds [53].

Electropolymerization is an attractive way to immobilize enzyme, since it provides a one-step route to make sensors, good control over the film thickness and allows the selection of different nano-scale materials [54, 55]. The enzyme mixes with a monomer which is electropolymerized at a GCE electrode, and thus the enzyme becomes embedded into the polymer matrix. Conducting polymers have gained much interest as a suitable matrix for enzyme immobilization because they have been proven effective in connecting the FAD of the enzyme and CNTs to the underlying electrode. Conducting polymers including polypyrrole (PPy) and its derivatives, polyaniline and polythiophene, have been the major sources of electrochemically generated enzyme electrodes. In this case, amperometric enzyme electrodes were produced by a one step method based on the co-immobilization of Carbon Nanotube (CNT) dopants and Glucose Oxidase (GOx) within an electropolymerized polypyrrole film. The CNT dopant retains its electrocatalytic activity toward hydrogen peroxide to impart high sensitivity upon entrapment within the PPy network [56].

Encapsulation is another immobilization method to encapsulate enzymes in hydrogel or sol-gel materials. Joshi and co-workers showed us two different methods for this purpose. i) the sensors were fabricated by casting a CNT film onto a GCE electrode, and then the redox hydrogel and sol-gel composite containing the enzymes is deposited on top of CNT-coated electrode; ii) the CNTs can be incubated with the enzyme solution, followed by incorporation into the hydrogel or sol-gel matrix to provide a support material for the CNTs [57].

Covalent coupling provides a direct connection between the enzyme and carbon structure, and always enables direct electron transfer to the FAD (Flavin Adenine Dinucleotide) center of the enzyme [58, 59]. It has been shown that chemical shortening of the SWNTs by strong acids leads to the formation of carboxylic (and phenolic) groups at the nanotube ends and sidewall defect sites, which allows the covalent immobilization of the SWNTs. Patolsky *et al.* [59] tried to align the enzyme (glucose oxidase) on the electrodes by using SWNTs as the electrical connectors between the enzyme redox centers and the electrodes. The SWNTs acted as conductive nanoneedles to electrically wire the enzyme redox-active site to the transducer surface. In the process, FAD was first covalently attached to the SWNT ends and then GOx was reconstituted at the immobilized FAD, as shown in Fig. **8**. The results show the electrode surface was linked to the aligned reconstitution of a redox flavoenzyme (glucose oxidase) on the edge of the carbon nanotubes, and the SWNTs acted as nanoconnectors that electrically contact the active site of the enzyme and the electrode. The electrons were transported along distances greater than 150 nm and the rate of electron transport is controlled by the length of the SWNTs.

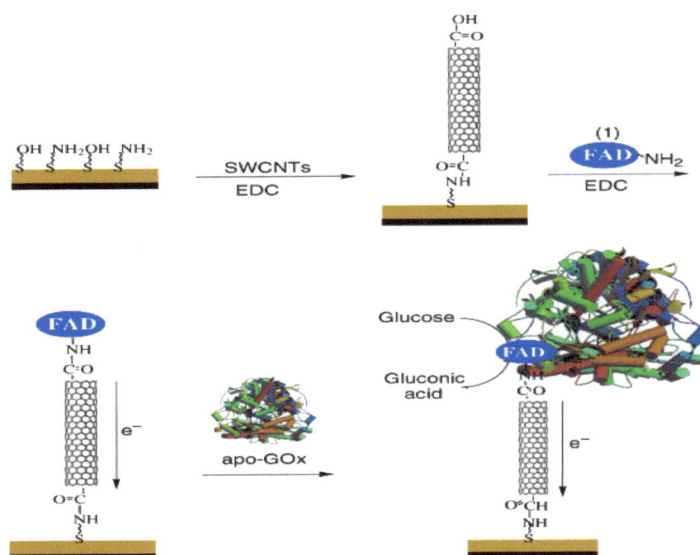

Figure 8. Assembly of the SWCNT electrically contacted glucose oxidase electrode. A 2-thioethanol/ cystamine mixed monolayer (3:1 ratio) was assembled on an Au electrode and the length fractionalized SWCNTs were coupled to the surface in the presence of the coupling reagent 1-ethyl-3-(3-dimethylaminopropyl)carbodiimide hydrochloride (EDC). The amino derivative of the FAD cofactor (1), was then coupled to the carboxy groups at the free edges of the standing SWCNTs Finally, Apo–glucose oxidase, apo–GOx, was then reconstituted on the FAD units linked to the ends of the standing SWCNTs. Reprinted with permission from [59].

It is easier to increase enzyme loading by using chemical cross-linking agents such as glutaraldehyde [51, 60]. The chemical cross-linking layer may be much thicker than the electropolymerized layer, which is usually 100nm thick, and has a large enzyme capacity. Among various methods, electrochemical co-deposition is frequently used in implantable biosensors because it can provide reproducible and site-selective immobilization of the enzyme [50].

1.2.5. SWNTs Transistors

There are two advantages of Field Effect Transistors (FETs) based biosensors over traditional electrochemical biosensors. FETs detect the electrical signal when the resistance changes due to the absorption of molecules on the FETs surface. Also, they can provide microscale and even nanoscale devices, which can measure the enzymatic activity at the molecular level and are suitable for integration with small chips [61]. Besteman *et al.* firstly introduced SWNTs into FET biosensing [62]. Controlled attachment of Glucose Oxidase (GOx) to the nanotube sidewall was achieved through a linking molecule, which on one side bonded to the SWNT through van der Waals coupling with a pyrene group and on the other side covalently bonded the enzyme through an amide bond, as depicted in Fig. **9**. A PH sensor with large and reversible changes in conductance upon changes in pH was achieved by this enzyme-coated tube. As for glucose detection, a steplike response can be monitored in real time after immobilization of GOx.

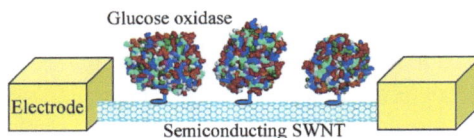

Figure 9. Schematic picture of two electrodes connecting a semiconducting SWNT with GOx enzymes immobilized on its surface. Reprint with permission from [62].

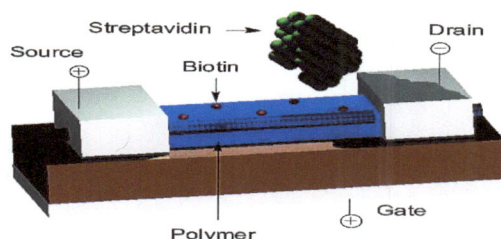

Figure 10. Schematic of the Nanotube Field Effect Transistor (NTFET). A polymeric functional layer, which coats the nanotube, is functionalized with a molecular receptor, biotin, a protein that recognizes a biomolecule, streptavidin. Reprinted with permission from [63].

Star *et al.* have used nanoscale field effect transistor devices with carbon nanotubes as the conducting channel to detect protein–receptor interactions (using biotin–streptavidin binding as an example), as depicted in Fig. **10** [63]. A PEI/PEG polymer coating layer has been employed to avoid nonspecific binding, which has several advantages: i) the polymer was used to attach molecular receptor molecules to the sidewalls of nanotubes; ii) polymer coatings have been shown to modify the characteristics of nanotube FET devices, and thus the coating process can be readily monitored; iii) the polymer coating could be used to prevent nonspecific binding of proteins. It was confirmed that Streptavidin, in which the biotin-binding sites were blocked by reaction with excess biotin, produced essentially no change in device characteristic of the biotinylated polymer-coated devices.

Single-cell analysis has become a highly attractive tool for studying cellular information. Sudibya *et al.* [64] proved that biocompatible glycosylated nanotube devices could interface with single living cells, and electronically detected biomolecules with high temporal resolution and sensitivity. This work was a significant development in the field of cell biology and also opened up new avenues to advance our fundamental understanding of dynamic secretion of bio-molecules from single cell.

1.3. Current Developments of CNTs Based Biosensors

1.3.1. Glucose Biosensor

Diabetes is a group of metabolic diseases affecting more than 200 million people in the world, and is one of the leading causes of death and disability, such as blindness, nerve degeneration and kidney failure. Electrochemical biosensors play a critical role in the diagnosis of diabetes. The dramatic decrease in the overpotentials of hydrogen peroxide as well as the direct electron transfer of glucose oxidase observed at CNT-modified electrodes show great promise for use in a glucose biosensor [65].

Wang *et al.* found that increasing the Nafion content (from 0.1 to 5 wt %) resulted in dramatic enhancement of the solubility of both types of CNT, and thus the use of Nafion as a solubilizing agent for CNTs overcomes a major obstacle for creating CNT-based biosensing devices [66]. The CNT/Nafion-coated electrode offered a marked decrease in the overvoltage for the hydrogen peroxide reaction to allow convenient low-potential amperometric detection (-0.05 V vs. Ag/AgCl). More experiments based on this work have shown palladium nanoparticles on a Nafion-solubilized Carbon Nanotube (CNT) film retained their biocatalytic activity and offered an efficient oxidation and reduction of the enzymatically liberated H_2O_2, allowing for fast and sensitive glucose quantification. The combination of Pd–GO_x electrodeposition with Nafion-solubilized CNTs enhances the storage time and performance of the sensor. An extra Nafion coating was also used to eliminate common interferent, such as uric and ascorbic acids. The fabricated Pd–GOx–Nafion CNT glucose biosensors exhibited a linear response up to 12 mM glucose and a detection limit of 0.15 mM (S/N = 3) [67].

Lin *et al.* investigated aligned carbon nanotubes arrays with an interspacing of more than several micrometers. They developed these Nanoelectrode Arrays (NEAs) into mediator-free and membrane-free glucose biosensors [41, 42]. Glucose oxidase was covalently immobilized on CNT NEAs *via* carodimmide chemistry by forming amide linkages between their amine residues and carboxylic acid groups on the CNT tips. A linear amperometric response was achieved over psychological levels from 2 – 30 mM, as shown in Fig. **11**. The signal response curve is effective at low detection limits at an attractive low-potential point, - 0.2V. The limit of detection can be as low as 0.08 mM based on a signal-to-noise ratio of 3.

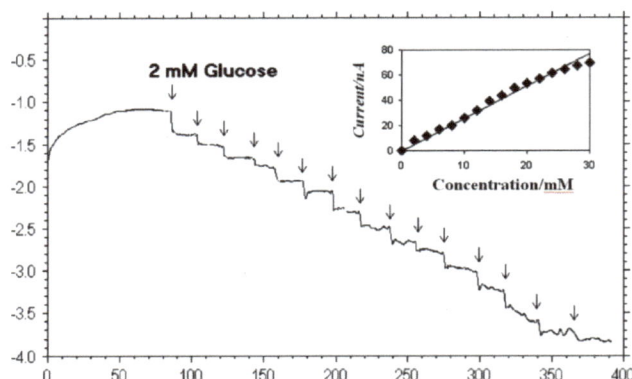

Figure 11. Amperometric responses of a NEE biosensor to successive additions of 2 mM glucose. Reprinted with permission from [41].

NTFETs, consisting of individual semiconducting SWNTs as a versatile biosensor, has been demonstrated by Besteman *et al.* [62] to measure device reactions at the molecular level. Controlled attachment of the glucose oxidase (GOx) to the nanotube sidewall was achieved through a linking molecule, and the oxidation of b-D-glucose ($C_6H_{12}O_6$) and D-glucono-1,5-lactone ($C_6H_{10}O_6$) were studied. The redox enzymes went through a catalytic reaction cycle, where groups in the enzyme temporarily change their charge state and conformational changes occur in the enzyme, which could be detected by the NTFET devices. A step-like response can be monitored in real time after immobilization of GOx in NTFETs.

1.3.2. Immunosensors

CNT based electroanalytical techniques, like electrochemical impedance spectroscopy and cyclic voltammetry provide a simple, cheap and reliable way to detect proteins for point-of-care when compared with traditional Enzyme-Linked Immunosorbent Assay (ELISA), radioimmunoassay and electrophoretic immunoassay [68]. Label-free methods are extremely attractive since they remove all additional steps of labelling with secondary antibodies fluorescence or radioactive materials. There are two main approaches for utilizing CNTs; aligned CNTs arrays and CNT FET electrodes.

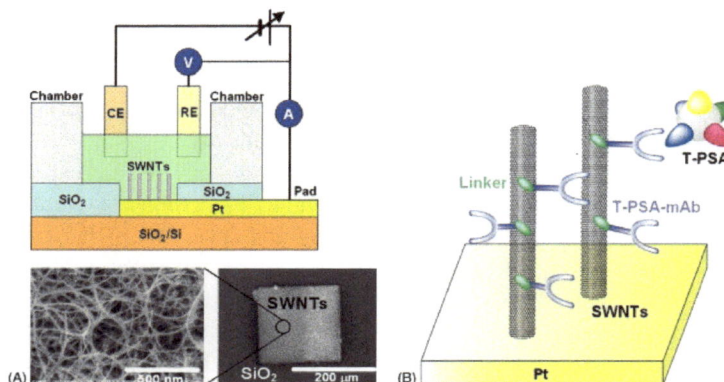

Figure 12. (A) Illustration for the experimental set-up with Single-Walled Carbon Nanotube (SWNT)-modified Pt microelectrode as the working electrode, Pt wire as the Counter Electrode (CE), and the miniaturized reference electrode (RE, Ag/AgCl) with the Scanning Electron Microscopy (SEM) images of a SWNT-modified microelectrode; **(B)** illustration for the label-free electrochemical immunosensor design. Monoclonal antibodies against total prostate-specific antigen (T-PSA-mAb) were covalently anchored onto the SWNTs using 1-pyrenebutanoic acid succinimidyl ester (Linker). Reprinted with the permission from [69].

A label-free electrochemical immunosensor using microelectrode arrays modified with Single-Walled Carbon Nanotubes (SWNTs) was achieved by Okuno *et al.* [69], as shown in the Fig. **12**. SWNTs were synthesized on Pt electrodes by thermal CVD using a catalyst, which contained $Fe(NO_3)_3 \cdot 9H_2O$, $MoO_2(acac)_2$ and alumina nanoparticles in the liquid phase. Afterwards, the substrate was heated up to 900 °C under Ar, and ethanol vapor was supplied for 10 min. The diameters of SWNTs were estimated to be 1–2 nm and an array of 30 microelectrodes with SWNT modification was formed on a single substrate. The current signals, derived from the oxidation of tyrosine (Tyr), and tryptophan (Trp) residues, increased with the interaction between T-PSA on T-PSA-mAb covalently immobilized on SWNTs. The selectivity of this biosensor was tested using bovine serum albumin (BSA) as the target protein. The detection limit for the T-PSA was determined to be 0.25 ng/mL, which is much lower than the cut-off limit of T-PSA between prostate hyperplasia and cancer (4 ng/mL).

Aptamers are synthetic oligonucleotides that can be generated to recognize amino acids, drugs, and proteins with high specificity. Aptamers have several advantages compared with antibodies, which are the most used capture agents providing high selectivity in biosensing, such as relatively cost-effective mass production, less temperature dependency and long shelf-lives. Maehashi *et al.* developed aptamer-modified Carbon Nanotube Field-Effect Transistors (CNT-FETs) for the detection of immunoglobulin E (IgE), as depicted in Fig. **13**. After the covalent immobilization of 5'-amino-modified 45-mer aptamers on the CNT channels, the electrical properties of the CNT-FETs were monitored in real time. The amount of the net source-drain current before and after IgE introduction on the aptamer-modified CNT-FETs increased as a function of IgE concentration. The detection limit for the IgE was determined as 250 pM. Furthermore, they proved that aptamers displayed a better performance for the detection of IgE than the monoclonal antibodies did under similar conditions [70].

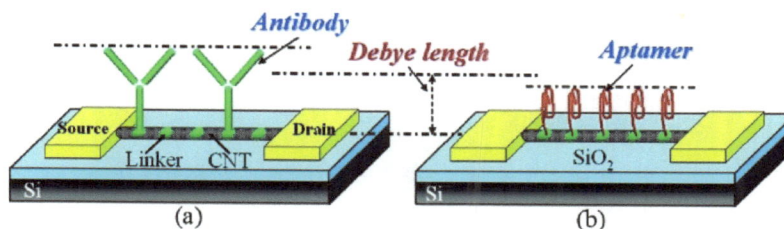

Figure 13. Schematic representation of label-free protein biosensors based on CNT-FETs: **(a)** antibody-modified CNT-FET; **(b)** aptamer-modified CNT-FET. Reprinted with permission from [70].

1.3.3. DNA Biosensors

DNA biosensors, based on nucleic acid recognition processes, are rapidly being developed towards the goal of rapid, simple and inexpensive testing of genetic and infectious disease. Electrochemical hybridization biosensors rely on the immobilization of a single-stranded DNA probe onto the transducer surface that convert the duplex formation into a useful electrical signal [40, 71-73].

Carbon Nanotube Paste Electrodes (CNTPE) were selected by Pedano and Ravios [74] to measure trace levels of nucleic acids. The influence of surface pretreatments, paste composition, nature of the nucleic acid, and accumulation conditions on the adsorption and further electrooxidation of different oligonucleotides and polynucleotides on the CNTPE have been observed. The electroactivity inherent to carbon nanotubes resulted in a large enhancement of the guanine oxidation signal compared to that obtained at its analogue carbon (graphite) paste electrode (CPE). Trace (μg/l) levels of the oligonucleotides and polynucleotides can be readily detected following short accumulation periods with a detection limit of 2.0 μg/l for 21 bases oligonucleotide, and 170 μg/l for calf thymus dsDNA.

A nanoelectrode array based on vertically aligned multiwall carbon nanotubes (MWNTs) embedded in SiO_2 was used for ultrasensitive DNA detection. The hybridization of less than a few attomoles of oligonucleotide targets can be easily detected combining the CNT nanoelectrode array with the $Ru(bpy)_3^{2+}$ mediated guanine oxidation method. The sensitivity can be further improved down to thousands of target

DNAs after optimization, which could provide faster, cheaper, and simpler solutions for molecular diagnosis [43].

A carbon nanotube transistor array was used to detect DNA hybridization, The polymer poly (methylmethacrylate$_{0.6}$-co-poly(ethyleneglycol)methacrylate$_{0.15}$-co-N-succinimidyl ethacrylate$_{0.25}$) was synthesized and bonded noncovalently to the nanotube. This polymer was well adsorbed to the walls of the CNTs and carried activated succinimidyl ester groups used to fix the NH$_2$-ssDNA probes. Aminated single-strand DNA was then attached covalently to the polymer. DNA hybridization produced statistically significant changes in the threshold voltages reflecting the charge trapping character of hybridized DNA [75].

2. INTRODUCTION TO CNT FIBRES

In biosensing applications, CNTs demonstrate faster response time and higher sensitivity than traditional electrodes, as was discussed in the previous sections. However, challenges remain for the development of robust, practical and stable sensors that are able to maximally exhibit the exceptional properties of individual CNTs [60]. In addition, their integration into biosensing electrodes has been challenging and concerns are often raised regarding toxicity of the nano-sized CNTs leaching from an implanted biosensor [76-80].

The alignment of nanotubes in a fibre is an effective way to exploit the exceptional anisotropic properties of individual CNTs and transform them for micro-/macroscale applications, as shown in Fig. **14** [81]. Apart from excellent mechanical properties, the fibre inherits the advantages of the high surface area and good electrocatalytic properties of CNTs, while avoiding the potential toxicity caused by the size of individual CNTs during implantation. To fully utilize CNT fibre, several production routes were developed to fabricate continuous CNT microfibres, including wet spinning of CNTs from polymer dispersions or acid dispersions [81-84], and direct spinning from CVD reactions [85-88]. Here, we aim to introduce the synthesis processes for CNT fibre and also discuss the recent development of CNT fibres in biosensing areas.

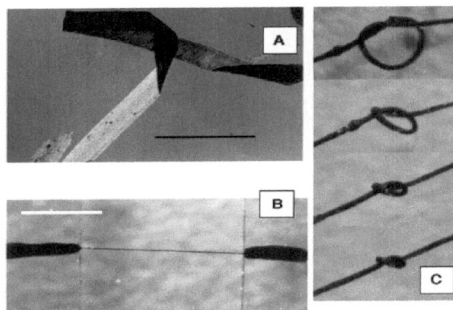

Figure 14. Optical micrographs of nanotube ribbons and fibres. (**A**) A single folded ribbon between horizontal and vertical crossed polarizers. (**B**) A freestanding nanotube fibre between two glass substrates. (scale bar = 1 mm). (**C**) Tying knots reveals the high flexibility and resistance to torsion of the nanotube microfibres. The fibre shown in the pictures has a radius of about 15 μm. Reprinted with permission from [81].

2.1 Synthesis and Processing of CNT Fibres

2.1.1 Particle-Coagulation Spinning (PCS) Process

Vigolo *et al* firstly invented the CNT fibre by the PCS process, involving the injection of a homogeneous and relatively concentrated aqueous CNT suspension in a flowing coagulating bath [81]. Flow-induced alignment of the nanotubes took place at the tip of the capillary, and thus the ribbons could be drawn in the third dimension to form a helical structure when the polymer solution was slowly pumped out from the bottom of the container. The injection rate of the SWNT dispersion was varied from 10 to 100 ml/hour. The capillary tip from which the nanotube solution was extruded was located at about 3.5 cm from the rotation axis of the polymer solution. The needle or the capillary tube was oriented so that the SWNT injection was tangential to the circular trajectory of the polymer solution. The velocity of the polymer solution at the tip of the capillary could thus be varied from 6.6 to 33 m/min. By pumping the polymer solution from the bottom, meter-long ribbons can be easily drawn.

Preliminary four-probe electrical measurements indicated that resistivity at room temperature is about 0.1 Ω^{-1} cm^{-1} and showed nonmetallic behavior when the temperature was decreased. Unlike classical carbon fibres, the nanotube fibres can be significantly bent without breaking (Fig. **14**). The elastic modulus obtained is 10 times higher than the modulus of high-quality bucky paper.

2.1.2 Wet spinning Method

Starting with purified SWNTs, Ericson *et al.* produced well-aligned continuous macroscopic fibres, without any supporting surfactant or polymer structure [82]. Fibres were made from concentrated dispersions of SWNTs in 102% sulfuric acid (2 wt% excess SO$_3$) *via* an industrially viable wet spinning technique. An 8 wt% dispersion of purified SWNTs in 102% sulfuric acid mixture was transferred to the mixing apparatus *via* a stainless steel syringe. Two alternating pneumatic pistons extensively mixed the SWNTs by pushing them back and forth through an actively rotating shear cell. When the viscosity reached a steady state, the SWNTs material was extruded through a small capillary tube (<125 μm in diameter) into a coagulation bath. Fibres were produced under a variety of conditions, including different doping temperatures (0° to 100°C), coagulants (diethyl ether, 5 wt% aqueous sulfuric acid, or water), and coagulation bath temperatures (0°C and room temperature). In the case of aqueous coagulants, fibres were washed for several hours before collection onto a Teflon drum. To remove water and residual acid, water-coagulated fibres were dried in a vacuum oven at 100°C, followed by annealing in a flow of H$_2$/Ar (1:1) at 1 atm and 850°C for 1 hour. The wet-spinning process produced the neat SWNT fibres with good mechanical properties (Young's modulus = 120 ± 10 GPa and a tensile strength =116 ± 10 MPa).

2.1.3 Chemical Vapor Deposition (CVD) method

Windle's group in Cambridge spun CNT fibres and ribbons directly using CVD synthesis and a liquid source of carbon and an iron nanocatalyst [85, 86]. The key requirements for continuous spinning are the rapid production of high-purity nanotubes to form an aerogel in the furnace hot zone and the forcible removal of the product from the reaction by a continuous wind-up. As illustrated in Fig. **15**, the ethanol containing ferrocene and thiophene was injected into the furnace hot zone with H$_2$ carrier gas. The nanotubes in this condition formed an areogel and stretched into form of a stock *via* the gas flow. As the areogel reaches the cool end of the furnace tube, it attaches to the wall and forms a diaphanous membrane across the tube that thickened with time. Due to the convection in the tube, a portion of the nanotube aerogel hangs down to form a fibre along the furnace axis and continues to grow when additional nanotubes adhere to it. The aerogel could be continuously drawn from the hot zone by winding it onto a rotating rod, and the geometry of the spindle affects the properties of the fibre. The best electrical conductivity measured along a fibre was 8.3 x 10^5 Ω^{-1} m^{-1} which is much higher than that value when using the PCS process (0.1 Ω^{-1}cm^{-1}) to produce fibre. Preliminary mechanical measurements indicate that the fibres have a range of strengths, dependent on process conditions. They lie between 0.05 N/Tex and 0.5 N/Tex (equivalent to 0.10 and 1.0 GPa, assuming a density of 2.0 g/cc, which is within the range of typical carbon fibres).

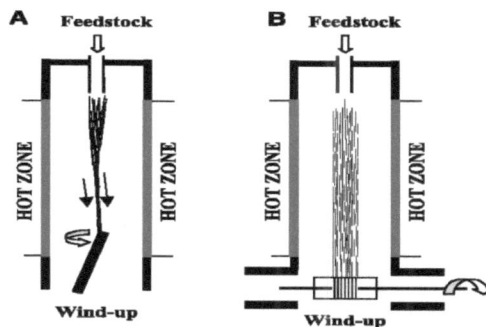

Figure 15. (**A**) Schematic of the CVD spinning process. The liquid feedstock, in which small quantities of ferrocene and thiophene are dissolved, is mixed with hydrogen and injected into the hot zone, where an aerogel of nanotubes forms. This aerogel is captured and wound out of the hot zone continuously as a fibre or film. Here, the wind-up is by an offset rotating spindle. (**B**) Schematic of the wind-up assembly that operates at a lower temperature, outside the furnace hot zone. Reprinted with the permission from [86].

2.2. Current Development of CNT Fibres for Biosensors

Wang *et al.* first introduced wet-spun CNT fibres as microelectrodes for electrochemical sensors and demonstrated the possibility of detection of NADH, hydrogen peroxide and dopamine in 2003 [89]. CNT fibres were activated at 300 °C for 60 min and subsequently inserted into pulled glass capillaries to produce CNT fibre microelectrodes (18 and 34 μm in diameter and 5 mm in length, for SWNTs and MWNT respectively). As depicted in Fig. **16**, the treated MWNT fibre electrodes respond to NADH over most of the potential range, with significant oxidation currents starting at +0.1 V and levelling off above +0.4 V (Fig. **16E**), which reflects the marked acceleration of the NADH redox process. A highly stable signal is observed over the entire operation upon using the CNT microelectrode, while a rapid loss of its activity is observed when carbon fibre is used according to the right insert of Fig. **16**. The electrocatalytic action of CNT fibre electrodes (34 μm) facilitates low-potential amperometric measurements (+0.2 V) of NADH over the entire concentration range (Fig. **16**, left inset). Moreover, the CNT fibre offers a marked decrease in the overvoltage for dopamine, and hydrogen peroxide.

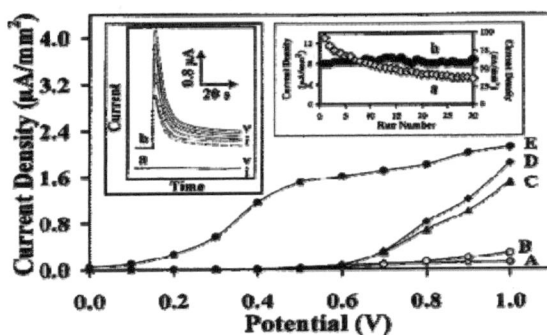

Figure 16. (**A**) Hydrodynamic voltammograms (current density vs applied potential) for 1 mM NADH at carbon (A,D) and CNT (B,C,E) fibre microelectrodes (untreated (A,B) and heat treated (C,D,E) fibres). Left inset: chronoamperometric signals measured at +0.2 V of NADH from 1.0 (i) to 5.0 (v) mM at the carbon (a) and CNT (b) fibre microelectrodes. Dashed line indicates the blank solution. Right inset: stability of the chronoamperometric response for repetitive measurements at +0.7 V of 3 mM NADH at the carbon (a, right axis) and CNT (b, left axis) fibre electrodes. Electrolyte, phosphate buffer (0.05 M, pH 7.4). Reprinted with permission from [89].

A CNT fibre microelectrode with proper mediators (2,4,7-trinitro-9-fluorenone) on the surface and various pre-treatments were reported by Viry *et al.* to assembly a biosensor [90, 91]. The possibility of increasing the average alignment of the tubes inside the fibres by stretching the fibre along the main axis offers an opportunity to exploit more efficiently their electronic properties. A glucose sensing electrode was built by adsorption of a mediator on the surface of a CNT fibre microelectrode. Electrocatalytic oxidation of analytes *via* a dehydrogenase works efficiently at 0V, which is a key point in developing such bioanalytical tools. This work shows that the CNT fibre microelectrodes produce more versatile tools in the biosensing area than conventional carbon microelectrodes.

Wet-spun fibres from polymer solution are actually composite fibres consisting of non-conductive polymer as spun, or carbon residue after thermal treatment, showing relatively low strength and low conductivity compared to pure CNT fibres. Recently, a wide variety of continuous yarns of CNTs were fabricated by direct spinning of pure CNT fibres from an aerogel formed during a CVD process using ethanol and acetone as the carbon source [85-88]. The resulting CNT micro-fibre has the potential to address the electrode design and toxicity concerns of CNT based sensors for long-term implantable biosensor applications. During the CVD process, the CNTs formed and self-assembled in the gas flow by van der Waals interactions at high temperature, and then were spun into nano-yarns along the fibre axis. Thus the direct spinning of the pure CNT fibre by the CVD method helped to achieve the best properties in terms of strength, stiffness, toughness, as well as electrical and thermal conductivities, in comparison with the CNT fibres spun from other methods. The specific fibre used was composed of double-walled CNTs that are compacted into concentric layers of CNT bundles organized as nano-yarns [60, 88], as shown in Fig. **17**. The CNT fibre resembles an electric wire, relying on nano-scale surface topography and porosity, which can facilitate molecular-scale interactions with agents like enzymes to efficiently capture and promote electron transfer reactions.

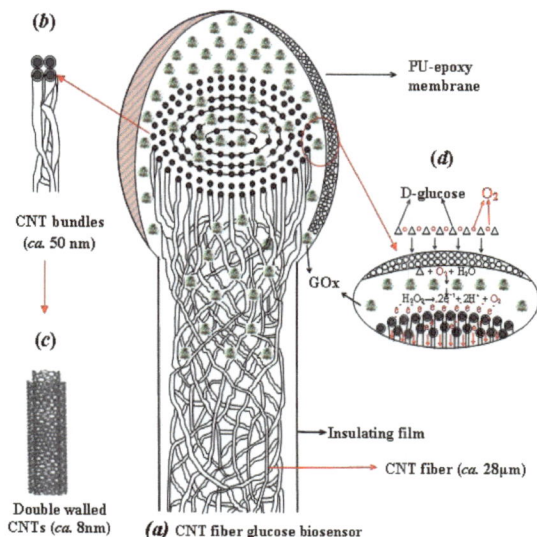

Figure 17. Schematic diagram showing **(a)** CNT fibre based glucose biosensor, **(b)** CNT bundles, **(c)** DWNT and **(d)** working principle of biosensor. The CNT fibre (ca. 28 μm) is made of bundles (ca. ~50 nm) of DWNTs (ca. 8-10nm) entangled to form concentrically compacted multiple layers of nano-yarns along the CNT fibre axis, as illustrated in (a). GOx enzyme is immobilized at the brush-like end of the CNT fibre and the enzyme layer is encapsulated by the epoxy-polyurethane (EPU) semi-permeable membrane. D-glucose and oxygen pass through the EPU membrane, GOx converts them into gluconic acid and hydro peroxide. The peroxide is converted to electrons, hydrogen and oxygen at a potential of 700mV. The electrons are captured by the fibre, which signal is used to measure glucose concentrations [60].

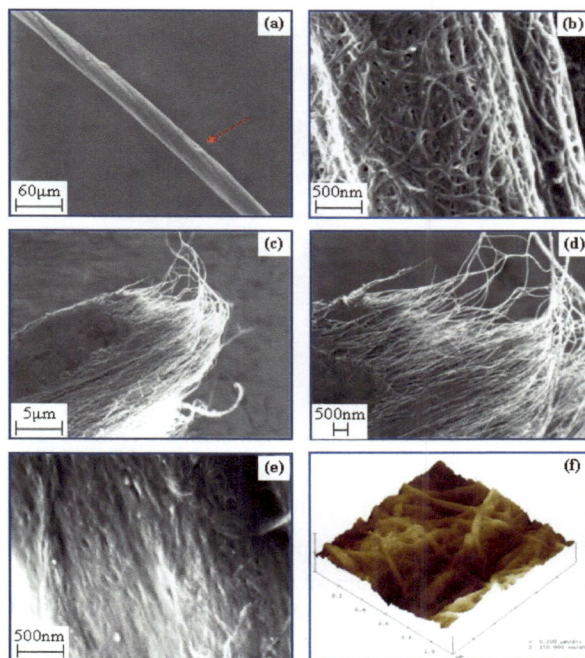

Figure 18. SEM photographs of **(a)** a twisted CNT fibre, diameter is *ca.* 28 μm; **(b)** unique nano-yarn and nano-porous membrane structure of the fibre surface, magnified (x100k) from arrow area in (a); **(c)** engineered brush-like nano-yarn tip **(d)** Magnified image from (c) (x25k) show the diameter of nano-yarn are ca. 50nm, as illustrated in Figure.1(b); **(e)** enzyme (glucose oxidase) directly immobilised on CNT fibre surface; **(f)** a AFM topology image in tapping-mode shows the porous structure of CNT fibre surface [60].

The SEM images illustrated in Fig. **18** show the morphology of the CNT fibre. The fibre has an average diameter of 28 µm and is slightly twisted along the fibre axis (Fig. **18a**). The slight twisting could be due to the spinning process. At a higher magnification (Fig. **18b**), the fibre's surface nano-topography is evident. CNT bundles of about 50 nm in diameter are entangled to form a nano-porous mesh-like surface topography. Brunauer-Emmett-Teller (BET) and Barrett-Joyener-Halenda (BJH) measurements respectively demonstrated that the surface of the CNT yarn has a high specific surface area of 194 m^2 g^{-1} and an average pore size around 2.4 nm. The CNT fibre has an electrical conductivity, 5.0×10^5 S m^{-1}, two orders of magnitude higher than that of wet-spun CNT fibres [$(1–2) \times 10^3$ S m^{-1}] [66, 81].

To realize the full potential of the CNT fibre as a biosensor, it was essential to unwind the CNT bundles (nano-yarns) at the ends of the fibres. The resulting brush-like nano-structure resembles a scaled-down electrical 'flex' and the individual nano-yarns within the brush-like end (Fig. **18c** and **d**) would act as multi-nano-electrodes, reminiscent of a dendrite-type nerve cell, which provides many nano-channels to promote and speed up the electron transfer as well as increase the surface area for enzyme immobilization. The whole process of assembling a CNT fibre based biosensor is depicted as Fig. **17**.

Fig. **19** shows the typical amperometric response of the CNT fibre-based glucose biosensor. Fast response of the biosensor towards glucose can be seen in that the sensor response current reaches dynamic equilibrium within tens of seconds (response time) of each addition of glucose, generating a near steady-state current signal. This indicates a fast electron transfer between the redox center of the enzyme and the CNT fibre, as described by Cai *et al.* and Guiseppi-Elie *et al.*[3, 92-94]. The linear calibration curves for the annealed and as-spun CNT fibre glucose biosensors are presented in the insert of Fig. **19**. Both the sensors showed a linear increase (R^2 of 0.991 and 0.997 respectively for annealed and as-spun CNT fibres) in the sensor response current covering the range 2-30 mM of glucose concentration as is needed to cover the whole range of physiological blood glucose levels. These results suggest that the miniature CNT electrode (~1mm) performs better than the reported ~10 mm long Pt coil electrode when tested as a glucose biosensor [50, 51].

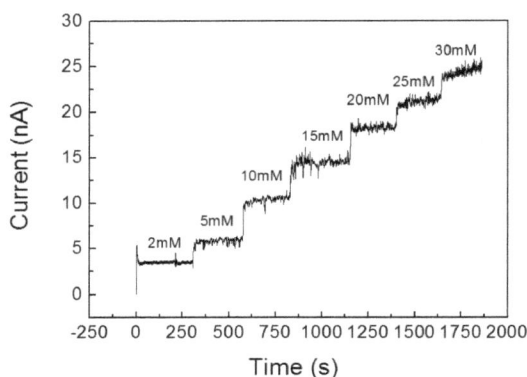

Figure 19. Amperometric responses of CNT fibre biosensors. Numbers in the chart represent the corresponding glucose concentration of the solution (n=3). Calibration equation: y = 2.9133+0.795x, R^2= 0.996872, linearity range: 2-30mM [60].

Furthermore, thermal annealing of the as-spun CNT fibre was shown to increase the sensitivity of the CNT fibre based glucose biosensor by 7.5 times; while the sensitivity of the sensors was demonstrated to be stable up to 70 days. Although coating the CNT fibre's with gold did not significantly affect the sensor's sensitivity, it resulted in the extension of the lower glucose detection limit from 2 mM to as low as 25 µM [60].

3. OUTLOOK AND SUMMARY

For biosensing applications, CNTs demonstrate faster response time and higher sensitivity than traditional electrodes at extremely low working potentials. Meanwhile, the use of CNT based molecular wires offers great promise for achieving efficient electron transfer from electrode surfaces to the redox sites of enzymes. However, better control of the chemical and physical properties of CNT based biosensors, like the separation process for different type of CNTs; the miniaturization of the sensor, the possibility of toxicity, *in-vivo* stability *etc.* still need to be addressed to meet the future requirements. In this section, how to develop the next generation CNT-based biosensors suit for mainstream applications will be demonstrated, by addressing the above issues.

Cost-effective, large scale fabrication of CNT Nanoelectrode Arrays (NEAs) is one attractive direction [41-43, 95, 96]. Such arrays of nanoelectrodes can produce a much higher current than a single nanoelectrode. This would avoid the need for expensive electronic devices and improve the signal to noise ratio, leading to ultrasensitive electrochemical sensors for chemical and biological sensing. Tu *et al.* reported a low-site density carbon nanotube based nanoelectrode array, and epoxy resin was spin-coated to create the electrode passivation layer effectively reducing the electrode capacitance and current leakage [95]. Yun and co-workers reported that gold nano particles enhanced long aligned MWNTs array electrodes for novel label-free immunosensing [97]. Further improvement in these promising CNT-NEAs for biosensor applications requires tunable and predictable assembly with well-ordered structures. Novel nano-technology such as soft lithography, nanoimprint lithography and highly ordered porous anodising alumina template can help achieve these.

Another promising approach for the direct detection of biological species is Field-Effect Transistors (FETs) which have possible advantages, like forming a conducting channel using SWNTs; absorption of molecules on the FETs surface and manipulation at the molecular level with cells. Single cell analysis has become a highly attractive tool for investigating cellular contents. Villamizaar *et al.* revealed a fast, sensitive and label-free biosensor based on a network of SWNTs which act as the conducting channel for the selective determination of salmonella Infantis [98]. Since the semiconductor properties of SWNTs are critical to form FETs for biosensing, a continuing problem is that the carbon materials produced by all CNT synthesis methods contain both semiconducting and metallic nanotubes. Recent reports addressing the separation and purification of semiconducting and metallic CNTs are undoubtedly a landmark advance, but higher yield and purity is still desired [99-101].

A very important issue related to the integration of CNTs into biological cells and tissues is the need to study their cytotoxicity towards biological species. Contradictory results have been reported in recent research on the toxicity of CNTs. Poland *et al.* demonstrated that asbestos-like pathogenic behaviour was associated with CNTs, they shown that the toxicity depends on length and thus suggested the use of commercially long CNTs [79]. Pantarotto *et al.* have shown that SWNTs will be toxic to mammalian cells beyond 10 μmol/L [78]. Meanwhile, Kam *et al.* reported that SWNTs are nontoxic up to a high concentration, 0.05mg/mL [102]. Therefore, an in-depth systematic and long-term study of the effect of CNTs on human cells and tissues as well as information related to safety issues have not yet been fully characterized. Much work is still required in this field.

Regardless of the contradictory results on toxicity of CNTs as discussed above, development of CNTs in non-particulate forms such as continuous CNT fibres is a safe way to avoid the potential risk of CNT leaching, especially when used in implantable electrodes for in-vivo testing. Although CNT fibres have high strength (1.8-3.0 GPa), stiffness (330GPa) and good electrical conductivity (8.3 x 105 s m-1), the development of CNT fibres into the biosening area is still at early stage. Wang *et al.* and Viry *et al.* have proven the possibility of detecting H_2O_2 and NADH by assembling CNT fibres as a microelectrode. Zhu *et al.* systemically exploit CVD synthesized CNT fibres into enzymatic glucose biosensors. Highly porous network structures of fibres and unique brush-like nanostructure fibre ends lead to short response times in amperometric test and fast electron transfer between the redox center of the enzyme and CNT fibre [60]. Thermal annealing of the CNT fibre will result in 7.5 fold increase in glucose sensitivity and long-term stability of this sensor was tested over 3 months. More work must be focused on the improvement of surface conductivity of the fibre; device design to fully utilize the physical and chemical CNT fibre properties and the surface functionalized biological agents for various biosensing applications. CNT fibres open a door for the safe use of CNT in the biosensing area when dealing with cells and tissue, especially for implantable biosensors. However, it is fair to say that there is still a long way to the full utilization of CNT fibres for biosensing applications.

ACKNOWLEDGMENTS

The author would like to thank the financial support by the Royal Society grant (RG 2009/R1) and the grant from EU Framework Programme 7 (CORONA, CP-FP 213969-2).

REFERENCES

[1]   Iijima S. Helical Microtubules of Graphitic Carbon. *Nature*, **1991**, *354*, 56-58.

[2]   Iijima S.; Ichihashi T. Single-Shell Carbon Nanotubes of 1-Nm Diameter. *Nature*, **1993**, *363*, 603-605.

[3]   Guiseppi-Elie A.; Lei C. H.; Baughman R. H. Direct electron transfer of glucose oxidase on carbon nanotubes. *Nanotech.*, **2002**, *13*, 559-564.

[4]   Heller I.; Janssens A. M.; Mannik J.; Minot E. D.; Lemay S. G.; Dekker C. Identifying the mechanism of biosensing with carbon nanotube transistors. *Nano Lett.*, **2008**, *8*, 591-595.

[5]   Martel R.; Schmidt T.; Shea H. R.; Hertel T.; Avouris P. Single- and multi-wall carbon nanotube field-effect transistors. *Appl. Phys. Lett.*, **1998**, *73*, 2447-2449.

[6]   Baughman R. H.; Zakhidov A. A.; De Heer W. A. Carbon nanotubes--the route toward applications. *Science*, **2002**, *297*, 787-792.

[7]   Shengli Z.; Shumin Z.; Jinyou L.; Minggang X. Helicity energy of a straight single-wall carbon nanotube. *Phys. Rev. B*, **2000**, *61*, 12693-12696.

[8]   Smith B. W.; Benes Z.; Luzzi D. E.; Fischer J. E.; Walters D. A.; Casavant M. J.; Schmidt J.; Smalley R. E. Structural anisotropy of magnetically aligned single wall carbon nanotube films. *Appl. Phys. Lett.*, **2000**, *77*, 663-665.

[9]   Hirsch A. Functionalization of single-walled carbon nanotubes. *Angew. Chem. Int. Ed. Engl.*, **2002**, *41*, 1853-1859.

[10]   Gooding J. J. Nanostructuring electrodes with carbon nanotubes: A review on electrochemistry and applications for sensing. *Electrochim. Acta*, **2005**, *50*, 3049-3060.

[11]   Sano M.; Kamino A.; Okamura J.; Shinkai S. Self-organization of PEO-graft-single-walled carbon nanotubes in solutions and Langmuir-Blodgett films. *Langmuir.*, **2001**, *17*, 5125-5128.

[12]   Wang J.; Musameh M. Carbon nanotube/teflon composite electrochemical sensors and biosensors. *Anal. Chem.*, **2003**, *75*, 2075-2079.

[13]   Wang J. X.; Li M. X.; Shi Z. J.; Li N. Q.; Gu Z. N. Electrocatalytic oxidation of norepinephrine at a glassy carbon electrode modified with single wall carbon nanotubes. *Electroanal.*, **2002**, *14*, 225-230.

[14]   Lin Y.; Taylor S.; Li H. P.; Fernando K. A. S.; Qu L. W.; Wang W.; Gu L. R.; Zhou B.; Sun Y. P. Advances toward bioapplications of carbon nanotubes. *J. Mater. Chem.*, **2004**, *14*, 527-541.

[15]   Popov V. N. Carbon nanotubes: properties and application. *Mater. Sci. & Eng. R*, **2004**, *43*, 61-102.

[16]   Bethune D. S.; Klang C. H.; De Vries M. S.; Gorman G.; Savoy R.; Vazquez J.; Beyers R. Cobalt-catalysed growth of carbon nanotubes with single-atomic-layer walls. *Nature*, **1993**, *363*, 605-607.

[17]   Guo T.; Nikolaev P.; Thess A.; Colbert D. T.; Smalley R. E. Catalytic growth of single-walled manotubes by laser vaporization. *Chem. Phys. Lett.*, **1995**, *243*, 49-54.

[18]   Scott C. D.; Arepalli S.; Nikolaev P.; Smalley R. E. Growth mechanisms for single-wall carbon nanotubes in a laser-ablation process. *Appl. Phys. A - Mater Sci. Process.*, **2001**, *72*, 573-580.

[19]   Jose-Yacaman M.; Mikiyoshida M.; Rendon L.; Santiesteban J. G. Catalytic Growth of Carbon Microtubules with Fullerene Structure. *Appl. Phys. Lett.*, **1993**, *62*, 202-204.

[20]   Ivanov V.; Nagy J. B.; Lambin P.; Lucas A.; Zhang X. B.; Zhang X. F.; Bernaerts D.; Vantendeloo G.; Amelinckx S.; Vanlanduyt J. The Study of Carbon Nanotubules Produced by Catalytic Method. *Chem. Phys. Lett.*, **1994**, *223*, 329-335.

[21]   Meyyappan M.; Delzeit L.; Cassell A.; Hash D. Carbon nanotube growth by PECVD: a review. *Plasma Sources Sci. T*, **2003**, *12*, 205-216.

[22]   Wei Y. Y.; Eres G.; Merkulov V. I.; Lowndes D. H. Effect of catalyst film thickness on carbon nanotube growth by selective area chemical vapor deposition. *Appl. Phys. Lett.*, **2001**, *78*, 1394-1396.

[23]   Lee C. J.; Park J. Growth model of bamboo-shaped carbon nanotubes by thermal chemical vapor deposition. *Appl. Phys. Lett.*, **2000**, *77*, 3397-3399.

[24]   Ren Z. F.; Huang Z. P.; Xu J. W.; Wang J. H.; Bush P.; Siegal M. P.; Provencio P. N. Synthesis of large arrays of well-aligned carbon nanotubes on glass. *Science*, **1998**, *282*, 1105-1107.

[25] Bower C.; Zhu W.; Jin S. H.; Zhou O. Plasma-induced alignment of carbon nanotubes. *Appl. Phys. Lett.*, **2000**, *77*, 830-832.

[26] Bower C.; Zhou O.; Zhu W.; Werder D. J.; Jin S. H. Nucleation and growth of carbon nanotubes by microwave plasma chemical vapor deposition. *Appl. Phys. Lett.*, **2000**, *77*, 2767-2769.

[27] Chhowalla M.; Teo K. B. K.; Ducati C.; Rupesinghe N. L.; Amaratunga G. A. J.; Ferrari A. C.; Roy D.; Robertson J.; Milne W. I. Growth process conditions of vertically aligned carbon nanotubes using plasma enhanced chemical vapor deposition. *J. Appl. Phys.*, **2001**, *90*, 5308-5317.

[28] Delzeit L.; Mcaninch I.; Cruden B. A.; Hash D.; Chen B.; Han J.; Meyyappan M. Growth of multiwall carbon nanotubes in an inductively coupled plasma reactor. *J. Appl. Phys.*, **2002**, *91*, 6027-6033.

[29] Boskovic B. O.; Stolojan V.; Khan R. U. A.; Haq S.; Silva S. R. P. Large-area synthesis of carbon nanofibres at room temperature. *Nature Mater.*, **2002**, *1*, 165-168.

[30] Baddour C.; Briens C. Carbon nanotube synthesis: A review. *Inter. J..Chem. Reactor Eng.*, **2005**, *3*, R3.

[31] Rubianes M.; Rivas G. A. Enzymatic Biosensors Based on Carbon Nanotubes Paste Electrodes. *Electroanal.*, **2005**, *17*, 73-78.

[32] Prasek J.; Hubalek J.; Adamek M.; Kizek R. Carbon nanotubes paste versus graphite working electrodes in electrochemical analysis. *IEEE Sensors* **2007**, 1257-1260.

[33] Rubianes M. D.; Arribas A. S.; Bermejo E.; Chicharro M.; Zapardiel A.; Rivas G. Carbon nanotubes paste electrodes modified with a melanic polymer: Analytical applications for the sensitive and selective quantification of dopamine. *Sensor Actuat. B. - Chem*, **2010**, *144*, 274-279.

[34] Abdollah S.; Rahman H.; Gholam-Reza K. Amperometric Detection of Morphine at Preheated Glassy Carbon Electrode Modified with Multiwall Carbon Nanotubes. *Electroanal.*, **2005**, *17*, 873-879.

[35] Wu F.-H.; Zhao G.-C.; Wei X.-W. Electrocatalytic oxidation of nitric oxide at multi-walled carbon nanotubes modified electrode. *Electrochem. Commun.*, **2002**, *4*, 690-694.

[36] Wu K.; Sun Y.; Hu S. Development of an amperometric indole-3-acetic acid sensor based on carbon nanotubes film coated glassy carbon electrode. *Sensors and Actuators B: Chemical*, **2003**, *96*, 658-662.

[37] Musameh M.; Wang J.; Merkoci A.; Lin Y. Low-potential stable NADH detection at carbon-nanotube-modified glassy carbon electrodes. *Electrochem. Commun.*, **2002**, *4*, 743-746.

[38] Arun Prakash P.; Yogeswaran U.; Chen S.-M. Direct electrochemistry of catalase at multiwalled carbon nanotubes-nafion in presence of needle shaped DDAB for H_2O_2 sensor. *Talanta*, **2009**, *78*, 1414-1421.

[39] Chen S. Z.; Ye F.; Lin W. M. Carbon nanotubes-Nafion composites as Pt-Ru catalyst support for methanol electro-oxidation in acid media. *Journal of Natural Gas Chemistry*, **2009**, *18*, 199-204.

[40] Wang J. Carbon-nanotube based electrochemical biosensors: A review. *Electroanal.*, **2005**, *17*, 7-14.

[41] Lin Y. H.; Lu F.; Tu Y.; Ren Z. F. Glucose biosensors based on carbon nanotube nanoelectrode ensembles. *Nano Lett.*, **2004**, *4*, 191-195.

[42] Tu Y.; Lin Y.; Ren Z. F. Nanoelectrode Arrays Based on Low Site Density Aligned Carbon Nanotubes. *Nano Lett.*, **2003**, *3*, 107-109.

[43] Li J.; Ng H. T.; Cassell A.; Fan W.; Chen H.; Ye Q.; Koehne J.; Han J.; Meyyappan M. Carbon Nanotube Nanoelectrode Array for Ultrasensitive DNA Detection. *Nano Lett.*, **2003**, *3*, 597-602.

[44] Masuda H.; Fukuda K. Ordered Metal Nanohole Arrays Made by a Two-Step Replication of Honeycomb Structures of Anodic Alumina. *Science*, **1995**, *268*, 1466-1468.

[45] Chong A. S. M.; Tan L. K.; Deng J.; Gao H. Soft Imprinting: Creating Highly Ordered Porous Anodic Alumina Templates on Substrates for Nanofabrication. *Adv. Funct. Mater.*, **2007**, *17*, 1629-1635.

[46] Lee W.; Schwirn K.; Steinhart M.; Pippel E.; Scholz R.; Gosele U. Structural engineering of nanoporous anodic aluminium oxide by pulse anodization of aluminium. *Nature Nanotechnol.*, **2008**, *3*, 234-239.

[47] Musselman K. P.; Mulholland G. J.; Robinson A. P.; Schmidt-Mende L.; Macmanus-Driscoll J. L. Low-Temperature Synthesis of Large-Area, Free-Standing Nanorod Arrays on ITO/Glass and other Conducting Substrates. *Adv. Mater.*, **2008**, *20*, 4470-4475.

[48] Bao J.; Tie C.; Xu Z.; Suo Z.; Zhou Q.; Hong J. A Facile Method for Creating an Array of Metal-Filled Carbon Nanotubes. *Adv. Mater.*, **2002**, *14*, 1483-1486.

[49] Maschmann M. R.; Franklin A. D.; Amama P. B.; Zakharov D. N.; Stach E. A.; Sands T. D.; Fisher T. S. Vertical single- and double-walled carbon nanotubes grown from modified porous anodic alumina templates. *Nanotech.*, **2006**, *17*, 3925-3929.

[50] Yu B.; Moussy Y.; Moussy F. Coil-type implantable glucose biosensor with excess enzyme loading. *Front Biosci*, **2005**, *10*, 512-520.

[51] Yu B.; Long N.; Moussy Y.; Moussy F. A long-term flexible minimally-invasive implantable glucose biosensor based on an epoxy-enhanced polyurethane membrane. *Biosens. Bioelectron.*, **2006**, *21*, 2275-2282.

[52] Yu B.; Wang C.; Ju Y. M.; West L.; Harmon J.; Moussy Y.; Moussy F. Use of hydrogel coating to improve the performance of implanted glucose sensors. *Biosens. Bioelectron.*, **2008**, *23*, 1278-1284.

[53] Zhao Q.; Guan L. H.; Gu Z. N.; Zhuang Q. K. Determination of phenolic compounds based on the tyrosinase-single walled carbon nanotubes sensor. *Electroanal.*, **2005**, *17*, 85-88.

[54] Eftekhari A. Electropolymerization of aniline onto passivated substrate and its application for preparation of enzyme-modified electrode. *Synth Metals*, **2004**, *145*, 211-216.

[55] Fu Y. C.; Chen C.; Xie Q. J.; Xu X. H.; Zou C.; Zhou Q. M.; Tan L.; Tang H.; Zhang Y. Y.; Yao S. Z. Immobilization of enzymes through one-pot chemical preoxidation and electropolymerization of dithiols in enzyme-containing aqueous suspensions to develop biosensors with improved performance. *Anal. Chem.*, **2008**, *80*, 5829-5838.

[56] Wang J.; Musameh M. Carbon-nanotubes doped polypyrrole glucose biosensor. *Anal. Chim. Acta*, **2005**, *539*, 209-213.

[57] Joshi P. P.; Merchant S. A.; Wang Y. D.; Schmidtke D. W. Amperometric biosensors based on redox polymer-carbon nanotube-enzyme composites. *Anal. Chem.*, **2005**, *77*, 3183-3188.

[58] Liu J. Q.; Chou A.; Rahmat W.; Paddon-Row M. N.; Gooding J. J. Achieving direct electrical connection to glucose oxidase using aligned single walled carbon nanotube arrays. *Electroanal.*, **2005**, *17*, 38-46.

[59] Patolsky F.; Weizmann Y.; Willner I. Long-range electrical contacting of redox enzymes by SWCNT connectors. *Angew. Chem. Int. Ed. Engl.*, **2004**, *43*, 2113-2117.

[60] Zhu Z.; Song W.; Burugapalli K.; Moussy F.; Li Y.-L.; Zhong X.-H. Nano-yarn carbon nanotube fiber based enzymatic glucose biosensor. *Nanotech.*, **2010**, *21*, 165501.

[61] Yang W. R.; Ratinac K. R.; Ringer S. P.; Thordarson P.; Gooding J. J.; Braet F. Carbon Nanomaterials in Biosensors: Should You Use Nanotubes or Graphene? *Angew. Chem. Int. Ed. Engl.*, **2010**, *49*, 2114-2138.

[62] Besteman K.; Lee J. O.; Wiertz F. G. M.; Heering H. A.; Dekker C. Enzyme-coated carbon nanotubes as single-molecule biosensors. *Nano Lett.*, **2003**, *3*, 727-730.

[63] Star A.; Gabriel J. C. P.; Bradley K.; Gruner G. Electronic detection of specific protein binding using nanotube FET devices. *Nano Lett.*, **2003**, *3*, 459-463.

[64] Sudibya H. G.; Ma J. M.; Dong X. C.; Ng S.; Li L. J.; Liu X. W.; Chen P. Interfacing Glycosylated Carbon-Nanotube-Network Devices with Living Cells to Detect Dynamic Secretion of Biomolecules. *Angew. Chem. Int. Ed. Engl.*, **2009**, *48*, 2723-2726.

[65] Wang J. Electrochemical glucose biosensors. *Chem. Rev.*, **2008**, *108*, 814-825.

[66] Wang J.; Musameh M.; Lin Y. H. Solubilization of carbon nanotubes by Nafion toward the preparation of amperometric biosensors. *J. Am. Chem. Soc.*, **2003**, *125*, 2408-2409.

[67] Lim S. H.; Wei J.; Lin J. Y.; Li Q. T.; Kuayou J. A glucose biosensor based on electrodeposition of palladium nanoparticles and glucose oxidase onto Nafion-solubilized carbon nanotube electrode. *Biosens. Bioelectron.*, **2005**, *20*, 2341-2346.

[68] Veetil J. V.; Ye K. M. Development of immunosensors using carbon nanotubes. *Biotechnol. Progr.*, **2007**, *23*, 517-531.

[69] Okuno J.; Maehashi K.; Kerman K.; Takamura Y.; Matsumoto K.; Tamiya E. Label-free immunosensor for prostate-specific antigen based on single-walled carbon nanotube array-modified microelectrodes. *Biosens. Bioelectron.*, **2007**, *22*, 2377-2381.

[70] Maehashi K.; Katsura T.; Kerman K.; Takamura Y.; Matsumoto K.; Tamiya E. Label-free protein biosensor based on aptamer-modified carbon nanotube field-effect transistors. *Anal. Chem.*, **2007**, *79*, 782-787.

[71] Niu S. Y.; Zhao M.; Hu L. Z.; Zhang S. S. Carbon nanotube-enhanced DNA biosensor for DNA hybridization detection using rutin-Mn as electrochemical indicator. *Sensor Actuat. B-chem*, **2008**, *135*, 200-205.

[72] Wang S. G.; Wang R. L.; Sellin P. J.; Chang S. Carbon Nanotube Based DNA Biosensor for Rapid Detection of Anti-Cancer Drug of Cyclophosphamide. *Current Nanoscience*, **2009**, *5*, 312-317.

[73] Zhu N. N.; Gao H.; Xua Q.; Lin Y. Q.; Su L.; Mao L. Q. Sensitive impedimetric DNA biosensor with poly(amidoamine) dendrimer covalently attached onto carbon nanotube electronic transducers as the tether for surface confinement of probe DNA. *Biosens. Bioelectron.*, **2010**, *25*, 1498-1503.

[74] Pedano M. L.; Rivas G. A. Adsorption and electrooxidation of nucleic acids at carbon nanotubes paste electrodes. *Electrochem. Commun.*, **2004**, *6*, 10-16.

[75]   Martinez M. T.; Tseng Y. C.; Ormategui N.; Loinaz I.; Eritja R.; Bokor J. Label-Free DNA Biosensors Based on Functionalized Carbon Nanotube Field Effect Transistors. *Nano Lett.*, **2009**, *9*, 530-536.

[76]   Hussain M. A.; Kabir M. A.; Sood A. K. On the cytotoxicity of carbon nanotubes. *Curr. Sci. India*, **2009**, *96*, 664-673.

[77]   Monteiro-Riviere N. A.; Nemanich R. J.; Inman A. O.; Wang Y. Y. Y.; Riviere J. E. Multi-walled carbon nanotube interactions with human epidermal keratinocytes. *Toxicol. Lett.*, **2005**, *155*, 377-384.

[78]   Pantarotto D.; Briand J. P.; Prato M.; Bianco A. Translocation of bioactive peptides across cell membranes by carbon nanotubes. *Chem. Commun.*, **2004**, 16-17.

[79]   Poland C. A.; Duffin R.; Kinloch I.; Maynard A.; Wallace W. A. H.; Seaton A.; Stone V.; Brown S.; Macnee W.; Donaldson K. Carbon nanotubes introduced into the abdominal cavity of mice show asbestos-like pathogenicity in a pilot study. *Nat. Nanotech.*, **2008**, *3*, 423-428.

[80]   Schipper M. L.; Nakayama-Ratchford N.; Davis C. R.; Kam N. W. S.; Chu P.; Liu Z.; Sun X. M.; Dai H. J.; Gambhir S. S. A pilot toxicology study of single-walled carbon nanotubes in a small sample of mice. *Nature Nanotechnol.*, **2008**, *3*, 216-221.

[81]   Vigolo B.; Penicaud A.; Coulon C.; Sauder C.; Pailler R.; Journet C.; Bernier P.; Poulin P. Macroscopic fibers and ribbons of oriented carbon nanotubes. *Science*, **2000**, *290*, 1331-1334.

[82]   Ericson L. M.; Fan H.; Peng H. Q.; Davis V. A.; Zhou W.; Sulpizio J.; Wang Y. H.; Booker R.; Vavro J.; Guthy C.; Parra-Vasquez A. N. G.; Kim M. J.; Ramesh S.; Saini R. K.; Kittrell C.; Lavin G.; Schmidt H.; Adams W. W.; Billups W. E.; Pasquali M.; Hwang W. F.; Hauge R. H.; Fischer J. E.; Smalley R. E. Macroscopic, neat, single-walled carbon nanotube fibers. *Science*, **2004**, *305*, 1447-1450.

[83]   Jiang K. L.; Li Q. Q.; Fan S. S. Nanotechnology: Spinning continuous carbon nanotube yarns - Carbon nanotubes weave their way into a range of imaginative macroscopic applications. *Nature*, **2002**, *419*, 801-801.

[84]   Zhang M.; Atkinson K. R.; Baughman R. H. Multifunctional carbon nanotube yarns by downsizing an ancient technology. *Science*, **2004**, *306*, 1358-1361.

[85]   Koziol K.; Vilatela J.; Moisala A.; Motta M.; Cunniff P.; Sennett M.; Windle A. High-performance carbon nanotube fiber. *Science*, **2007**, *318*, 1892-1895.

[86]   Li Y. L.; Kinloch I. A.; Windle A. H. Direct spinning of carbon nanotube fibers from chemical vapor deposition synthesis. *Science*, **2004**, *304*, 276-278.

[87]   Motta M.; Moisala A.; Kinloch I. A.; Windle A. H. High performance fibres from 'Dog bone' carbon nanotubes. *Adv. Mater.*, **2007**, *19*, 3721-3726.

[88]   Zhong X. H.; Li Y. L.; Liu Y. K.; Qiao X. H.; Feng Y.; Liang J.; Jin J.; Zhu L.; Hou F.; Li J. Y. Continuous Multilayered Carbon Nanotube Yarns. *Adv. Mater.*, **2010**, *22*, 692-696.

[89]   Wang J.; Deo R. P.; Poulin P.; Mangey M. Carbon nanotube fiber microelectrodes. *J. Am. Chem. Soc.*, **2003**, *125*, 14706-14707.

[90]   Viry L.; Derre A.; Garrigue P.; Sojic N.; Poulin P.; Kuhn A. Optimized carbon nanotube fiber microelectrodes as potential analytical tools. *Anal. Bioanal. Chem.*, **2007**, *389*, 499-505.

[91]   Viry L.; Derre A.; Garrigue P.; Sojic N.; Poulin P.; Kuhn A. Optimized carbon nanotube fiber microelectrodes as potential analytical tools. *Anal. Bioanal. Chem.*, **2007**, *389*, 499-505.

[92]   Cai C. X.; Chen J. Direct electron transfer of glucose oxidase promoted by carbon nanotubes. *Anal. Biochem.*, **2004**, *332*, 75-83.

[93]   Ruifang G.; Jianbin Z. Amine-terminated ionic liquid functionalized carbon nanotube-gold nanoparticles for investigating the direct electron transfer of glucose oxidase. *Electrochem. Commun.*, **2009**, 608-611.

[94]   Zhao H. Z.; Sun J. J.; Song J.; Yang Q. Z. Direct electron transfer and conformational change of glucose oxidase on carbon nanotube-based electrodes. *Carbon*, **2010**, *48*, 1508-1514.

[95]   Tu Y.; Lin Y. H.; Yantasee W.; Ren Z. F. Carbon nanotubes based nanoelectrode arrays: Fabrication, evaluation, and application in voltammetric analysis. *Electroanal.*, **2005**, *17*, 79-84.

[96]   Yan Y. H.; Chan-Park M. B.; Zhang Q. Advances in carbon-nanotube assembly. *Small*, **2007**, *3*, 24-42.

[97]   Yun Y. H.; Dong Z.; Shanov V. N.; Doepke A.; Heineman W. R.; Halsall H. B.; Bhattacharya A.; Wong D. K. Y.; Schulz M. J. Fabrication and characterization of carbon nanotube array electrodes with gold nanoparticle tips. *Sensor Actuat. B - Chem.*, **2008**, *133*, 208-212.

[98]   Villamizar R. A.; Maroto A.; Rius F. X.; Inza I.; Figueras M. J. Fast detection of Salmonella Infantis with carbon nanotube field effect transistors. *Biosens. Bioelectron.*, **2008**, *24*, 279-283.

[99]   Park T. J.; Banerjee S.; Hemraj-Benny T.; Wong S. S. Purification strategies and purity visualization techniques for single-walled carbon nanotubes. *J. Mater. Chem.*, **2006**, *16*, 141-154.

[100] Yu A.; Bekyarova E.; Itkis M. E.; Fakhrutdinov D.; Webster R.; Haddon R. C. Application of Centrifugation to the Large-Scale Purification of Electric Arc-Produced Single-Walled Carbon Nanotubes. *J. Am. Chem. Soc.*, **2006**, *128*, 9902-9908.

[101] Lu F.; Wang X.; Meziani M. J.; Cao L.; Tian L.; Bloodgood M. A.; Robinson J.; Sun Y.-P. Effective Purification of Single-Walled Carbon Nanotubes with Reversible Noncovalent Functionalization. *Langmuir.*, **2010**, *26*, 7561-7564.

[102] Kam N. W. S.; Jessop T. C.; Wender P. A.; Dai H. J. Nanotube molecular transporters: Internalization of carbon nanotube-protein conjugates into mammalian cells. *J. Am. Chem. Soc.*, **2004**, *126*, 6850-6851.

CHAPTER 9

Biosensors Based on Selected Gold Nanoparticles

Ram Singh[1*], Geetanjali[2], Vinita Katiyar[3] and S. Bhanumati[4]

[1]*Department of Applied Chemistry, Delhi Technological University, Bawana Road, Delhi - 110 042, India;* [2]*Department of Chemistry, Kirori Mal College, University of Delhi, Delhi - 110 007, India;* [3]*Civil Engineering Department, Indian Institute of Technology, Delhi - 110 016, India and* [4] *Department of Chemistry, Gargi College, University of Delhi, Srifort Road, New Delhi - 110 049, India*

Abstract: Biosensors are devices which comprise of a biological recognition element and a transducer that is a physicochemical detector system. In other words, a biosensor is a sophisticated tool that combines a biological component; a bioreceptor, with a physicochemical detector component; the transducer, for the purpose of detection of an analyte. The transducer converts the information into a measurable effect such as an electric signal which can be measured and quantified. The sensitivity of biosensors can be improved and enhanced by changing the nano material in the construction. This field of research is fast developing. A conceptual background of biosensors based on selected Gold nanoparticles, their properties, synthesis, characterization and their associated uses has been described in this chapter.

Keywords: Gold nanoparticles; biosensors; smart nanoparticles; selectivity.

1. INTRODUCTION

A revolutionary development in nanotechnology has taken place through the integration of electronics, material sciences, and biology; put together as nanobiotechnology that caters to several fields like medicine, agriculture, industry and defence services. The domain of nanobiotechnology is concerned with the interfacing of naturally or synthetic biological materials with inorganic electronic materials, components or systems. This has led to the development of two broad categories of devices; one to transduce biochemical signals generated by biological components into electrical signals and the second to transduce electronically generated signals into biochemical signals. The first category of devices permits the monitoring of living cells, the second, enables control of cellular processes. Based on this, many devices like microbial bio-fuel cells (produce electricity), bioelectric reactors (control of cellular metabolism), living cell biosensors (detections) and devices that permit monitoring and control of mammalian physiology have been manufactured.

1.1. Biosensors

Biosensors are devices that contain sensors for quantitative measurement of samples and convert these into signals that can be displayed on a mechanical device. They produce measurable responses to change in physical conditions or chemical concentrations. The basic components of this include a sensing element and a signal transducer. Their very high sensitivity, fast response and low cost, make them an important unit of biological and medical applications [1, 2]. There are certain sensors that consist of biological recognition elements, often called bioreceptors [3] and are rightly given a name with prefix bio as biosensors [4, 5]. This was first reported in 1962 [6].

1.1.1. Principal Features of Biosensors

Biosensors have two basic differences from the conventional chemical sensors: (i) they use biological structures like cells, enzymes, or nucleic acids as sensing elements; and (ii) they are used to measure biological processes or changes (Fig. **1**). In other words, a biosensor is a device that detects, records, and

---

Address correspondence to Ram Singh: Department of Applied Chemistry, Delhi Technological University, Bawana Road, Delhi - 110 042, India; E-mail: singh_dr_ram@yahoo.com

Songjun Li, Yi Ge and He Li (Eds)
All rights reserved - © 2012 Bentham Science Publishers

transmits information regarding a physiological change or detects the presence of various chemical or biological materials present in the environment. Available in different sizes and shapes, they are used to monitor changes in environmental conditions. They can detect and measure concentrations of specific bacteria or hazardous chemicals and also measure acidity levels (pH). The search for novel devices that offer higher sensitivity, greater analyte discrimination, and lower operating costs, is a demanding exercise. Intensive research and consistent efforts are being put to improve the sensing and transducing performance of a biosensor in a big way.

1.1.2. Design of Biosensors

Biosensors consist of three parts: (i) The sensitive biological element such as tissue, micro organisms, organelles, cell receptors, enzymes, antibodies, nucleic acids, *etc.*, (ii) The transducer or the detector element that works in a physicochemical way. They transforms the signal resulting from the interaction of the analyte with the biological element into another signal, and (iii) An associated electronics or signal processors that are primarily responsible for the display of the results in a user-friendly way [7].

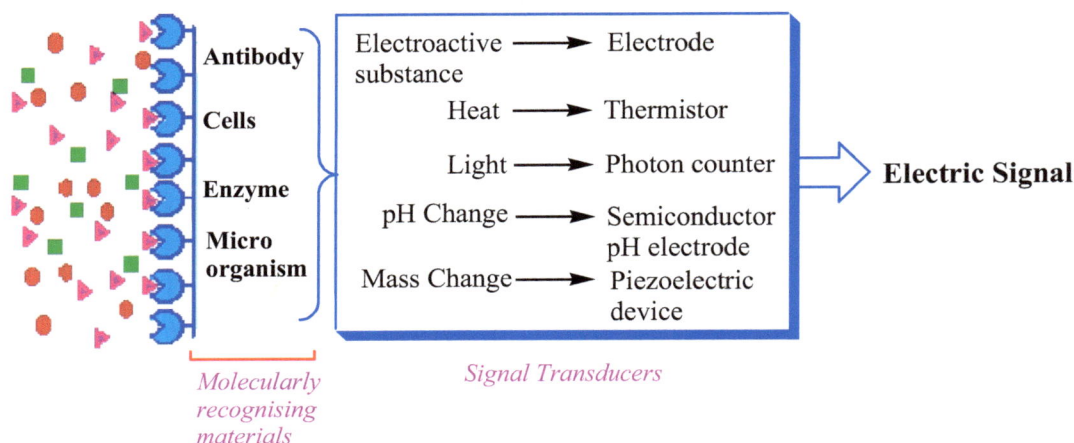

Figure 1. Principal features of a Biosensor.

1.1.3. Evolution of Biosensors

With the advent of nanotechnology, biosensing has entered in a new era [8-10]. The reason is the development of advanced sensors that can detect low level concentrations of analytes using a portable device that was impossible in the past [11-13]. New nano-materials have high strength, good electrical conductivity, nano scale size, and are compatible with biological molecules. Hence, they are ideal for developing biosensors with a low detection limit [11-13]. Many sensors operate through the variation of a surface parameter, like surface conductivity, with analyte concentration. The effective surface area of the device, *i.e.* the area actually interacting with the analyte, determines the sensitivity. For increase in the surface area, nanoparticles provide an easy answer. In recent years, a wide variety of nanoparticles with different properties have found broad application in biosensors. Owing to their small size (normally in the range of 1-100 nm), nanoparticles exhibit unique chemical, physical, and electronic properties that is different from those of bulk materials [14, 15]. The high surface-to-volume ratio of nanoparticles has been exploited for improving the performance of biosensors.

1.1.4. Types of Biosensors

Basically, biosensors can be divided into two main types:

(i)   Single molecule or molecular complexes (proteins: enzymes, antibodies, DNA, RNA, *etc.*

(ii)   Cell based biosensors (whole cells, tissues, whole organisms).

Based on the use, they can be classified as:

(i)   calorimetric.

(ii) colorimetric.

(iii) potentiometric.

(iv) amperometric.

(v)   optical and acoustic wave biosensors.

1.1.5. *Nanoparticles*

The interest in nanoparticles is due to the fact that, owing to the small size of the building blocks and the high surface to volume ratio, these particles demonstrate unique mechanical, optical, electronic and magnetic properties and find wide range of applications (Fig. **2**).

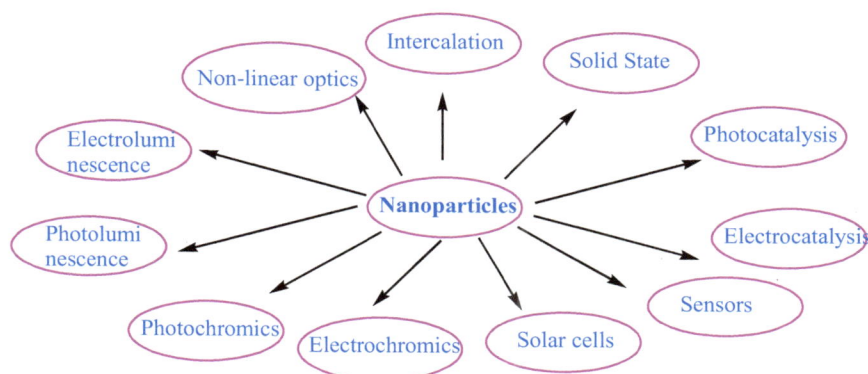

Figure 2. Application of nanoparticles.

Many kinds of nanoparticles find their applications in the construction of biosensors such as metal nanoparticles, oxide nanoparticles, semiconductor nanoparticles, and even nanodimensional conducting polymers. The molecular tool box is being constantly improved in the hope of revolutionizing devices and equipments that would judiciously blend biomolecules to chemically synthesized nano materials. For example, several groups have reported the use of gold and silver nanoparticles, or silver-silica hybrid nanostructures that are used in the construction of biosensor substrates [16-18]. Some oxide nanoparticles and semiconductor nanoparticles such as MnO_2 nanoparticles [19] and CdS nanoparticles [20] are also applied to construct biosensors. Due to different unique properties of nanoparticles, they always play different roles in different sensing systems. Usually, metal nanoparticles are always used as components of electronic wires. Oxide nanoparticles are often applied to immobilize biomolecules, while semiconductor nanoparticles are often used as labels or tracers [21].

Among the nano materials used as component in biosensors, gold nanoparticles have received greatest interest because they have several kinds of intriguing properties [22, 23]. Gold Nanoparticles (GNPs) with the diameter of 1-100 nm, have high surface-to-volume ratio and high surface energy to provide a stable immobilization of a large amount of biomolecules retaining their bioactivity. Moreover, they have an ability to permit fast and direct electron transfer between a wide range of electro active species and electrode materials. In addition, the light-scattering properties and extremely large enhancement ability of the local electromagnetic field enables GNPs to be used as signal amplification tags in diverse biosensors. Tiny gold nanoparticles linked to DNA could detect free floating target DNA.The gold bound complementary strands bind to the target strands and create a simple colour based detector created by the GNPs. This results in rapid diagnosis of diseases.

2. SMART NANOPARTICLES

The smart materials are the next advanced level of nanotechnology. They are defined as materials with properties engineered to change in a controlled manner under the influence of external stimuli [24]. These external influences can include temperature, force, moisture, electric charge, magnetic fields and pH. Many of the current examples of smart materials are biomimetic, since nature employs and depends on dynamic and rapid switching for critical functions such as vision, camouflage, and ion channel regulations [25]. Smart nonmaterials use nano-scale engineering and system integration of existing materials to continuously develop better materials and better products [26]. Existing smart materials are already an intrinsic part of modern society. These materials include: piezoelectric materials, thermoresponsive materials, shape memory alloys, polychromic, chromogenic or halochromic materials etc [27]. The smart nanomaterials are also expected to make their presence strongly felt in areas like healthcare, environment, energy generation and conservation, security and terrorism, textiles and surveillance [28-31].

3. SYNTHESIS AND CHARACTERIZATION OF NANOPARTICLES

3.1. Synthesis of Nanoparticles

Methods for the synthesis of nanoparticles are usually grouped into two categories: top-down and bottom-up. The first involves division of a massive solid into smaller portions. This approach may involve milling or attrition, chemical methods, and volatilization of a solid followed by condensation of the volatilized components. The second, bottom-up, method of nanoparticle fabrication involves condensation of atoms or molecular entities in a gas phase or in solution. The latter approach is far more popular in the synthesis of nanoparticles [32].

The gas-phase and sol-gel methods have been used for the preparation of nanoparticles with diameters ranging from 1 to 10 nm with consistent crystal structure, surface derivatization, and a high degree of monodispersity. The initial size variances were about 20%, but for measurable enhancement of the quantum effect, this must be reduced to less than 5% [33]. The evaporation and condensation in a sub-atmospheric inert-gas environment was initially used to prepare nanoparticles [34, 35]. Different aerosol processing techniques were reported to improve the production yield of nanoparticles [36, 37]. Some of these methods included the use of combustion flame [38, 39], plasma [40], spray pyrolysis [41] *etc.* Sol-gel processing is a wet chemical synthetic approach that can be used to generate nanoparticles by gelation, precipitation, and hydrothermal treatment [42]. Some other methods include sonochemical processing [43], cavitation processing [44], microemulsion processing [45, 46], and high-energy ball milling [47, 48]. Process for the preparation of monodispersed nanoparticles without the use of a size classification procedure was also developed. A monodispersed gold colloidal nanoparticle with diameters of about 1 nm was prepared by reduction of metallic salt with UV irradiation in the presence of dendrimers [49]. Poly(amidoamine) dendrimers with surface amino groups of higher generations have spherical 3-D structures, which may have an effective protective action for the formation of gold nanoparticles. The advent of carbon-based nanotubes has created yet another way to fabricate nanometer fibers and tubes. These nanotubes were used as templates for the fabrication of carbide and oxide nanotubes [50, 51].

Gold nanoparticles have attracted extensive interest due to their potential applications in electronics, optics, catalysts, and biosensors. Various procedures have been reported for the synthesis of hybrid gold nanoparticles modified by polymer [52]. These methods are classified as grafting from, grafting to, and post-modification techniques. The 'grafting from' method requires tedious synthesis and purification, while the 'grafting to' technique does not have a high grafting density. However, the post-modification technique can overcome the above drawbacks [52]. The post-modification method has been further simplified by using click chemistry and controlled free radical polymerization [53].

3.2. Characterization of Nanoparticles

Nanoparticle characterization is necessary to establish understanding and control of nanoparticles synthesis and applications. Characterization is done by using a variety of different techniques, mainly drawn from materials sciences. Common Techniques are Electron Microscopy (TEM, SEM), atomic force microscopy (AFM), Dynamic Light Scattering (DLS), X-ray Photoelectron Spectroscopy (XPS), X-ray diffraction (XRD), Fourier Transform Infrared Spectroscopy (FTIR), Matrix-Assisted Laser Desorption/Ionization Time-of-Flight Mass Spectrometry (MALDI-TOF), ultraviolet-visible spectroscopy, dual polarization interferometer and nuclear magnetic resonance (NMR) spectroscopy. It is difficult to determine one single best technique for all situations. Determining the best technique for a particular situation requires knowledge of the particles being analyzed, the ultimate application of the particles, and the limitations of techniques being considered. Depending on the application of interest, a number of techniques can be used to analyze and characterize nanoparticles. A partial list of commercially available techniques for particle measurement is given in Table **1**. The details of few important methods are discussed below:

3.2.1. Quasi-Elastic Light Scattering (QELS) Studies

QELS is particularly suited in determining small changes in mean diameter such as those due to adsorbed layers on the particle surface. This technique is used to determine the size distribution profile of small particles in suspension. In this method, a solution containing the particles is placed in the path of a monochromatic beam of light and the temporal fluctuations of the scattered light due to the Brownian motion of the particles are determined [54, 55].

Table 1. Partial list of commercially available techniques for particle measurement.

X-ray Diffraction (XRD)	Transmission Electron Microscopy (TEM)
Dynamic Light Scattering (DLS) or Photon Correlation Spectroscopy (PCS) or Quasi-elastic Light Scattering (QELS)	Gas Absorption Surface Area Analysis (e.g. BET)
Tapered Element Oscillating Microbalance (TEOM)	Electrokinetic Sonic Amplitude
Sieving	Electroacoustic Spectroscopy
Sedimentation (Gravitational & Sentrifugal)	Electrical Zone Sensing (Coulter Counting)
Scanning Electron Microscopy (SEM)	Differential Mobility Analysis (DMA)
Microelectrophoresis	Condensation Nucleus Counter (CNC)
Aerosol Mass Spectroscopy (Aerosol MS)	Cascade Impaction Light Microscopy or Optical Imaging
Laser Doppler Velocimetry (LDV)	Laser Light Diffraction or Static Light Scattering
Acoustic Attenuation Spectroscopy	

3.2.2. Transmission Electron Microscope (TEM)

The size of nanoparticles can be confirmed and also their shapes can be visualized by using the Transmission Electron Microscope (TEM). In TEM, a beam of high energy electrons irradiate a sample and the resulting image can be seen on a fluorescent screen. The sample should be thin, approximately 500Å. When the irradiated electron beam passes through the sample, a transmitted as well as a diffracted beam is observed. The image is found by interference between the transmitted and the diffracted beam. This permits a very high resolution of the order of 2Å [56].

3.2.3. Scanning Electron Microscope (SEM)

The Scanning Electron Microscope (SEM) is a microscope that uses electrons rather than light to form an image. There are many advantages to using the SEM instead of a light microscope. The SEM has a large depth of field, which allows a large amount of the sample to be in focus at one time. The SEM also produces images of high resolution, which means that closely spaced features can be examined at a high magnification. Preparation of the samples is relatively easy since most SEM only require the sample to be conductive. The combination of higher magnification, larger depth of focus, greater resolution, and ease of sample observation makes the SEM one of the most heavily used instruments in nanoparticle research areas today [57, 58].

3.2.4. Atomic Force Microscope (AFM)

AFM is based on the distance between tip and sample. If the distance between tip and sample is higher than 1nm, this is considered as non-contact mode. In this case, van der Waals, electrostatic, magnetic or capillary forces produce images. In contact mode, the leading role is played by ionic repulsive forces [59]. The distance between sample and tip is adjusted with the help of piezoelectric elements. The image is taken by scanning the sample relative to probing tip and measuring the deflection of the cantilever as a function of lateral position. The non-contact mode has advantage over the contact mode.

3.2.5. X-Ray Diffraction (XRD)

X-ray Diffraction (XRD) is a versatile, non-destructive technique that reveals detailed information about the chemical composition and crystallographic structure of natural and manufactured materials [60, 61]. These techniques are based on observing the scattered intensity of an X-ray beam hitting a sample as a function of incident and scattered angle, polarization, and wavelength or energy. Identification is achieved by comparing the X-ray diffraction pattern, or 'diffractogram', obtained from the unknown sample with an internationally recognized database containing reference patterns for more than 70,000 phases. Modern computer-controlled diffractometer systems use automatic routines to measure, record and interpret the unique diffractograms produced by individual constituent in even highly complex mixtures.

3.2.6. Zeta Potential

Zeta potential is a physical property that is exhibited by any particle in suspension. The formulations of suspensions and emulsions can be optimized with this potential. The description about this can reduce the time needed for trial formulations. It is also useful in producing long term stability [62]. The magnitude of the zeta potential gives an indication of the potential stability of the colloidal system.

The development of a net charge at the particle surface affects the distribution of ions in the surrounding interfacial region. This leads to an increased concentration of counter ions close to the surface causing an electrical double layer around each particle [63]. The liquid layer surrounding the particle exists as two parts; an inner region, called the Stern layer, where the ions are strongly bound and an outer diffuse region where they are less firmly attached. Within the diffuse layer there is a notional boundary inside which the ions and particles form a stable entity. When a particle moves (*e.g.* due to gravity), ions within the boundary move with it, but any ions beyond the boundary do not travel with the particle. This boundary is called the surface of hydrodynamic shear or slipping plane. The potential that exists at this boundary is known as the Zeta potential [63].

4. APPLICATION OF GOLD NANOPARTICLES (GNPs) AS BIOSENSORS

Gold is a chemical element with the symbol Au and an atomic number of 79. It has been a highly sought-after precious metal for coinage, jewellery, and other articles since the beginning of recorded history. The use of this metal as nanoparticles is well known. The unique properties of GNPs have stimulated interest in the application of GNPs in interfacing biological recognition events with signal transduction and in designing biosensing devices exhibiting novel functions [64]. Examples of the selected GNP based biosensors are discussed below.

4.1. Gold Nanoparticle (GNP)-Based Electrochemical Biosensors

Electrochemical biosensors convert the biological bindings into useful electrical signals. They have fast, simple and low-cost detection capabilities. The excellent biocompatibility, conductivity and catalytic properties make GNPs as a potential candidate for electrochemical biosensing. They are also used to amplify the electrode surfaces, enhance the electron transfer between redox centers in proteins, and as catalysts to increase electrochemical reactions [65].

4.1.1. GNPs Act as Immobilization Platform

The adsorption of biomolecules directly onto naked surfaces of bulk materials usually result in their denaturation and loss of bioactivity. The GNPs are excellent candidates for the immobilization platform. Due to the biocompatibility and the high surface free energy of GNPs, they provide an amazing platform for the adsorption of biomolecules onto the surfaces and still can retain their bioactivity and stability [66-68]. A sensitive and reagent less electrochemical glucose biosensor based on surface-immobilized periplasmic glucose receptors on GNPs was developed [69]. The sensor was fabricated by immobilization of genetically engineered periplasmic glucose receptors to the GNPs, and showed selective detection of glucose in the micromolar concentration range, with a detection limit of 0.18 µM [69]. Some other biosensors based on the immobilization of different proteins with GNPs include horseradish peroxidase [70], microperoxidase-11 [71], tyrosinase [72] and human serum albumin [73]. The biosensing elements are attached to the gold surface *via* thiol linkers [74], and some are immobilized through covalent bonds [75], or amine groups [76]. Thiol groups are the most widely used groups for DNA and gold linkages [77, 78]. GNPs are also conjugated with other nanomaterials to improve their binding capacity [79].

4.1.2. GNPs Act as Electron Transfer "Electron Wires"

The electron transfer between the redox-protein and the electrode surface is an important bioelectrochemical reaction in the living system. The difficult aspect is the surrounding environment. The active centres of most oxidoreductases are surrounded by considerable thick insulating protein shells, resulting in the blockage of electron transfer between electrodes and the active centres. This lead to poor analytical performance of electrochemical biosensing without electron transfer mediators. The conductivity properties of GNPs enhance the electron transfer between the active centres of proteins and electrodes and thus they act as electron transfer electron wires [80-82]. The studies showed that the excellent electron transfer ability of GNPs are due to an environment similar to that of redox proteins in native systems and give the protein molecules more freedom in orientation. GNPs dispersed in polymeric matrices are also used to construct electrochemical biosensors with increased stability, improved process ability, reusability and solubility in a variety of solvents [83]. The nanocomposite of GNPs and biopolymer has also been employed as an excellent matrix for fabricating novel biosensors. A nanocomposite composed of carboxymethyl chitosan and GNPs have been used for H_2O_2 bioelectrochemical sensing [84]. GNP modified silicon nanowire is also suitable for the detection of biomolecules. The wire is functionalized with cysteamine and used as a working electrode in a three-electrode system for cyclic voltammetric detection of glutathione in solution. The sensor showed a wide range of linear concentration dependence with high sensitivity [85].

4.1.3. GNPs Act as Electrocatalyst

The GNPs exhibit extraordinary catalytic activity, which has been widely utilized in electrochemical biosensing [86]. The studies on the effect of GNPs diameter and supported material on the catalytic activity of GNPs showed that very small gold entities (~1.4 nm) derived from 55-atom gold clusters and support materials are efficient and robust catalysts for the selective oxidation of styrene by dioxygen [87]. A sharp size threshold in catalytic activity, where particles with diameters of 2 nm and above are completely inactive, was also determined [87]. An electrocatalytic oxidation of NADH by GNPs has been studied [88]. The biosensor was fabricated by self assembling of GNPs on a (3-mercaptopropyl)-trimethoxysilane (MPTS) modified polycrystalline gold (polygold) electrode, which showed a decrease in overpotential of 780mV in the presence of NADH without any electron transfer mediators [88]. Nitric oxide electrochemical sensors fabricated on GNPs modified platinum microelectrode [89] or dense GNP film modified electrode [90] was also reported to be based on the catalytic oxidation of GNPs. A glucose biosensor has been developed utilizing the electrodeposited biocomposites of GNPs and glucose oxidase enzyme [91]. Also, the GNPs could be utilized to fabricate enzyme-free biosensors [92]. The catalysis by GNPs also exhibits selectivity that enables selective electrochemical analysis [93].

4.2. GNP-Based Piezoelectric Biosensors

Piezoelectric biosensors measure the mass change arising from the biological recognition process. Quartz has been used in piezoelectric analysis and hence sensitivity of Quartz Crystal Microbalance (QCM) was the prime focus because for biosensing, the target detection molecules are always in trace quantity. As the high density and high surface-to-volume ratio of GNPs can amplify the mass change on the crystals during the analysis, numerous research groups focus on improving the analytical sensitivity by coupling GNPs with the QCM sensing process. Few research groups have reported amplified DNA microgravimetric sensor by GNPs [94, 95]. Studies have also been done with the immobilization of GNPs onto the gold surface of QCM, followed by immobilization of 17mer-oligonucleotide probes onto the GNPs [94]. GNPs are also successfully used as amplification tags in some other DNA piezoelectric biosensors, such as a DNA point mutation detection method based on DNA ligase reaction and GNP amplification [96]. The work has also been performed for immuno-sensing or ligand-sensing [97]. The conjugation of GNPs with other nanoparticles also gains much interest in piezoelectric biosensors. For example, the employment of GNP/HA [hydroxyapatite $Ca_5(PO_4)_3(OH)$] hybrid nanomaterial in QCM endowed an α-fetoprotein immunosensor with a remarkably higher sensitivity than that used only GNP or HA [98]. The incorporation of γ-Fe_2O_3, a magnetic material that allows magnetic separation, into the GNP/HA hybrid nanomaterial, a reusable piezoelectric immunosensor using antibody-adsorbed magnetic nanocomposite was also developed [99].

4.3. GNP-Based Optical Biosensors

Optical biosensors generally measure changes in light or photon output. Out of several optical sensing modalities for GNPs, the Surface Plasmon Resonance (SPR) has been widely explored. SPR is an optical phenomenon arising from the interaction between an electromagnetic wave and the conduction electrons in a metal. This resonance is a consistent oscillation of the surface conduction electrons excited by electromagnetic radiation. The binding of specific molecules onto the surface of metallic films can induce a variation in the dielectric constant, which can cause a change in the reflection of laser light from a metal-liquid surface (Fig. 3) [100]. The basic application of SPR includes probing and characterization of physicochemical changes of thin films on metal surface [100]. GNPs have the ability to amplify the SPR signals. A fiber-based biosensor for organophosphorous pesticide determination has been developed utilizing the Localized Surface Plasmon Resonance (LSPR) effect on GNPs [101]. The comparative study of the fiber sensor with and without GNPs suggested that the GNPs coated on optical fiber can substantially enhance the sensitivity of the sensor [101].

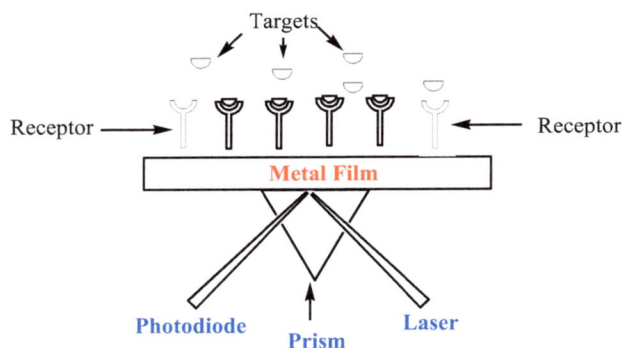

Figure 3. Surface Plasmon Resonance Detection Unit.

A sensing method for the detection of DNA hybridization by GNP-enhanced SPR has also been demonstrated [102]. Similarly, Li *et al.* developed a sensitive method for the analysis of single nucleotide polymorphisms (SNPs) in genomic DNA using GNP-enhanced Surface Plasmon Resonance Imaging (SPRI) [103]. In addition to SPR-based biosensors, GNPs are also incorporated into other optical structures, like interferometer-based biosensors [104], Raman scattering (SERS) [105] *etc.* A multiplexed detection method of oligonucleotide targets used nanoparticles functionalized with oligonucleotides and Raman-active dyes [106].

4.4. GNP-Based Nucleic Acid Biosensors

Various Nucleic Acid Biosensors (NAB) in connection with different transducers are available [107]. The sensitivities of NAB have been enhanced by the use of nanomaterial labels and novel signal amplification strategies labels but most of them have not been applied for routine use in the research laboratories or in clinical diagnosis applications because of their less accuracy, reproducibility, and complex operations, such as multiple incubation and washing steps. Hence, further modification becomes important. For the nucleic acid biosensors to be applicable at the near patient (point of care or POC), or near process level, simple, easy-to-use, cost competitive systems are required. The emergence of Lateral Flow Nucleic Acid (LFNA) test strips, also called dipstick test strips, offer a promising approach to realize POC nucleic acid detection. The use of lateral flow immunoassays for the detection of DNA amplification products has been studied [108]. A study has also been done on a dipstick-type LFNA biosensors with nanomolar detection limit based on dye-encapsulating liposome labels [109]. The biosensors have been applied to detect Dengue virus in blood samples [110] and viable Escherichia coli in drinking water [111]. Most of the formats of the reported LFNA test strips limit their infield and POC applications. Hence, a dry-reagent strip biosensor based on DNA aptamer functionalized GNP probes for qualitative (visual)/ quantitative detection of protein within minutes has been developed. Recently, a Disposable Nucleic Acid Biosensor (DNAB) based on the oligonucleotide functionalized GNPs and lateral flow device for fast, sensitive, and POC detection of nucleic acid samples has been developed [112].

5. CONCLUSIONS AND FUTURE PROSPECTS

The role of sensors in medicine and individual medicine is becoming increasingly essential. If a sensor can detect a specific protein called a biomarker or signaling molecule in either the inside or outside of cells, early diagnosis of diseases is possible. Last decade has seen the work in the area of intracellular signaling transduction which regulates intracellular activities and results in the secretion of cytokines and growth factors for tissue organization. Errors in a specific intracellular pathway can develop into a certain disease. A biosensor has to play a leading role in this detection.

Nanotechnology is playing an important role in the development of sensors. As to a biosensor, the vital parameters are its sensitivity and detection ranges. In recent years, nanoparticles relating to biosensors show significant maturation and hence have very high-impact applications. The major functions of nanoparticle-based biosensors include glucose biosensors, electrochemical biosensors, enzyme biosensors, optical biosensors, and amperometric biosensors. Nanoparticles play different roles under different environment for example one type of nanoparticle can play different roles in different biosensor systems, and it can also play more than one role in the same biosensor system. Various types of nanoparticles are associated with biosensors to increase their applicability in different aspects of life.

Among the nanoparticles used as component in biosensors, Gold Nanoparticles (GNPs) have received maximum attention because of their several exciting properties. GNPs, with the diameter of 1-100 nm, have high surface-to-volume ratio and high surface energy to provide stable immobilization of a large amount of biomolecules preserving their bioactivity. Moreover, GNPs have an ability to permit fast and direct electron transfer between a wide range of electroactive species and electrode materials. In addition, the light-scattering properties and extremely large enhancement ability of the local electromagnetic field enables GNPs to be used as signal amplification tags in diverse biosensors. Major efforts have been undertaken to create increasingly sophisticated GNPs that start to mimic biological materials with respect to precision, architecture, and functionality. Combinations of theoretical and experimental methods significantly widen the design parameters available for smart GNPs. With the advent of novel technologies, the control of the shape and dimensions of gold based nano-objects and the design of specific properties are no longer unapproachable challenges.

GNPs-based biosensors are an inter-discipline research area. There is a need to bring cooperative research network. There must be more frequent conferences to bring together the biomedical researchers with the chemists, materials scientists, and the physicists, to share and disseminate knowledge and information on the enhancement of various biosensors derived from gold nanoparticles.

ACKNOWLEDGMENTS

We are thankful to Dr. Rupesh Kumar and Dr. Amit Saxena, University of Delhi, India for providing us with several literature references.

REFERENCES

[1]   Ferrari, M. Cancer nanotechnology: opportunities and challenges. *Nat. Rev. Cancer*, **2005**, *5*, 161-171.

[2]   Nishiyama, N.; Kataoka, K. Current state, achievements, and future prospects of polymeric micelles as nanocarriers for drug and gene delivery. *Pharmacol. Ther.*, **2006**, *112*, 630-648.

[3]   Vo, Dinh, T.; Cullum, B. Biosensors and biochips: Advances in biological and medical diagnostics. *Fresenius' J. Anal. Chem.*, **2000**, *366*, 540-551.

[4]   International Union of Pure and Applied Chemistry. "biosensor". Compendium of Chemical Terminology, Internet edition.

[5]   Cavalcanti, A.; Shirinzadeh, B.; Zhang, M.; Kretly, L.C. Nanorobot hardware architecture for medical defense. *Sensors*, 2008, *8*, 2932-2958.

[6]   Clark, Jr. L.C.; Lyons, C. Electrode systems for continuous monitoring in cardiovascular surgery. *Ann. N.Y. Acad. Sci.*, **1962**, *102*, 29-45.

[7]   Velasco, M.N. Optical biosensors for probing at the cellular level: A review of recent progress and future prospects. *Semin Cell Dev. Biol.*, **2009**, *20*, 27-33.

[8]   Shipway, A.N.; Katz, E.; Willner, I. Nanoparticle arrays on surfaces for electronic, optical, and sensor applications. *Chem. Phys. Chem.*, **2008**, *1*, 18-52.

[9]   Chen, J.R.; Miao, Y.Q.; He, N.Y.; Wu, X.H.; Li, S.J. Nanotechnology and biosensors. *Biotechnol. Adv.*, **2004**, *22*, 505-518.

[10]   Luo, X.L.; Morrin, A.; Killard, A.J.; Smyth, M.R. Application of nanoparticles in electrochemical sensors and biosensors. *Electroanalysis*, **2006**, *18*, 319-326.

[11]   You, C.; Bhagawati, M.; Brecht, A.; Piehler, J. Affinity capturing for targeting proteins into micro and nanostructures. *Anal. Bioanal. Chem.*, **2009**, *393*, 1563-1570.

[12]   Fan, X.; White, I.M.; Shopova, S.I.; Zhu, H.; Suter J.D.; Sun, Y. Sensitive optical biosensors for unlabeled targets: A review. *Anal. Chim. Acta*, **2008**, *620*, 8-26.

[13]   Khanna, V.K. New-generation nano-engineered biosensors, enabling nanotechnologies and nanomaterials. *Sens. Rev.*, **2008**, *28*, 39-45.

[14]   Luo, X.L.; Morrin, A.; Killard, A.J.; Smyth, M.R. Application of nanoparticles in electrochemical sensors and biosensors. *Electroanalysis*, **2006**, *18*, 319-326.

[15]   Hamley, I.W. Nanotechnology with soft materials. *Angew. Chem. Int. Ed. Eng.*, **2003**, *42*, 1692-1712.

[16]   Xiao, Y.; Patolsky, F.; Katz, E.; Hainfeld, J.F.; Willner, I. Plugging into enzymes: Nanowiring of redox enzymes by a gold nanoparticle. *Science*, **2003**, *299*, 1877-1881.

[17]   Schierhorn, M.; Lee, S.J.; Boettcher, S.W.; Stucky, G.D.; Moskovits, M. Metal–Silica hybrid nanostructures for surface-enhanced raman spectroscopy. *Adv. Mater.*, **2006**, *18*, 2829-2832.

[18]   Cai, H.; Xu, Y.; Zhu, N.; He, P.; Fang, Y. An electrochemical DNA hybridization detection assay based on a silver nanoparticle label. *Analyst*, **2002**, *127*, 803-808.

[19]   Luo, X.L.; Xu, J.J.; Zhao, W.; Chen, H.Y. A novel glucose ENFET based on the special reactivity of MnO_2 nanoparticles. *Biosens. Bioelectron.*, **2004**, *19*, 1295-1300.

[20]   Wang, J.; Liu, G.; Polsky, R.; Merkoçi, A. Electrochemical stripping detection of DNA hybridization based on cadmium sulfide nanoparticle tags. *Electrochem. Commun.*, **2002**, *4*, 722-726.

[21]   Luo, X.; Morrin, A.; Killard, A.J. Smyth, M.R. Application of nanoparticles in electrochemical sensors and biosensors. *Electroanalysis*, **2006**, *18*, 319-326.

[22]   Wang, J.; Polsky, R.; Xu, D. Silver-enhanced colloidal gold electrochemical stripping detection of DNA hybridization. *Langmuir*, **2001**, *17*, 5739-5741.

[23]   Wang, J.; Xu, D.; Polsky, R. Magnetically-Induced solid-state electrochemical detection of DNA hybridization. *J. Am. Chem. Soc.*, **2002**, *124*, 4208-4209.

[24]   Yoshida, M.; Lahann, J. Smart nanomaterials. *ACS Nano*, **2008**, *2*, 1101-1107.

[25]   Langer, R.; Tirrell, D.A. Designing materials for biology and medicine. *Nature*, **2004**, *428*, 487-492.

[26]   Ballauff, M.; Lu, Y. Smart nanoparticles: Preparation, characterization and applications, *Polymer*, **2007**, *48*, 1815-1823.

[27]   http://www.azonano.com/Details.asp?ArticleID=1877 (Accessed June 20, 2010).

[28]   http://www.nanotech-now.com (Accessed June 20, 2010).

[29]   Space Mission Analysis and Design, W. J. Larson and James Wertz, Space Technology Library, published by Microcosm, Inc., 2nd ed., **1992**.

[30]   Loh, K.S.; Lee, Y.H.; Musa, A.; Salmah, A.A.; Zamri, I. Use of Fe_3O_4 Nanoparticles for enhancement of biosensor response to the herbicide 2,4-dichlorophenoxyacetic acid. *Sensors*, **2008**, *8*, 5775-5791.

[31]   Lia, S.; Tiwari, A.; Gec, Y.; Fei, D. A pH-responsive, low crosslinked, molecularly imprinted insulin delivery system. *Adv. Mat. Lett.*, **2010**, *1*, 4-10

[32]   Gopalakrishnan J. Chimie Douce approaches to the synthesis of metastable oxide materials. *Chem. Mater.* **1995**, *7*, 1265-1275.

[33]   Murray, C.B.; Norris, D.J.; Bawendi, M.G. Synthesis and characterization of nearly monodisperse CdE (E = sulfur, selenium, tellurium) semiconductor nanocrystallites. *J. Am. Chem. Soc.*, **1993**, *115*, 8706-8715.

[34]   Gleiter, H. Nanocrystalline materials. *Prog. Mater. Sci.*, **1989**, *33*, 223-315.

[35]   Siegel, R.W. *Physics of new materials*. F.E. Fujita (Ed.), Springer Series in Materials Science, *27*, Berlin: Springer, **1994**.

[36]   Friedlander, S.K.; Jang, H.D.; Ryu, K.H. Elastic behavior of nanoparticle chain aggregates. *Appl. Phy. Lett.*, **1998**, *72*, 173-175.

[37]   Uyeda, R. Studies of ultrafine particles in Japan: crystallography. *Prog. in Mater. Sci.*, **1991**, *35*, 1-96.

[38]   Axelbaum, R.L. Developments in sodium/halide flame technology for the synthesis of unagglomerated non-oxide nanoparticles. In Proc. of the Joint NSF-NIST Conference on Nanoparticles: Synthesis, Processing into Functional Nanostructures and Characterization, May 12-13, Arlington, VA, **1997**.

[39]   Pratsinis, S.E. Precision synthesis of nanostructured particles. *In Proc. of the Joint NSF-NIST Conference on Nanoparticles: Synthesis, Processing into Functional Nanostructures and Characterization*, May 12-13, Arlington, VA, **1997**.

[40]   Rao, N.P.; Tymiak, N.; Blum, J.; Neuman, A.; Lee, H.J.; Girshick, S.L.; McMurry, P.H.; Heberlein, J. Nanostructured materials production by hypersonic plasma particle deposition. *In Proc. of the Joint NSF-NIST Conference on Nanoparticles: Synthesis, Processing into Functional Nanostructures and Characterization*, May 12-13, Arlington, VA, **1997**.

[41]   Messing, G.L.; Zhang, S.; Selvaraj, U.; Santoro, R.J.; Ni, T. Synthesis of composite particles by spray pyrolysis. *In Proc. of the Joint NSF-NIST Conference on Nanoparticles: Synthesis, Processing into Functional Nanostructures and Characterization,* May 12-13, Arlington, VA, **1997**.

[42]   Kung, H.H.; Ko, E.I. Preparation of oxide catalysts and catalysts supports - a review of recent advances, *Chem. Eng. J.*, **1996**, *64*, 203-214.

[43]   Suslick, K.S.; Hyeon, T.; Fang, F. Nanostructured materials generated by high-intensity ultrasound: sonochemical synthesis and catalytic studies. *Chem. Mater.*, **1996**, *8*, 2172-2179.

[44]   Sunstrom, J.E. IV; Moser, W.R.; Marshik-Guerts, B. General route to nanocrystalline oxides by hydrodynamic cavitation. *Chem. Mater.*, **1996**, *8*, 2061-2067.

[45]   Hopwood, J.; Mann, S. Synthesis of barium sulfate nanoparticles and nanofilaments in reverse micelles and microemulsions. *Chem. Mater.*, **1997**, *9*, 1819-1828.

[46]   Pillai, V.; Kumar, P.; Hou, M.J.; Ayyub, P.; Shah, D.O. Preparation of nanoparticles of silver halides, superconductors and magnetic materials using water-in-oil microemulsions as nano-reactors. *Adv. Colloid Interf. Sci.*, **1995**, *55*, 241-269.

[47]   Leslie-Pelecky, D.L.; Reike, R.D. Magnetic properties of nanostructured materials *Chem. Mater.*, **1996**, *8*, 1770-1783.

[48]   Koch, C.C. Materials synthesis by mechanical alloying. *Ann. Rev. Mater. Sci.*, **1989**, *19*, 121-143.

[49]   Esumi, A.; Suzuki, A.; Aihara, N.; Uswi, K.; Torigoe, K. Preparation of gold colloids with UV irradiation using dendrimers as stabilizer. *Langmuir*, **1998**, *14*, 3157-3159.

[50]   Dai, H.; Wong, E.W.; Lu, Y.Z.; Fan, S.; Leiber, C.M. Synthesis and characterization of carbide nanorods. *Nature*, **1995**, *375*, 769-771.

[51]   Kasuga, T.; Hiramatsu, M.; Hoson, A.; Sekino, T.; Niihara, K. Formation of titanium oxide nanotube. *Langmuir*, **1998**, *14*, 3160-3163.

[52]   Shan, J.; Tenhu, H. Recent advances in polymer protected gold nanoparticles: synthesis, properties and applications. *Chem. Commun.,* **2007**, 4580-4598.

[53]   Zhang, T.; Zheng, Z.; Ding, X.; Peng, Y. Smart surface of GNPs fabricated by combination of RAFT and click chemistry. *Macromol. Rapid Commun.*, **2008**, 29, 1716-1720.

[54]   Chu B.; *Laser Light Scattering*, Academic Press, New York, **1994**.

[55]   Berne, B.J.; Pecora, R. *Dynamic Light Scattering*; Wiley: New York, **1976**.

[56]   http://nobelprize.org/educational/physics/microscopes/tem/index.html ((Accessed July 03, 2010).

[57] Suzuki, E. High-resolution scanning electron microscopy of immunogold-labelled cells by the use of thin plasma coating of osmium. *J. Microsc.*, **2002**, *208*, 153-157.

[58] Goldstein, G. I.; Newbury, D.E.; Echlin, P.; Joy, D.C.; Fiori, C.; Lifshin, E. *Scanning electron microscopy and x-ray microanalysis.* New York: Plenum Press, **1981**.

[59] Waser, R. *Nanoelectronics and information technology: Advanced Electronic Materials and Nobel devices*, Wiley VCH, Second Edition, **2005**.

[60] Azároff, L.V.; Kaplow, R.; Kato, N.; Weiss, R.J.; Wilson, A.J.C.; Young, R.A. *X-ray diffraction*, McGraw-Hill, **1974**.

[61] Glatter, O.; Kratky, O. *Small Angle X-ray Scattering.* Academic Press, **1982**.

[62] Lyklema, J. *Fundam. Interface. Colloid. Sci.*, 2, page 3.208, **1995**.

[63] www.nbtc.cornell.edu/facilities/./Zetasizer%20chapter%2016.pdf (assessed on july 10, **2010**.

[64] Lia, Y.; Schluesener, H.J.; Xua, S. Gold nanoparticle-based biosensors. *Gold Bulletin*, **2010**, *43*, 29-41.

[65] Katz, E.; Willner, I.; Wang, J. Electroanalytical and bioelectroanalytical systems based on metal and semiconductor nanoparticles. *Electroanalysis*, **2004**, *16*, 19-44.

[66] Gole, A.; Dash, C.; Ramakrishnan, V.; Sainkar, S.R.; Mandale, A.B.; Rao, M.; Sastry, M. Pepsin–Gold colloid conjugates: preparation, characterization, and enzymatic activity. *Langmuir*, **2001**, *17*, 1674-1679.

[67] Gole, A.; Vyas, S.; Phadtare, S.; Lachke, A.; Sastry, M. Studies on the formation of bioconjugates of *Endoglucanase* with colloidal gold. *Colloid. Surface. B*, **2002**, *25*, 129-138.

[68] Tang, L.; Zeng, G.M.; Shen, G.L.; Li, Y.P.; Zhang, Y.; Huang, D.L. Rapid detection of picloram in agricultural field samples using a disposable immunomembrane-based electrochemical sensor. *Environ. Sci. Technol.*, **2008**, *42*, 1207-1212.

[69] Andreescu, S.; Luck, L.A. Studies of the binding and signaling of surface-immobilized periplasmic glucose receptors on gold nanoparticles: A glucose biosensor application. *Anal. Biochem.*, **2008**, *375*, 282-290.

[70] Luo, X.L.; Xu, J.J.; Zhang, Q.; Yang, G.J.; Chen, H.Y. Electrochemically deposited chitosan hydrogel for horseradish peroxidase immobilization through gold nanoparticles self-assembly. *Biosens. Bioelectron.*, **2005**, *21*, 190-196.

[71] Patolsky, F.; Gabriel, T.; Willner, I. Controlled electrocatalysis by microperoxidase-11 and Au-nanoparticle superstructures on conductive supports. *J. Electroanal. Chem.*, **1999**, *479*, 69-73.

[72] Liu, Z.M.; Wang, H.; Yang, Y.; Yang, H.F.; Hu, S.Q.; Shen, G.L.; Yu, R.Q. Amperometric tyrosinase biosensor using enzyme-labeled Au colloids immobilized on cystamine/chitosan modified gold surface. *Anal. Lett.*, **2004**, *37*, 1079-1091.

[73] Yin, T.; Wei, W.; Yang, L.; Liu, K.; Gao, X. Kinetics parameter estimation for the binding process of salicylic acid to human serum albumin (HSA) with capacitive sensing technique. *J. Biochem. Biophys. Methods*, **2007**, *70*, 587-593.

[74] Zhao, S.; Zhang, K.; Bai, Y.; Yang, W.; Sun, C. Glucose oxidase/colloidal gold nanoparticles immobilized in Nafion film on glassy carbon electrode: Direct electron transfer and electrocatalysis. *Bioelectrochemistry*, **2006**, *69*, 158-163.

[75] Zhang, S.; Wang, N.; Yu, H.; Niu, Y.; Sun, C. Covalent attachment of glucose oxidase to an Au electrode modified with gold nanoparticles for use as glucose biosensor. *Bioelectrochemistry*, **2005**, *67*, 15-22.

[76] Li, X.; Yuan, R.; Chai, Y.; Zhang, L.; Zhuo, Y.; Zhang, Y. Amperometric immunosensor based on toluidine blue/nano-Au through electrostatic interaction for determination of carcinoembryonic antigen. *J. Biotechnol.*, **2006**, *123*, 356-366.

[77] Jin, R.; Wu, G.; Li, Z.; Mirkin, C.A.; Schatz, G.C. What controls the melting properties of DNA-linked gold nanoparticle assemblies? *J. Am. Chem. Soc.*, **2003**, *125*, 1643-1654.

[78] Kang, J.; Li, X.; Wu, G.; Wang, Z.; Lu, X. A new scheme of hybridization based on the Au$_{nano}$–DNA modified glassy carbon electrode. *Anal. Biochem.*, **2007**, *364*, 165-170.

[79] Wang, H.; Wang, X.; Zhang, X.; Qin, X.; Zhao, Z.; Miao, Z.; Huang, N.; Chen, Q. A novel glucose biosensor based on the immobilization of glucose oxidase onto gold nanoparticles-modified Pb nanowires. *Biosens. Bioelectron.*, **2009**, *25*, 142-146.

[80] Brown, K.R.; Fox, A.P.; Natan, M.J. Morphology-Dependent electrochemistry of cytochrome *c* at Au colloid-modified SnO$_2$ electrodes. *J. Am. Chem. Soc.*, **1996**, *118*, 1154-1157.

[81] Liu, S.; Peng, L.; Yang, X.; Wu, Y.; He, L. Electrochemistry of cytochrome P450 enzyme on nanoparticle-containing membrane-coated electrode and its applications for drug sensing. *Anal. Biochem.*, **2008**, *375*, 209-216.

[82] Zhang, L.; Yuan, R.; Chai, Y.; Li, X. Investigation of the electrochemical and electrocatalytic behavior of positively charged gold nanoparticle and l-cysteine film on an Au electrode. *Anal. Chim. Acta*, **2007**, *596*, 99-105.

[83] Shenhar, R.; Norsten, T.B.; Rotello, V.M. Polymer-mediated nanoparticle assembly: structural control and applications. *Adv. Mater.*, **2005**, *17*, 657-669

[84] Xu, Q.; Mao, C.; Liu, N.N.; Zhu, J.J.; Sheng, J. Direct electrochemistry of horseradish peroxidase based on biocompatible carboxymethyl chitosan–gold nanoparticle nanocomposite. *Biosens. Bioelectron.*, **2006**, *22*, 768-773.

[85] Yang, K.; Wang, H.; Zou, K.; Zhang, X. Gold nanoparticle modified silicon nanowires as biosensors. *Nanotechnology*, **2006**, *17*, S276-S279.

[86] Kang, Q.; Yang, L.; Cai, Q. An electro-catalytic biosensor fabricated with Pt–Au nanoparticle-decorated titania nanotube array. *Bioelectrochemistry*, **2008**, *74*, 62-65.

[87] Turner, M.; Golovko, V.B.; Vaughan, O.P.; Abdulkin, P.; Berenguer-Murcia, A.; Tikhov, M.S.; Johnson, B.F.; Lambert, R.M. Selective oxidation with dioxygen by gold nanoparticle catalysts derived from 55-atom clusters. *Nature*, **2008**, *454*, 981-983.

[88] Jena, B.K.; Raj, C.R. Electrochemical biosensor based on integrated assembly of dehydrogenase enzymes and gold nanoparticles, *Anal. Chem.*, **2006**, *78*, 6332-6339.

[89] Zhu, M.; Liu, M.; Shi, G.; Xu, F.; Ye, X.; Chen, J.; Jin, L.; Jin, J. Novel nitric oxide microsensor and its application to the study of smooth muscle cells. *Anal. Chim. Acta*, **2002**, *455*, 199-206.

[90] Yu, A.; Liang, Z.; Cho, J.; Caruso, F. Nanostructured electrochemical sensor based on dense gold nanoparticle films. *Nano. Lett.*, **2003**, *3*, 1203-1207.

[91] Bharathi, S.; Nogami, M. A glucose biosensor based on electrodeposited biocomposites of gold nanoparticles and glucose oxidase enzyme. *Analyst*, **2001**, *126*, 1919–1922.

[92] Jena, B.K.; Raj, C.R. Amperometric Sensing of Glucose by Using Gold Nanoparticles. *Chem. - A Eur. J.*, **2006**, *12*, 2702–2708.

[93] Raj, C.R.; Okajima, T.; Ohsaka, T. Gold nanoparticle arrays for the voltammetric sensing of dopamine. *J. Electroanal. Chem.*, **2003**, *543*, 127–133.

[94] Lin, L.; Zhao, H.; Li, J.; Tang, J.; Duan, M.; Jiang, L. Study on colloidal Au-Enhanced DNA sensing by quartz crystal microbalance. *Biochem. Biophys. Res. Commun.*, **2000**, *274*, 817–820.

[95] Zhou, X.C.; O'Shea, S.J.; Li, S.F.Y. Amplified microgravimetric gene sensor using Au nanoparticle modified oligonucleotides. *Chem. Commun.*, **2000**, *11*, 953–954.

[96] Pang, L.; Li, J.; Jiang, J.; Shen, G.; Yu, R.; DNA point mutation detection based on DNA ligase reaction and nano-Au amplification: A piezoelectric approach. *Anal. Biochem.*, **2006**, *358*, 99–103.

[97] Chu, X.; Zhao, Z.L.; Shen, G.L.; Yu, R.Q. Quartz crystal microbalance immunoassay with dendritic amplification using colloidal gold immunocomplex. *Sens. Actuators B*, **2006**, *114*, 696–704.

[98] Ding, Y.; Liu, J.; Wang, H.; Shen, G.; Yu, R. A piezoelectric immunosensor for the detection of α-fetoprotein using an interface of gold/hydroxyapatite hybrid nanomaterial. *Biomaterials*, **2007**, *28*, 2147–2154.

[99] Zhang, Y.; Wang, H.; Yan, B.; Zhang, Y.; Li, J.; Shen, G.; Yu, R. A reusable piezoelectric immunosensor using antibody-adsorbed magnetic nanocomposite. *J. Immunol. Methods*, **2008**, *332*, 103–111.

[100] Daniel, M.C.; Astruc, D. Gold Nanoparticles: Assembly, Supramolecular Chemistry, Quantum-Size-Related Properties, and Applications toward Biology, Catalysis, and Nanotechnology. *Chem. Rev*, **2004**, *104*, 293–346.

[101] Lin, T.J.; Huang, K.T.; Liu, C.Y. Determination of organophosphorous pesticides by a novel biosensor based on localized surface plasmon resonance. *Biosens. Bioelectron*, **2006**, *22*, 513–518.

[102] He, L.; Musick, M.D.; Nicewarner, S.R.; Salinas, F.G.; Benkovic, S.J.; Natan, M.J.; Keating, C.D. Colloidal Au-Enhanced Surface Plasmon Resonance for Ultrasensitive Detection of DNA Hybridization. *J. Am. Chem. Soc.*, **2000**, *122*, 9071–9077.

[103] Li, Y.; Wark, A.W.; Lee, H.J.; Corn, R.M. Single-Nucleotide Polymorphism Genotyping by Nanoparticle-Enhanced Surface Plasmon Resonance Imaging Measurements of Surface Ligation Reactions. *Anal Chem*, **2006**, *78*, 3158–3164.

[104] Tseng, Y.T.; Chuang, Y.J.; Wu, Y.C.; Yang, C.S.; Wang, M.C.; Tseng, F.G. A gold-nanoparticle-enhanced immune sensor based on fiber optic interferometry *Nanotechnology*, **2008**, *19*, 345–501.

[105] Hossain, M.K.; Huang, G.G.; Kaneko, T.; Ozaki, Y. Characteristics of surface-enhanced Raman scattering and surface-enhanced fluorescence using a single and a double layer gold nanostructure. *Phys. Chem. Chem. Phys.*, **2009**, *11*, 7484–7490.

[106] Cao, Y.C.; Jin, R.; Mirkin, C.A. Nanoparticles with Raman Spectroscopic Fingerprints for DNA and RNA Detection. *Science*, **2002**, *297*, 1536–1540.

[107] Odenthal, K.J.; Gooding, J.J. An introduction to electrochemical DNA biosensors. *Analyst* **2007**, *132*, 603–610.

[108] Fong, W.K.; Modrusan, Z.; Mcnevin, J. P.; Marostenmaki, J.; Zin, B.; Bekkaoui, F. Rapid Solid-Phase Immunoassay for Detection of Methicillin-Resistant *Staphylococcus aureus* Using Cycling Probe Technology. *J. Clin. Microbiol.,* **2000**, *38*, 2525–2529.

[109] Baeumner, A.J.; Pretz, J.; Fang, S. A Universal Nucleic Acid Sequence Biosensor with Nanomolar Detection Limits. *Anal. Chem.,* **2004**, *76*, 888–894.

[110] Baeumner, A.J.; Schlesinger, N.A.; Slutzki, N.S.; Romano, J.; Lee, E.M.; Montagna, R.A. Biosensor for Dengue Virus Detection: Sensitive, Rapid, and Serotype Specific. *Anal. Chem.,* **2002**, *74*, 1442–1448.

[111] Baeumner, A.J.; Cohen, R.N.; Miksic, V.; Min, J. RNA biosensor for the rapid detection of viable *Escherichia coli* in drinking water. *Biosens. Bioelectron.,* **2003**, *18*, 405–413.

[112] Mao, X.; Ma, Y.; Zhang, A.; Zhang, L.; Zeng, L.; Liu, G. Disposable Nucleic Acid Biosensors based on gold nanoparticle probes and lateral flow strip. *Anal. Chem.,* **2009**, *81,* 1660–1668.

CHAPTER 10

1D Nanostructures for Sensing Purposes

Alessio Giuliani[1,2*] and Yi Ge[2]

[1]SINTEA PLUSTEK Srl, Via E. Fermi 44, 20090 Assago (MI), Italy and [2]Cranfield Health, Vincent Building, Cranfield University, Bedfordshire, MK43 0AL, UK

Abstract: Nanowires, Nanorods and Nanotubes are well-known monodimensional (1D) nanostructures. They exhibit a very high surface to volume ratio as well as unique magnetic and electrical properties which make them ideal candidates for new and innovative sensing devices. One of the key aspects in new biosensor developments is the alignment of monodimensional nanostructures that can be obtained by employing Langmuir Blodgett techniques as well as electrical and magnetic external fields. There recently have been many successful applications of 1D nanostructures in high performance sensors, such as field biomedical sensors that are able to detect very low concentrations of DNA and novel gas sensors for the detection of ultra-low concentration of toxic gases. The well-improved sensors employing 1D-nanostructures have a profound impact on healthcare and safety, thereby playing a more and more important role in the near future.

Keywords: Monodimentional (1D) nanostructures; nanomaterials; sensors.

1. INTRODUCTION

Monodimensional nanostructures are metallic or semiconducting nanoparticles some tens of microns long, characterized by a cross section << 1nm. Nanowires (NWs), Nanorods And Nanotubes (NTs) (Fig. **1-3**) are the most studied mono dimensional (1D) nanostructures. In particular, there are two types of Carbon Nanotubes (CNTs): Single Walled Carbon Nanotubes (SWCNTs) which are rolled up graphene sheets characterized by covalently bound carbon atoms and Multi Walled Carbon Nanotubes (MWCNTs) that are simply a coaxial assembly of SWCNTs.

Monodimensional nanostructures exhibit a very high surface–to–volume ratio, and their tunable electron transfer and many other distinguished electronic, magnetic and optical properties make them ideal candidates for sensing devices [1-3]. The great interest in 1D nanostructures arises from their ability to more effectively transfer electrons than 2D and 3D nanomaterials: the electron transport in 1D nanostructures is directly in contact with the surrounding environment, [7-8] so that they could be applied for sensors with an-ultra high sensitivity. For example, even if a very small biomolecule binds to a SWCNT, the rate of electron transport between the biomolecule and the nanotube would be altered. Thus, the binding could be sensed and detected [9]. Generally speaking, the basic mechanism exploited by nanostructures based sensors is the variation of the electrical conductivity in response to the interaction between the material and the analytes that should be detected. In addition, their very large surface–to–volume ratio enhances their ability to bind and detect biomolecules.

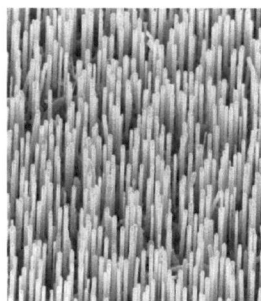

Figure 1. Image of nanowires [4]. Reproduced with permission.

*****Address correspondence to Alessio Giuliani:** SINTEA PLUSTEK Srl, Via E. Fermi 44, 20090 Assago (MI), Italy; E-mail: giuliani@sinteaplustek.com

Songjun Li, Yi Ge and He Li (Eds)
All rights reserved - © 2012 Bentham Science Publishers

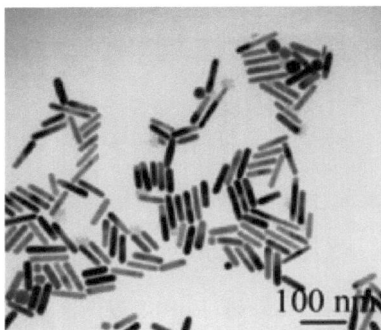

Figure 2. Image of nanorods [5]. Reproduced with permission.

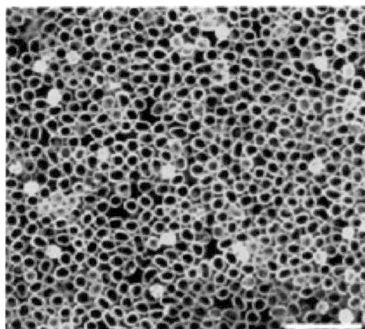

Figure 3. Image of nanotubes. [6] Reproduced with permission.

The properties of NWs largely depend on the method adopted in their synthesis as demonstrated by Xu *et al.* [10], who produced, in a non-acqueous bath, CdSe uniformly oriented NWs with the c-axis normal to the substrate. These results were further agreed by some other researchers [11] while on the contrary Peng *et al.* [12], showed that NWs deposited in an aqueous bath were oriented randomly.

Field Emission Transistor (FET) has attracted considerable interests and investigations in the last decade. It is characterized by some drain electrodes and a p-type silicon semiconductor connected to a metal source. A gate electrode coupled by a thin dielectric, allows the on/off switching of conductance between the other two electrodes. For example, the application of a positive gate voltage leads to carriers depletion and thus results the reduction of conductance. This mechanism has been exploited successfully in biomedical sensors, where the binding of charged molecules to the gate causes an electrical field analogous to the application of a voltage. NWs-based FET sensors are characterized by the accumulation of carriers in the bulk of the nanomaterial [13] and they show a much higher sensitivity enabling the detection of even single viruses [14]. It's worth noting that Zheng and his co-workers [15] successfully developed a silicon NWs based sensor platform with an objective to elucidate the role of these structures in sensing devices. They demonstrated the importance of such nanostructures by showing all electrical signals obtained in the experiments were produced just by those phenomena occurred at the silicon NWs surface.

The sensors market is very big and growing fast especially due to the share and importance of biomedical sensing devices. In addition, there is a great demand for new and innovative devices that are able to effectively and simultaneously detect the high variety of pathogens and biomolecules presented in the environment. As a result, by taking the advantages of 1D nanostructures, 1D nanostructures based-sensors have make a considerable impact on patient recovery and public health monitoring [16].

2. SYNTHESIS OF 1D NANOSTRUCTURES

Nanomaterials can be produced by using either bottom-up or top-down techniques. The bottom-up approach starts from single molecules which then assemble together in order to obtain more complex structures. The top-down approach starts from a chunk of material and then removes parts of it until the final product, characterized by dimensions in the nanometer range, is obtained.

Among bottom-up approaches, the method of seed-mediated growth is widely applied for synthesizing colloidal gold nanorods. It offers many advantages such as simplicity, high quality, high yield and flexibility. The origin of this method dates back to 1989, when Wiesner and Wokaun [17] added gold nuclei to HAuCl₄ growth solutions and H_2O_2. In contrast, the current seed mediated chemical growth procedure has just been established recently where some gold nanorods are generated by adding citrate-capped small gold nanospheres to a $HAuCl_2$ growth solution. By using such method, the first produced gold nanorods are able to be used as seeds to allow the growth of the second and third stage products [18]. Unfortunately, this procedure implies a large production of nanospheres which in turn require a lot of time to be separated from the nanorods, whose yield, size and shape can be tuned by acting on some parameters such as seed and surfactant concentration, temperature and pH [19]. Orendorff and Murphy [20] reported that diffused $AuCl_2^-$ ions could present on CTAB (cetrimonium bromide) micelles by using an electric field. As a result, the spherical symmetry of nanospheres could be broken into a geometry with different facets, some of which were characterized by a faster deposition of silver ions which in turn allowed the particles to change their shape into a rod-like one.

Nanorods can also be prepared by using electrochemical methods. One of the most employed techniques was developed by Wang *et al.* in 1990s [21]. It was based on the use of a platinum cathode and a gold anode that were immersed in an electrolytic solution comprising CTAB as surfactant and Tetraoctylammonium Bromide (TOAB) as co-surfactant. The $AuBr4^-$ ions produced at the anode were complexed with CTAB and then deposited onto the cathode where they were changed into gold atoms. The concentration of ions produced during the redox reaction gave rise to nanorods which were subsequently separated from the cathode *via* ultrasonication. Nanorods produced by the above electrochemical methods are monocrystalline nanostructures without faults, twins and dislocations [22].

The photochemical reduction is another modern method for producing gold nanorods [23]. E. Leontidis *et al.* [24] reported that the rod formation could be obtained by a two-step aggregation process, during which nuclei and crystal aggregation led to primary particles growth and rod formation respectively. The addition of sodium chloride electrolytes influenced the length of nanorods while silver ions enhanced reaction yield and uniformity of the product. An important element of this process was the application of UV light which accelerated the growth of nanostructures. In particular, O. R. Miranda *et al.* [25], confronted 300 nm and 254nm UV light wavelengths and found out that only the first one (300nm) worked for the successful production of longer nanorods which were also characterized by a narrower distribution.

B. M. I. van der Zande *et al.* [26] demonstrated that the template-based methods, usually are employed for producing aligned nanocomposites, could also be used to produce gold nanorods, offering an opportunity to conciliate an easy surface chemistry with a very precise control of the nanomaterial geometry. AAO (Anodic Aluminium Oxide) and polycarbonate membranes are the most adapted templates. The advantage of this method is the possibility of exploiting it for the generation of nanowires made by different materials, while the main problem of it is related to the polycristallinity of the obtained nanostructures. Another research group [27] suggested that sol-gel electrophoretic deposition could be used to produce mesoporous silica nanorods. In details, an electrical field was applied to guide charged sol nanoclusters into pores of a template while counter ions moved in the opposite direction. The process continued until the pores were completely filled before nanorods were thermally treated (500–700 °C for 15-30 min) in order to achieve desired crystal structure. Capillarity is the only driving force in this process. Thus, in high density solutions of sol, the diffusion of nanoclusters in the small pores of the template is not easily achievable and this leads to an unsuccessful synthesis of nanorods with diameters less than 50nm [28]. On the other hand, it's worth noting that low concentrations of sol may cause serious shrinkage phenomena in the final nanostructures [29]. D. S. Xu *et al.* [30] showed that quasi-1D semiconducting nanowires could be produced by using Direct Current Electrochemical Deposition (DCED). As can be seen in Fig. **4**, the three-step method includes: **1)** the generation of a metal film onto the back of a template by sputtering; **2)** the potentiostatical and galvanostatical deposition of the semiconductor materials from a solution containing metal ions; and **3)** the removal of the template.

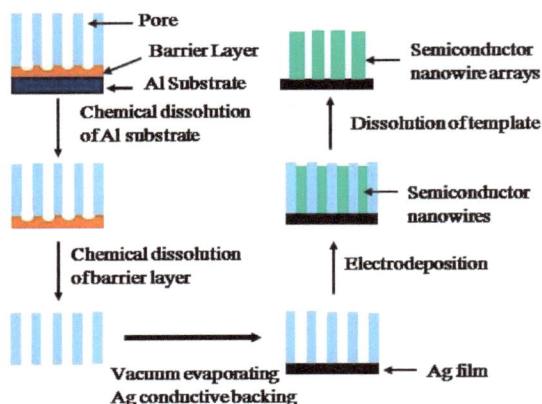

Figure 4. The DC electrochemical process. [30] Reproduced with permission.

The direct electrochemical deposition is useful to allow a precise control of the 1D nanostructures synthesis. For example, Xu *et al.* [29] prepared monocrystalline CdS nanowires starting from a 0.05 M $CdCl_2$ and 0.10 M thioacetamide bath. This process was carried out under a condition of 0.65V (potential) and 70 °C in a glass cell. After 8 hours, it led to uniform hexagonal nanowires with diameters smaller than 5 nm and lengths up to 10 μm.

Sol-gel method can also be applied to generate monocrystalline TiO_2 nanowires. For example, Miao Z. *et al.* [31] produced hydroxyl ions which caused the increase of pH at the surface of the electrode. This product is able to produce titanium oxyhydroxide gel, whose deposition occurred in the pores of template (Fig. **5**). Once the deposition was completed, template was removed in order to obtain desired TiO_2 nanostructures.

Furthermore, NWs can be produced by using VLS (Vapour Solid Liquid) process based on the use of a catalyst and a gas precursor; particularly when a nanometer droplet of the catalyst liquid, lying onto a surface, comes in contact with the gas, a component of interest is absorbed into the drop, until supersaturation arises. The absorbed component starts to precipitate at the level of the drop, leading to the growth of the nanowire whose diameter depends on the droplet dimension.

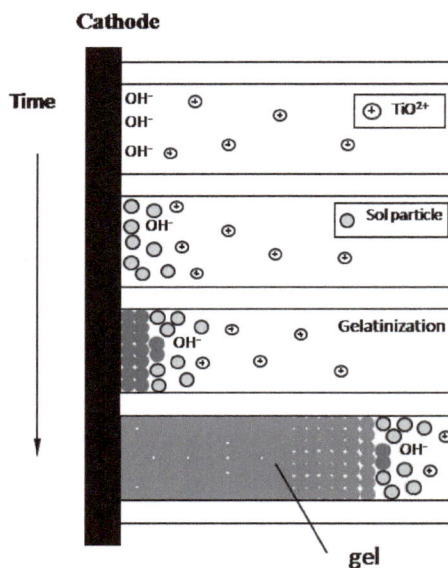

Figure 5. The Sol-gel technique for producing TiO_2 nanowires [31]. Reproduced with permission.

It is worth noting that the material of substrate has a great importance in the growth rate as well as its temperature, precursor composition and pressure. In particular, M. T. Borgstrom *et al.* [32] found out that the higher the concentration of the catalytic elements, the higher the nanowire growth rate.

A. I. Persson *et al.* [33] explained that the importance of these parameters was due to their influence on the size of the collection area which is the migration length of the adsorbed species during the vapour growth process. VLS can be employed also to produce ZnO NWs. The epitaxial relationship between the substrate and NWs, plays a very important role in the alignment of the final monodimensional nanostructures, while their growth depends mainly on their crystal structure, chamber pressure, oxygen partial pressure and thickness of the catalyst layer [34].

Hydrothermal synthesis is a broadly adopted method for the production of mono dimensional nanostructures; particularly it has been demonstrated that this process could be applied successfully to synthesize ZnO nanowires starting from ZnO thin films and ZnO nanoparticles [35-36]. It's worth noting that a group of researchers [37] have recently developed a new hydrothermal process which does not use ZnO seeds nanoparticles: firstly, the substrate is modified by using ultrasonication, baking and sputtering, secondly, the nutrient solution comprising zinc nitrate and hexamethylenetetramine (HMTA) in 1:1 proportion is prepared and finally, an annealing process is carried out to produce uniform crystalline layers. ZnO NWs derive from the reaction between Zn^{2+} ions contained in the Zinc nitrate salt and O^{2-} ions coming from water molecules contained in solution. During this process, the precursor concentration influences NWs density while the growth temperature and growth duration affect their aspect ratio.

Another interesting approach for the synthesis of mono dimensional nanostructures is electrospinning which is usually adopted for producing nanofibers. This method is characterized by the application of a high voltage to a capillary where a polymer/solvent mixture flows through; the fluid, jetting through a needle into an electrical field, is collected onto a counter-electrode giving rise to the desired nanofibers once the solute evaporates. Such fibers are usually organic-inorganic, hence, in order to obtain inorganic nanostructures, an oxidative process has to be used. For example, D. Li *et al.* [38] used this technique and injected a mixture of PVP and titanium tetraisopropoxide in a strong electrical field obtaining composite inorganic nanofibers made of PVP and TiO_2, whose diameters ranged from 20nm to 200nm.

The growth of mono dimensional oxide nanostructures, such as long, pure and uniform mono crystalline oxide nanobelts, can be achieved by using the Vapor Solid growth process which is characterized by the absence of a catalyst and by a high operating temperature [39]. It is based on the vaporization of a source powder material which subsequently condenses and generates the final product.

Carbon Nanotubes (CNTs) were discovered during an arc discharge process [40], though the first large production were realized by two researchers at NEC's Fundamental Research Laboratory in 1992 [41]. The arc discharge method is based on the use of an anode and a cathode, made of high purity graphite, under a helium atmosphere and a high voltage (necessary to establish a stable electric arc). During this process the anode is consumed and CNTs deposit on the cathode.

CNTs can be also obtained by the condensation, achieved through a pulsed laser, of a vaporized graphite target onto a water cooled surface [42-43]. It is also worth to mention the Chemical Vapour Deposition technique, in which carbon nanotubes derive from the diffusion and the deposition at the growth surface of carbon atoms dissolved into metal droplets, as demonstrated by W. Z. Li *et al.* [44] who applied this process by using Fe nanoparticles contained in mesoporous silica. The catalysts play a very important role since they are able to influence the diameter and the growth rate of the nanotubes [45]. CNTs can be employed also to produce metal nanotubes, in fact Xu *et al.* [46] developed a metal nanotubes array by depositing CNTs in the walls of an AAO template followed by Ni electrodeposition into the pores of the structure.

Another interesting approach for developing metal oxide NTs is to start by synthesizing NWs or nanorods and subsequently dissolve them along their c-axis. For example L. Xu *et al.*[47] used this approach starting from ZnO nanorods they developed. The resulting ZnO NTs have smooth tubular structure and walls characterized by a uniform thickness (Fig. **6**).

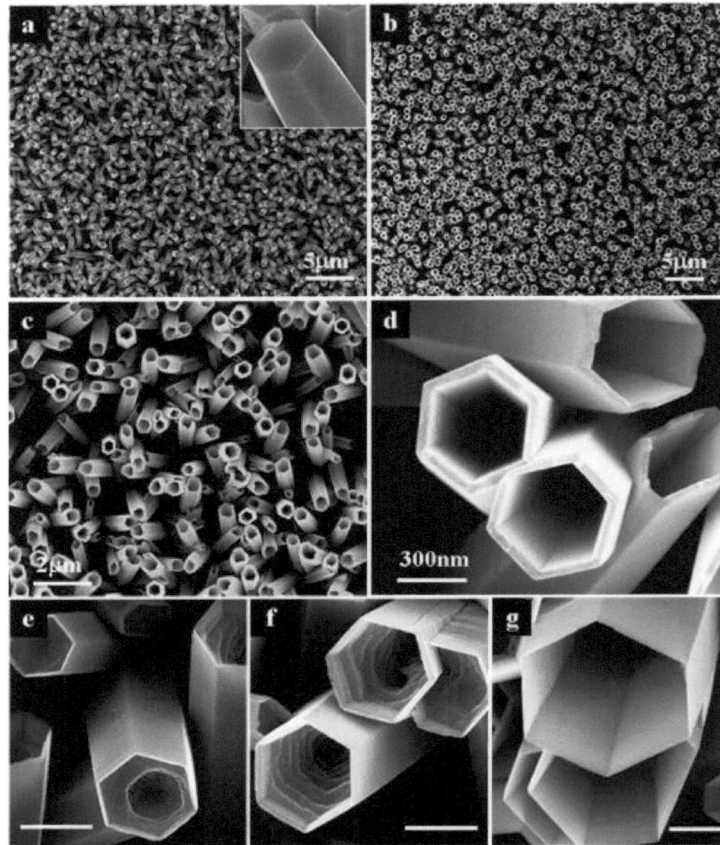

Figure 6. The ZnO nanorods **(a)** compared with ZnO NTs **(b-d)** obtained by dissolution of ZnO nanorods [47]. Reproduced with permission.

Top-down techniques, can be successfully exploited for large mass production of nanostructures and are very attractive because of the high yield, uniformity and alignment that they allow to achieve. Two are the most used top-down techniquesare: Electron Beam Lithography (EBL) and Anodization.

As shown by E. J. Smythe [48], EBL is characterized by an electron beam which dissolves an electron sensitive resist layer, leaving some nanometer openings where metal nanoparticles, for example made of gold, can be deposited. This method can be used to produce very complicated nanostructures with a resolution of tens of nanometers.

Nodozation is an electrochemical process in which the metal should be treated to form the anode of an electric circuit. This process allows the modification of the surface topography as well as its crystal structure. Many researchers [49-51] demonstrated that anodization is a powerful method for producing TiO$_2$ nanotubes and they particularly showed that the use of glycerol electrolytes allowed the fabrication of high regular and homogeneous structures while water content influenced NTs length. Additionally they reported that the applied potential, reaction duration and fluoride ions concentration, could influence TiO$_2$ morphology. Controlling such parameters is then a key aspect of this process, as demonstrated by S. P. Albu *et al.* [51-52[who succeeded in producing TiO$_2$ NTs characterized by a length of 250 nm and a double wall.

3. 1D NANOSTRUCTURES FUNCTIONALIZATION

Functonalization is a very important topic in biosensing. This term refers to the modification of a surface with both chemical functional groups and biological molecules. In particular, the binding of some biomolecules such as proteins or antibodies allows the production of sensors which can be specifically designed to capture only the molecules that need to be analyzed.

It is well-known that the Au-S bonds can be exploited to bind alkythiol-functionalized PEG [53] and DNA [54] molecules to gold nanorods. Biomolecules can be linked also by the use of some polymers. For example, it has been showed that antibodies could be bound to gold nanorods, via hydrophobic interactions, if they had been previously coated with anionic poly(sodium-4-styrenesulfonate) (PSS) polyelectrolytes [55-56]

In contrast, for silicon NWs, 3-(trimethoxysilyl) propyl aldehyde (TMSPA) is usually adopted for generating an aldehyde-rich surface allowing the binding of proteins and DNA molecules [15].

Carbon nanotubes can be functionalized with a two-step approach which was carried out for the first time at Stanford University by Dai *et al.* [57] They bound one side of the linker to SWCNTs and the other side to bioaffinitive agents, by using non covalent and covalent methods respectively. In their research work, non-covalent binding did not alter CNTs structure while the covalent functionalization, which was carried out at the open ends of the nanotube [58], implied the use of strong acids in order to form carboxylic groups which CNTs could be bound. The work of Matsui and his co-workers [59] is also worth mentioning: firstly, they coated CNTs sidewalls with Au nanocrystals, then they added avidin molecules and finally, they employed chemical etching to remove nanocrystals and molecules just from the sidewalls, leaving only avidin at the ends of the CNTs. Avidin molecules were not altered by this process and thus they were subsequently exploited to bind biotin (Fig. **7**).

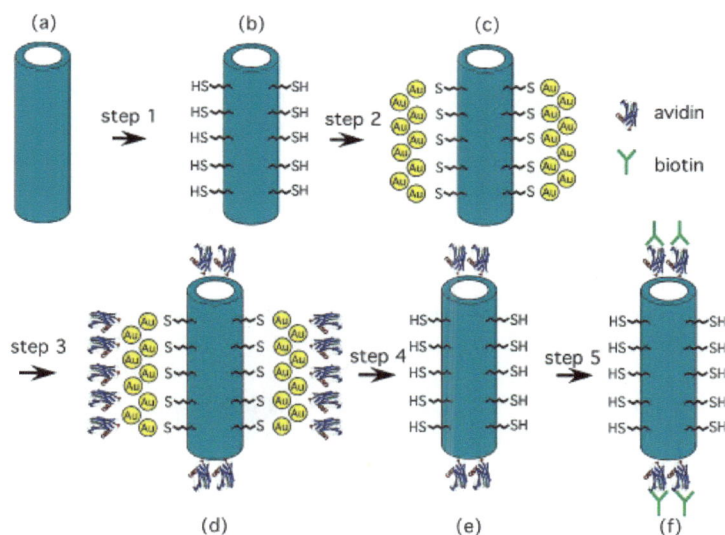

Figure 7. CNTs functionalization by Au nanocrystals. **a)** CNT; **b)** thiolation of CNT sidewall; **c)** Au coating on the CNT sidewall; **d)** incubation of avidin molecules; **e)** Au and avidin removal by chemical etching from the sidewall; **f)** biotin immobilization [59]. Reproduced with permission.

In order to prevent non-specific binding of some molecules, some researchers [60] employed PEO (polyethylene oxide) protein-resistant polymer and biotin to obtain a specific binding of streptavidin onto SWCNTs. Some other approaches [61-62] are characterized by the use of conducting polymers to entrap some target molecules, though these techniques may drastically reduce the activities of proteins and enzymes which are essential for the biosensor functionality.

4. 1D NANOSTRUCTURES ALIGNMENT

Assembly and alignment of 1D nanomaterials, are also key aspects that should be taken into account during the development of new nanostructures based sensors. It is worth noting that any technique employed for this scope should be affordable and compatible with a wide range of materials.

Many researchers [63-65] applied the Langmuir Blodgett technique to assemble silicon and silver nanowires. NWs initially were dispersed into a water solution where thay interacted with surfactans and aligned parallel to the barriers by varying the pressure of the solution. The monolayer was then obtained at the air water interface, followed by being transferred to a planar surface *via* a simple dipping procedure.

Y. Huang and his coworkers [66-67] reported that the alignment of 1D nanostructures could be induced also by letting NWs suspend in microfluidic channels, flowing on substrates. The alignment degree and their orientation depended on the flow rate and direction, while the surface coverage was influenced by the duration of the process. The substrate topography could be also modified by electron beam lithography in order to guide the NWs alignment.

Electrical fields can also be used for the alignment of NWs suspensions. In fact, depending on the field strength and the thermal energy, the application of an electrical potential is able to polarize NWs, allowing them to align parallel to the direction of the electrical field itself. Some researchers [68-69] have applied this technique to Au NWs while others employed it for assembling CNTs [70].

NWs exhibit magnetic properties due to the ferromagnetic elements, such as Co and Ni, which they originate from. NWs are dipoles characterized by shape anisotropic magnetic properties due to their large aspect ratios and this enables them to align parallel to the applied external field. It is worth noting that if this method needs to be used to align non-metallic NWs, metallic nanoparticles should be linked to the non metallic NWs ends. This finding was demonstrated by Crone *et al.* [71] who bound Ni nanoparticles to the end of CuSn NWs and oriented them by exploiting the magnetic field generated by two nickel electrodes.

Another very absorbing approach, named "Blown Bubble Film", is based on the use of a solution of polymers and NWs and it is characterized by an expanding bubble which forces NWs to align and to deposit onto rigid or flexible substrates [72]. The main advantage of this technique is its applicability to large areas (225-300 mm) without the need of using lithography, though the controllability of the NWs density onto the substrates needs to be optimized.

Finally, a very simple and effective approach was suggested by A. Javey *et al.* [73] who oriented some non-aligned NWs by sliding the surface containing them against another substrate. The result was very positive and it successfully demonstrated the capability of a direct and dry transfer of monodimensional nanostructures by using sheer forces.

5. APPLICATIONS OF 1D NANOSTRUCTURES TO SENSORS

One of the most important and interesting topics in medicine is the trace detection of biomolecules for diagnosing diseases in time in order to enhance the effectiveness of treatments at an early stage. Thus there is an increasing need for advanced sensors that are capable to measure/detect tiny amount of analytes/samples. As described and discussed previously, the monodimensional nanostructures have a great capability and potential to better serve the needs for advanced sensors.

Lieber's group [13] firstly reported the use of 1D nanostructures based sensors for proteins detection with a very high sensitivity. They functionalized p-Si NWs with biotin and after an exposure to a solution containing a streptavidin concentration of 10 pM, they found out the electrical conductance increasing to a stable value.

The immobilization of peptide nucleic acid has also been exploited successfully for the detection of DNA. Z. Li *et al.* [74] showed that the immobilization of single strand DNA allowed to detect up to 25 pM complementary single strand DNA while some others [75] reported that the PNA (Peptide Nucleic Acid) functionalization of Si NWs enabled the detection of 10 fM single strand DNA. It is worth noting that the sensor sensitivity is highly related to the distance between the DNA layer and the Si NWs surface. The higher the distance the weaker the electrical signal [76].

Employing PNA as a capture probe molecule seems a promising approach for the detection of DNA [77]. PNA backbone is neutral and this allows hybridization to occur at low ionic strength, without generating a high electrical field at the NWs surface which in turn reduces the background response and then the noise level. In addition the neutrality of PNA backbone could help to reduce repulsions between the bound strands and this results in a higher stability of the system.

DNA can also be detected by using SWCNTs-based FET sensors, reported by researchers at Nanomix [78]. The sensor was used to measure 12-bp ss-DNA firstly and then 10 nM complementary DNA solution. It was found out that the transconductance shifted progressively to more negative values. It was speculated that this effect was due to an electron doping phenomenon caused by the ss-DNA adsorption on the SWCNT sidewall.

Some other interesting researches of 1D nanostructures based FET sensors employed for proteins analysis were investigated by Byon and Choi [79], who succeeded in detecting both specific and non specific proteins interactions at 1 pM concentration; and by Abe *et al.* [80] who measured 5 nM Pig Serum Albumin (PSA) concentration by using a CNTs based FET transistor covered by an insulator.

The worldwide's concerns about bio-terrorism raise an increasing interest of public and scientific community on the development of new methods for the detection of viruses and bacteria and thus there is a growing demand of sensors able to measure those organisms in time and with a very higher sensitivity. Viruses could be characterized by some proteins on their surfaces so they can be bound to antibody-functionalized Si-NWs and thus be successfully detected by FET sensors [14] which further are able to distinguish one species from another.

H. M. So *et al.* [81] employed an aptamer-functionalized SWCNTs based FET sensor to detect Escherichia coli bacterium. They found out that the conductance dropped off by 50% with a nominal shift along the voltage axis once the bacterium bound to the surface, and could return to the base value when the organism was removed.

For the detection of viruses, Patolsky *et al.* [14] reported a solution with a 50 viruses/mL concentration to antihemagglutinin (protein-specific antibody) functionalized Si-NWs and they observed discrete and defined changes in the electrical conductance due to the binding and unbinding phenomena occurring at the sensor surface. In their experiments they employed two different NWs arrays: NW1 and NW2 which had been functionalized with adenovirus and Influenza A receptors respectively. They firstly delivered adenovirus particles to the sensor and only a positive change of NW1 conductance was observed. Influenza A was then analyzed and they recorded only a modification of NW2 conductance. Finally it was shown that NW1 and NW2 were unambiguously able to detect in parallel the two different viruses when they measured a mixture of adenovirus and Influenza A.

Interestingly, Reed *et al.* [82] demonstrated Si NWs based sensors could be employed successfully for immunoassaying and cellular immune response analysis in real time. In the first case they functionalized NWs with mouse immunoglobulin G (IgG) and immunoglobulin A (IgA) antibodies showing a more selective response than that obtained with PEG functionalized NWs that were adopted as a control. In the second case they added a mouse anti-CD3 to 6,000 T-lymphocyte mouse cells and they immediately observed a drop off of conductance, although they didn't discover any appreciable change when a control human anti-CD3 antibody were added to the solution that they were studying.

One of the most promising applications of nano-sensors in the biomedical field is the analysis of proteins for early detection of complex oncologic diseases. Li *et al.* [83] developed a biosensor comprising In_2O_3 NWs and SWCNTs, which proved to be able to measure up to 5 ng/ml concentration of Prostate Specific Antigen (PSA) - the biomarker of one of the most diagnosed diseases in men (prostate cancer).

It is worth mentioning that the oncologic pathologies heterogeneity requires the analysis of multiple cancer biomarkers in order to get all necessary clinical information to diagnose and treat disease effectively [84].

Biosensors which allow multiplex analysis of protein biomarkers are attracting a lot of attentions from industries to scientific community. They can be produced by modifying the sequence of metal segments[1] or the diameters of metallic wires [85] and silica nanotubes [86] allowing simultaneous and different biological assays *via* optical methods.

Humans and environment are constantly under the threat of toxic and combustible gaseous substances. Thus their early identification would be very important for health and safety. There is an increasing interest in developing new devices which are able to detect a wide range of gases within a certain area, with a higher sensitivity and selectivity. Wan *et al.* [87] succeeded in detecting an ethanol vapours concentration of 200 ppm by using ZnO NWs based sensor while some other researchers [88] employed ZnO nanorods which were mixed with an adhesion agent to develop a sensing device with a higher sensitivity (a few ppm concentration of other inflammable gases such as gasoline and liquid petroleum).

One of the most monitored combustible gases is NO_2, which is present in tropospheric ozone and has a great importance in smog formation. NO_2 can be detected by using SnO_2 commercial sensors which need to be operated at a very high temperature such as 300-500 C° [89]. Hence, by considering the explosive environment in which they work, there is a real risk to implement the detection. Some researchers have tried to resolve this issue by using SnO_2 nanoribbons based sensors which proved to be more effective for the detection of such kind of gas at room temperature and under UV light [90].

Du *et al.* [91] studied the applicability of NTs in gas sensing by particularly synthesizing In_2O_3 NTs with diameters ranging from 20 nm to 60 nm and then assembling them onto CNTs templates in order to develop a sensor which could have a higher sensitivity towards NH_3 at room temperature. Some similar results were obtained by Y. Li *et al.* [92] who deposited VO_x-NTs onto a ceramic chip to develop a novel gas sensor comprising Ag-Pd electrodes which showed a higher sensitivity towards NO molecules at room temperature. It is also worth noting that Zhang *et al.* [93] produced three types of metallic nanotubes ($NiCo_2O_4$, $CuCo_2O_4$, and $ZnCo_2O_4$) and then used them for developing the same number of sensing devices that are very sensitive and selective for the detection of ethanol and SO_2. In addition, A. Star *et al.* [94] further improved the detection of CO_2, one of the most common gases in the environment. They developed a SWNTs based sensor [95], which were functionalized with Polyethyleneimine (PEI) and starch. When such a device came in contact with CO_2, a decrease of the electrical resistance was detected due to the increased selective gas absorption of carbamates, which were generated by the reaction of polymer amine groups with the gas itself.

6. CONCLUSION

In conclusion, monodimensional (1D) nanostructures have been successfully developed and demonstrated to be functional and advantageous for producing new and innovative sensing devices with higher sensitivity and selectivity. As a result, precise diagnoses as well as early, rapid and effective detections of pathogens and other dangerous substances could be achieved, paving the way towards to a more personalized healthcare and a more effective treatment against diseases.

REFERENCES

[1]   Brunker, S. E.; Cederquist, K. B.; Keating, C. D. Metallic barcodes for multiplexed bioassays", *Nanomedicine*, **2007**, *2*, 695-710.

[2]   Patolsky, F.; Timko, B.P.; Zheng, G.; Lieber, C. M. Nanowire-based nanoelectronic devices in the life sciences. *MRS Bull.*, **2007**, *32*, 142-149.

[3]   Wang, J. Nanomaterial-based electrochemical biosensors. *Analyst*, **2005**, *130*, 421-426.

[4]   Wang, Z. L. ZnO nanowire and nanobelt platform for nanotechnology. *Mater. Sci. Eng. R: Rep.*, **2009**, *64*, 33-71.

[5]   Huang, X.; Neretina, S.; El-Sayed, M. A. Gold nanorods: From synthesis and properties to biological and biomedical applications. *Adv. Mater*, **2009**, *21*, 4880-4910.

[6]   Mu, C.; Yu, Y.; Liao, W.; Zhao, X.; Xu, D.; Chen, X. and Yu, D. "Controlling growth and field emission properties of silicon nanotube arrays by multistep template replication and chemical vapor deposition", *App. Phys. Lett.*, **2005**, *87*(11), pp. 1-3.

[7]   Kong, J.; Franklin, N.R.; Zhou, C.; Chapline, M.G.; Peng, S.; Cho, K.; Dai, H. Nanotube molecular wires as chemical sensors. *Science*, **2000**, *287*, 622-625.

[8]   Shim, M.; Javey, A.; Kam, N.W.S.; Dai, H. Polymer functionalization for air-stable n-type carbon nanotube field-effect transistors. *J. Am. Chem. Soc.*, **2001**, *123*, 11512-11513.

[9]   Star, A.; Gabriel, J.P.; Bradley, K.; Grüner, G. Electronic detection of specific protein binding using nanotube FET devices. *Nano Lett.*, **2003**, *3*, 459-463.

[10]   Xu, D.; Shi, X.; Guo, G.; Gui, L.; Tang, Y. Electrochemical Preparation of CdSe Nanowire Arrays. *J. Phys. Chem. B*, **2000**, *104*, 5061-5063.

[11]   Chen, R.; Xu, D.; Guo, G.; Gui, L. Silver telluride nanowires prepared by dc electrodeposition in porous anodic alumina templates. *J. Mater. Chem.*, **2002**, *12*, 2435-2438.

[12]   Peng, X.S.; Zhang, J.; Wang, X.F.; Wang, Y.W.; Zhao, L.X.; Meng, G.W.; Zhang, L.D. Synthesis of highly ordered CdSe nanowire arrays embedded in anodic alumina membrane by electrodeposition in ammonia alkaline solution. *Chem. Phys. Lett.*, **2001**, *343*, 470-474.

[13]   Cui, Y.; Wei, Q.; Park, H.; Lieber, C.M. Nanowire nanosensors for highly sensitive and selective detection of biological and chemical species. *Science*, **2001**, *293*, 1289-1292.

[14]   Patolsky, F.; Zheng, G.; Hayden, O.; Lakadamyali, M.; Zhuang, X.; Lieber, C.M. Electrical detection of single viruses. *Proc. Natl. Acad. Sci. USA*, **2004**, *101*, 14017-14022.

[15]   Zheng, G.; Patolsky, F.; Cui, Y.; Wang, W.U.; Lieber, C.M. Multiplexed electrical detection of cancer markers with nanowire sensor arrays. *Nat. Biotech.*, **2005**, *23*, 1294-1301.

[16]   Hinman, A.R. Global progress in infectious disease control. *Vaccine*, **1998**, *16*, 1116-1121.

[17]   Wiesner, J.; Wokaun, A. Anisometric gold colloids. Preparation, characterization, and optical properties. *Chem. Phys. Lett.*, **1989**, *157*, 569-575.

[18]   Murphy, C.J.; Sau, T.K.; Gole, A.M.; Orendorff, C.J.; Gao, J.; Gou, L.; Hunyadi, S.E.; Li, T. Anisotropic metal nanoparticles: Synthesis, assembly, and optical applications. *J. Phys. Chem.*, **2005**, *109*, 13857-13870.

[19]   Nikoobakht, B.; El-Sayed, M.A. Preparation and growth mechanism of gold nanorods (NRs) using seed-mediated growth method. *Chem. Mater.* **2003**, *15*, 1957-1962.

[20]   Huang, W.; El-Sayed, M.A. Photothermally excited coherent lattice phonon oscillations in plasmonic nanoparticles. *Eur. Phys. J. Spec. Top.*, **2008**, *153*, 325-333.

[21]   Chang, S.; Shih, C.; Chen, C.; Lai, W.; Wang, C.R.C. The Shape Transition of Gold Nanorods. *Langmuir*, **1999**, *15*, 701-709.

[22]   Wang, Z.L.; Mohamed, M.B.; Link, S.; El-Sayed, M.A. Crystallographic facets and shapes of gold nanorods of different aspect ratios. *Surf. Sci.*, **1999**, *440*, 809-814.

[23]   Esumi, K.; Matsuhisa, K.; Torigoe, K.; Preparation of rodlike gold particles by UV irradiation using cationic micelles as a template. *Langmuir*, **1995**, *11*, 3285-3287.

[24]   Leontidis, E.; Kleitou, K.; Kyprianidou-Leodidou, T.; Bekiari, V.; Lianos, P. Gold colloids from cationic surfactant solutions. 1. Mechanisms that control particle morphology. *Langmuir*, **2002**, *18*, 3659-3668.

[25]   Miranda, O.R.; Ahmadi, T.S. Effects of intensity and energy of CW UV light on the growth of gold nanorods. *J. Phys. Chem. B*, **2005**, *109*, 15724-15734.

[26]   Van Der Zande, B.M.I.; Böhmer, M.R.; Fokkink, L.G.J.; Schönenberger, C. Colloidal dispersions of gold rods: Synthesis and optical properties. *Langmuir*, **2000**, *16*, 451-458.

[27]   Limmer, S.J.; Hubler, T.L.; Cao, G. Nanorods of various oxides and hierarchically structured mesoporous silica by sol-gel electrophoresis. *J. Sol-Gel Sci. Tech.*, **2003**, *26*, 577-581.

[28]   Limmer, S.J.; Seraji, S.; Forbess, M.J.; Wu, Y.; Chou, T.P.; Nguyen, C.; Cao, G.Z. Electrophoretic growth of lead zirconate titanate nanorods. *Adv. Mat.*, **2001**, *13*, 1269-1272.

[29]   Xu, D.; Xu, Y.; Chen, D.; Guo, G.; Gui, L.; Tang, Y. reparation of CdS single-crystal nanowires by electrochemically induced deposition. *Adv. Mat.*, **2000**, *12*, 520-522.

[30]   Xu, D.; Chen, D.; Xu, Y.; Shi, X.; Guo, G.; Gui, L.; Tang, Y. Preparation of II-VI group semiconductor nanowire arrays by dc electrochemical deposition in porous aluminum oxide templates. *P. App. Chem.*, **2000**, *72*, 127-135.

[31]   Miao, Z.; Xu, D.; Ouyang, J.; Guo, G.; Zhao, X.; Tang, Y. Electrochemically Induced Sol-Gel Preparation of Single-Crystalline TiO2 Nanowires. *Nano Lett.*, **2002**, *2*, 717-720.

[32]   Borgström, M.T.; Immink, G.; Ketelaars, B.; Algra, R.; Bakkers, E.P.A.M. Synergetic nanowire growth. *Nat. Nanotech.*, **2007**, *2*, 541-544.

[33]   Persson, A.I.; Fröberg, L.E.; Jeppesen, S.; Björk, M.T.; Samuelson, L. Surface diffusion effects on growth of nanowires by chemical beam epitaxy. *J. App. Phys.*, **2007**, *101*, 034313-034313-6.

[34]   Wang, X.; Song, J.; Summers, C.J.; Ryou, J.H.; Li, P.; Dupuis, R.D.; Wang, Z.L. Density-controlled growth of aligned ZnO nanowires sharing a common contact: A simple, low-cost, and mask-free technique for large-scale applications. *J. Phys. Chem. B,* **2006**, *110*, 7720-7724.

[35]   Hsu, H.; Tseng, Y.; Cheng, H.; Kuo, J.; Hsieh, W. Selective growth of ZnO nanorods on pre-coated ZnO buffer layer. *J. Crys. Gr.,* **2004**, *261*, 520-525.

[36]   Ma, T.; Guo, M.; Zhang, M.; Zhang, Y.; Wang, X.; Density-controlled hydrothermal growth of well-aligned ZnO nanorod arrays. *Nanotechnology,* **2007**, *18*, 035605-035605-7.

[37]   Xu, S.; Lao, C.; Weintraub, B.; Wang, Z.L. Density-controlled growth of aligned ZnO nanowire arrays by seedles chemical approach on smooth surfaces. *J. Mat. Res.,* **2008**, *23*, 2072-2077.

[38]   Li, D.; Xia, Y. Fabrication of titania nanofibers by electrospinning. *Nano. Lett.,* **2003**, *3*, 555-560.

[39]   Wang, Z.L.; Pan, Z.W.; Dai, Z.R. Structures of oxide nanobelts and nanowires. *Micr. Microan.,* **2002**, *8*, 467-474.

[40]   Iijima, S. Helical microtubules of graphitic carbon. *Nature,* **1991**, *354*, 56-58.

[41]   Ebbesen, T.W.; Ajayan, P.M. Large-scale synthesis of carbon nanotubes. *Nature,* **1992**, *358*, 220-222.

[42]   Guo, T.; Nikolaev, P.; Thess, A.; Colbert, D.T.; Smalley, R.E. Catalytic growth of single-walled manotubes by laser vaporization. *Chem. Phys. Lett.,* **1995**, *243*, 49-54.

[43]   Guo, T.; Nikolaev, P.; Rinzler, A.G.; Tomànek, D.; Colbert, D.T.; Smalley, R.E. Self-assembly of tubular fullerenes. *J. Phys. Chem.,* **1995**, *99*, 10694-10697.

[44]   Li, W.Z.; Xie, S.S.; Qian, L.X.; Chang, B.H.; Zou, B.S.; Zhou, W.Y.; Zhao, R.A.; Wang, G. Large-scale synthesis of aligned carbon nanotubes. *Science,* **1996**, *274*, 1701-1703.

[45]   Choi, Y.C.; Shin, Y.M.; Lee, Y.H.; Lee, B.S.; Park, G.; Choi, W.B.; Lee, N.S.; Kim, J.M. Controlling the diameter, growth rate, and density of vertically aligned carbon nanotubes synthesized by microwave plasma-enhanced chemical vapor deposition. *App. Phys. Lett.,* **2000**, *76*, 2367-2369.

[46]   Mu, C.; Yu, Y.; Wang, R.; Wu, K.; Xu, D.; Guo, G. Uniform metal nanotube arrays by multistep template replication and electrodeposition. *Adv. Mat.,* **2004**, *16*, 1550-1553.

[47]   Xu, L.; Liao, Q.; Zhang, J.; Ai, X.; Xu, D. Single-crystalline ZnO nanotube arrays on conductive glass substrates by selective disolution of electrodeposited ZnO nanorods. *J. Phys. Chem. C,* **2007**, *111*, 4549-4552.

[48]   Smythe, E.J.; Cubukcu, E.; Capasso, F. Optical properties of surface plasmon resonances of coupled metallic nanorods. *Optics Express,* **2007**, *15*, 7439-7447.

[49]   Macak, J.M.; Tsuchiya, H.; Taveira, L.; Aldabergerova, S.; Schmuki, P. Smooth anodic TiO2 nanotubes. *Ang. Chem. Int. Ed.,* **2005**, *44*, 7463-7465.

[50]   Yoriya, S.; Grimes, C.A. Self-Assembled TiO_2 nanotube arrays by anodization of titanium in diethylene glycol: Approach to extended pore widening. *Langmuir,* **2010**, *26*, 417-420.

[51]   Albu, S.P.; Ghicov, A.; Macak, J.M.; Schmuki, P. 250 im long anodic TiO2 nanotubes with hexagonal self-ordering. *Phys. Stat. Sol. Rap. Res. Let.,* **2007**, *1*, R65-R67.

[52]   Albu, S.P.; Ghicov, A.; Aldabergenova, S.; Drechsel, P.; LeClere, D.; Thompson, G.E.; Macak, J.M.; Schmuki, P. Formation of double-walled TiO_2 nanotubes and robust anatase membranes. *Adv. Mat.,* **2008**, *20*, 4135-4139.

[53]   Niidome, T.; Yamagata, M.; Okamoto, Y.; Akiyama, Y.; Takahashi, H.; Kawano, T.; Katayama, Y.; Niidome, Y. PEG-modified gold nanorods with a stealth character for *in vivo* applications. *J. Contr. Rel.,* **2006**, *114*, 343-347.

[54]   Chen, C.; Lin, Y.; Wang, C.; Tzeng, H.; Wu, C.; Chen, Y.; Chen, C.; Chen, L.; Wu, Y. DNA-gold nanorod conjugates for remote control of localized gene expression by near infrared irradiation. *J. Am. Chem. Soc.,* **2006**, *128*, 3709-3715.

[55]   Huang, X.; El-Sayed, I.H.; Qian, W.; El-Sayed, M.A. Cancer cell imaging and photothermal therapy in the near-infrared region by using gold nanorods. *J. Am. Chem. Soc.,* **2006**, *128*, 2115-2120.

[56]   Durr, N.J.; Larson, T.; Smith, D.K.; Korgel, B.A.; Sokolov, K.; Ben-Yakar, A. Two-photon luminescence imaging of cancer cells using molecularly targeted gold nanorods. *Nano Lett.,* **2007**, *7*, 941-945.

[57]   Chen, R.J.; Zhang, Y.; Wang, D.; Dai, H. Noncovalent sidewall functionalization of single-walled carbon nanotubes for protein immobilization. *J. Am. Chem. Soc.,* **2001**, *123*, 3838-3839.

[58]   Bahr, J.L.; Tour, J.M. Covalent chemistry of single-wall carbon nanotubes. *J. Mat. Chem.,* **2002**, *12*, 1952-1958.

[59]   Banerjee, I.A.; Yu, L.; Matsui, H. Location-specific biological functionalization on nanotubes: Attachment of proteins at the ends of nanotubes using Au nanocrystal masks. *Nano Lett.,* **2003**, *3*, 283-287.

[60]   Chen, R.J.; Bangsaruntip, S.; Drouvalakis, K.A.; Wong Shi Kam, N.; Shim, M.; Li, Y.; Kim, W.; Utz, P.J.; Dai, H. Noncovalent functionalization of carbon nanotubes for highly specific electronic biosensors. *Proc. Natl. Acad. Sci. USA,* **2003**, *100*, 4984-4989.

[61] Coche-Guerente, L.; Deronzier, A.; Mailley, P.; Moutet, J. Electrochemical immobilization of glucose oxidase in poly(amphiphilic pyrrole) films and its application to the preparation of an amperometric glucose sensor. *Analytica Chimica Acta,* **1994**, *289*, 143-153.

[62] Cosnier, S.; Innocent, C. A novel biosensor elaboration by electropolymerization of an adsorbed amphiphilic pyrrole-tyrosinase enzyme layer. *J. Electr. Chem.,* **1992**, *328*, 361-366.

[63] Kim, F.; Kwan, S.; Akana, J.; Yang, P. Langmuir-Blodgett nanorod assembly. *J. Amer. Chem. Soc.,* **2001**, *123*, 4360-4361.

[64] Whang, D.; Jin, S.; Wu, Y.; Lieber, C.M. Large-scale hierarchical organization of nanowire arrays for integrated nanosystems. *Nano Lett.*, **2003**, *3*, 1255-1259.

[65] Tao, A.; Kim, F.; Hess, C.; Goldberger, J.; He, R.; Sun, Y.; Xia, Y.; Yang, P. Langmuir-Blodgett silver nanowire monolayers for molecular sensing using surface-enhanced Raman spectroscopy. *Nano Lett.*, **2003**, *3*, 1229-1233.

[66] Huang, Y.; Duan, X.; Wei, Q.; Lieber, C.M. Directed assembly of one-dimensional nanostructures into functional networks. *Science,* **2001**, *291*, 630-633.

[67] Lu, W.; Lieber, C. M. Nanoelectronics from the bottom-up. *Nature Materials,* **2007**, *6*, 841-850.

[68] Van Der Zande, B.M.I.; Koper, G.J.M.; Lekkerkerker, H.N.W. Alignment of rod-shaped gold particles by electric fields. *J. Phys. Chem. B,* **1999**, *103*, 5754-5760.

[69] Smith, P.A.; Nordquist, C.D.; Jackson, T.N.; Mayer, T.S.; Martin, B.R.; Mbindyo, J.; Mallouk, T.E. Electric-field assisted assembly and alignment of metallic nanowires. *App. Phys. Lett.,* **2000**, *77*, 1399-1401.

[70] Chen, X.Q.; Saito, T.; Yamada, H.; Matsushige, K. Aligning single-wall carbon nanotubes with an alternating-current electric field. *App. Phys. Lett.,* **2001**, *78*, 3714-3716.

[71] Bentley, A.K.; Trethewey, J.S.; Ellis, A.B.; Crone, W.C. Magnetic manipulation of copper-tin nanowires capped with nickel ends. *Nano Lett.,* **2004**, *4*, 487-490.

[72] Yu, G.; Cao, A.; Lieber, C.M. Large-area blown bubble films of aligned nanowires and carbon nanotubes. *Nat. Nanotech.,* **2007**, *2*, 372-377.

[73] Javey, A.; Nam, S.; Friedman, R.S.; Yan, H.; Lieber, C.M. Layer-by-layer assembly of nanowires for three-dimensional, multifunctional electronics. *Nano Lett.,* **2007**, *7*, 773-777.

[74] Li, Z.; Chen, Y.; Li, X.; Kamins, T.I.; Nauka, K.; Williams, R.S. Sequence-Specific Label-Free DNA Sensors Based on Silicon Nanowires. *Nano Lett.,* **2004**, *4*, 245-247.

[75] Gao, Z.; Agarwal, A.; Trigg, A.D.; Singh, N.; Fang, C.; Tung, C.; Fan, Y.; Buddharaju, K.D.; Kong, J. Silicon nanowire arrays for label-free detection of DNA. *An. Chem.,* **2007**, *79*, 3291-3297.

[76] Zhang, G.; Zhang, G.; Chua, J.H.; Chee, R.; Wong, E.H.; Agarwal, A.; Buddharaju, K.D.; Singh, N.; Gao, Z.; Balasubramanian, DNA sensing by silicon nanowire: Charge layer distance dependence. *Nano Lett.,* **2008**, *8*, 1066-1070.

[77] Barth, S.; Hernandez-Ramirez, F.; Holmes, J.D.; Romano-Rodriguez, A. Synthesis and applications of one-dimensional semiconductors. *Pr. Mat. Sci.,* **2010**, *55*, 563-627.

[78] Star, A.; Tu, E.; Niemann, J.; Gabriel, J.P.; Joiner, C.S.; Valcke, C. Label-free detection of DNA hybridization using carbon nanotube network field-effect transistors. *Proc. Natl. Acad. Sci. USA,* **2006**, *103*, 921-926.

[79] Byon, H.R.; Choi, H.C. Network single-walled carbon nanotube-field effect transistors (SWNT-FETs) with increased schottky contact area for highly sensitive biosensor applications. *J. Am. Chem. Soc.,* **2006**, *128*, 2188-2189.

[80] Abe, M.; Murata, K.; Kojima, A.; Ifuku, Y.; Shimizu, M.; Ataka, T.; Matsumoto, K. Quantitative detection of protein using a top-gate carbon nanotube field effect transistor. *J. Phys. Chem. C,* **2007**, *111*, 8667-8670.

[81] So, H.; Park, D.; Jeon, E.; Kim, Y.; Kim, B.S.; Lee, C.; Choi, S.Y.; Kim, S.C.; Chang, H.; Lee, J. Detection and titer estimation of Escherichia coli using aptamer-functionalized single-walled carbon-nanotube field-effect transistors. *Small,* **2008**, *4*, 197-201.

[82] Stern, E.; Klemic, J.F.; Routenberg, D.A.; Wyrembak, P.N.; Turner-Evans, D.B.; Hamilton, A.D.; LaVan, D.A.; Fahmy, T.M.; Reed, M.A. Label-free immunodetection with CMOS-compatible semiconducting nanowires. *Nature,* **2007**, *445*, 519-522.

[83] Li, C.; Curreli, M.; Lin, H.; Lei, B.; Ishikawa, F.N.; Datar, R.; Cote, R.J.; Thompson, M.E.; Zhou, C. Complementary detection of prostate-specific antigen using In$_2$O$_3$ nanowires and carbon nanotubes. *J. Am. Chem. Soc.,* **2005**, *127*, 12484-12485.

[84] Sidransky, D. Emerging molecular markers of cancer. *Nature Reviews Cancer,* **2002**, *2*, 210-219.

[85] Matthias, S.; Schilling, J.; Nielsch, K.; Müller, F.; Wehrspohn, R.B.; Gösele, U. Monodisperse diameter-modulated gold microwires. *Adv. Mat.,* **2002**, *14*, 1618-1621.

[86]   He, B.; Sang, J.S.; Sang, B.L. Suspension array with shape-coded silica nanotubes for multiplexed immunoassays. *An. Chem.,* **2007**, *79,* 5257-5263.

[87]   Wan, Q.; Li, Q.H.; Chen, Y.J.; Wang, T.H.; He, X.L.; Li, J.P.; Lin, C.L. Fabrication and ethanol sensing characteristics of ZnO nanowire gas sensors. *App. Phys. Lett.,* **2004**, *84,* 3654-3656.

[88]   Xu, J.; Chen, Y.; Li, Y.; Shen, J. Gas sensing properties of ZnO nanorods prepared by hydrothermal method. *J. Mat. Sci.,* **2005**, *40,* 2919-2921.

[89]   Barsan, N.; Schweizer-Berberich, M.; Göpel, W. Fundamental and practical aspects in the design of nanoscaled SnO2 gas sensors: A status report. *Fresenius J. Anal. Chem.,* **1999**, *365,* 287-304.

[90]   Law, M.; Kind, H.; Messer, B.; Kim, F.; Yang, P. Photochemical sensing of NO_2 with SnO_2 nanoribbon nanosensors at room temperature. *Angew. Chem. - Intl Ed.,* **2002**, *41,* 2405-2408.

[91]   Du, N.; Zhang, H.; Chen, B.; Xiangyang, M.; Liu, Z.; Wu, J.; Yang, D. Porous indium oxide nanotubes: Layer-by-layer assembly on carbon-nanotube templates and application for room-temperature NH_3 gas sensors. *Adv. Mat.,* **2007**, *19,* 1641-1645.

[92]   Yubo, L.; Zhiwei, H.; Shiquan, R. A vanadium oxide nanotube-based nitric oxide gas sensor. *Sensors and Materials,* **2006**, *18,* 241-249.

[93]   Zhang, G.; Guo, B.; Chen, J. MCo_2O_4 (M = Ni, Cu, Zn) nanotubes: Template synthesis and application in gas sensors. *Sens. Act., B: Chem.,* **2006**, *114,* 402-409.

[94]   Star, A.; Han, T.; Joshi, V.; Gabriel, J.P.; Grüner, G. Nanoelectronic carbon dioxide sensors. *Adv. Mat.,* **2004**, *16,* 2049-2052.

[95]   Dong, X.; Fu, D.; Ahmed, M.O.; Shi, Y.; Mhaisalkar, S.G.; Zhang, S.; Moochhala, S.; Ho, X.; Rogers, J.A.; Li, L. Heme-enabled electrical detection of carbon monoxide at room temperature using networked carbon nanotube field-effect transistors. *Chemistry of Materials,* **2007**, *19,* 6059-6061.

Index

Songjun Li, Yi Ge and He Li (Eds)
All rights reserved - © 2012 Bentham Science Publishers

S

Selectivity 54, 149, 172
Self-assembly 12-24, 82-84, 93-97
Smart nanomaterials 3
Smart nanoparticles 149, 152

T

Thermosensitive polymers 74-78

W

Wound healing 111-123

www.ingramcontent.com/pod-product-compliance
Lightning Source LLC
Chambersburg PA
CBHW041703210326
41598CB00007B/518